THE CITY

CITY, REGION AND REGIONALISM

URBAN GEOGRAPHY

CITY, REGION AND REGIONALISM

A Geographical Contribution to Human Ecology

ROBERT E. DICKINSON

Routledge
Taylor & Francis Group
LONDON AND NEW YORK

First published in 1947

This edition published in 2007
Routledge
2 Park Square, Milton Park, Abingdon, Oxon, OX14 4RN

Simultaneously published in the USA and Canada by Routledge
711 Third Avenue, New York, NY 10017
Routledge is an imprint of Taylor & Francis Group, an informa business

Transferred to Digital Printing 2007

First issued in paperback 2013

British Library Cataloguing in Publication Data
A CIP catalogue record for this book
is available from the British Library

City, Region and Regionalism

ISBN13: 978-0-415-41777-8 (volume hbk)
ISBN13: 978-0-415-41934-5 (subset)
ISBN13: 978-0-415-41318-3 (set)
ISBN13: 978-0-415-8045-1 (volume pbk)

Routledge Library Editions: The City

CITY REGION AND REGIONALISM

A GEOGRAPHICAL CONTRIBUTION TO HUMAN ECOLOGY

by

ROBERT E. DICKINSON

READER IN GEOGRAPHY UNIVERSITY OF LONDON

LONDON

KEGAN PAUL, TRENCH, TRUBNER & CO., LTD.

BROADWAY HOUSE: 68-74 CARTER LANE, E.C.4

First published 1947

THIS BOOK IS PRODUCED IN COMPLETE CONFORMITY WITH THE AUTHORIZED ECONOMY STANDARDS

Printed in Great Britain by Butler & Tanner Ltd., Frome and London

CONTENTS

LIST OF FIGURES

ACKNOWLEDGMENTS

We are indebted for the following permissions for the direct reproduction of maps.

Figs. 6, 7*a*, 7*b*, 9 to 15, to J. H. Kolb, Professor of Rural Sociology, University of Wisconsin, Madison, Wisconsin, U.S.A.

Figs. 30*a*, 30*b*, 39 to 47, to the Editor, *Geographical Review*, New York City, U.S.A.

Figs. 32 to 35, to Dr. Chauncy D. Harris, University of Chicago, U.S.A.

Fig. 48, to Professor C. B. Fawcett and the Royal Geographical Society.

Figs. 61 to 67, to Mr. E. W. Gilbert and the Royal Geographical Society.

Figs. 57 to 60, to the Association for Planning and Regional Reconstruction.

Figs. 68 to 75, to the United States Government Printing Office and the National Resources Committee, Washington, D.C., U.S.A.

Substantial portions of the book have appeared a s articles by the author in periodicals since 1929, and we are indebted for permissions for verbatim reproduction to the Editors of *Geography* (London), *The Geographical Review* (New York), *Sociological Review* (London), and *The Architects' Journal* (London).

PREFACE

This book is not about planning. It is concerned with certain aspects of the inherent spatial or geographical structure of society upon which planning must be based, and it insists that knowledge of the anatomy of society must precede the treatment of its defects. The theme of the book is the concept of the nucleated settlement, whether village, town or city or city sub-centre, as a focus of human activity and organization in the service of a surrounding tributary area. Such an area, through the welding force of its centre, acquires a degree of homogeneity which permits its being regarded as a natural social unit or community space-grouping. This space-grouping is a fundamental characteristic of society that has hitherto received no more than fleeting attention in this country, although its study has made great progress in recent years abroad.

This field of investigation is called "human ecology" by sociologists in the United States and, more commonly, "social morphology" on this side of the Atlantic. It is concerned with the structure of social groups and has two main aspects, the spatial or geographical aspect and the biological or demographic aspect. The field has recently been outlined and discussed by Professor M. Halbwachs in his *Morphologie Sociale* (Collection Armand Colin, 1938). It has obviously much ground in common with geography, sociology, economics and anthropology, and demands varied equipment in the individual worker and the co-operation of specialists. The neglect of such study in Britain has undoubtedly been due to the absence of such co-operation and of clear perception of the problems ; as well as to a reluctance to accept it as a legitimate sphere of sociological, or, for that matter, geographical, research, and, furthermore, to the scepticism with which "case studies" and the cartographic approach to the study of economic and social conditions are still regarded.

The approach of this book is from the standpoint of the geographer. I am fully aware that some of its material and argument is marginal to geography, which, as I agree (and have tried to argue elsewhere), should be the study of places rather than of men. But it would be futile and sheer frustration to circumscribe study in this field, as in so many other problems of contemporary society, by the arbitrarily fixed limits of particular disciplines. What matters is the problem. Nevertheless, I have sought to base the approach throughout on a logical application of the concept of geography as just stated to the interpretation of the urban settlement. The scope and techniques of urban geography, though still vaguely defined and little developed in Britain, are well established on the Continent and in the United States, and it is for this reason that most of the illustrations are drawn from foreign studies.

It is therefore important to undertake a closer study of the actual human space relations in the city, town and countryside. Quite apart from any question of defining new planning or political units, which must await the announcement of clear directives, these space relations lie at the root of a thorough understanding of the structure of an urbanized society, whose characteristics are based fundamentally on the fact of mobility. Such study is likely to be one of the most rewarding aspects of research in the social sciences. Significant advances have been made in Britain during the last five years by such bodies as the Association for Planning and Regional Reconstruction, the West Midland Group on Post-War Reconstruction and Planning, by Max Lock in his direction of the civic surveys of Hull and Middlesbrough, and by Thomas Sharp in surveys of Durham and Exeter. There is need, however, for more research workers with adequate training, for the development of sound techniques of observing, recording and mapping data, and for the standardization of terminology. In all these respects much can be learned from parallel researches in the United States during the past twenty-five years.

The volume falls into four parts. Part I deals with the nucleated settlement—the urban settlement as a regional centre. Part II deals with the general features of the structure of the city, both as a whole and with respect to its parts. Part III deals with the concept of the city as a regional focus and with the idea of the city-region; and Part IV with the subject of regionalism, with special reference to the relations between regionalism and the city-region. In Part IV a chapter on Germany has been omitted since it is dealt with more fully in my book on the *Regions of Germany* in this series. Eight chapters dealing with individual and comparative studies of European cities have been omitted and these, it is hoped, will finally appear as the basis of a second volume.

Since the work was ready in its first draft at the outbreak of war, the task of redrafting at long intervals when fully occupied with war-time duties has been especially difficult.

It is with sincere thanks that I acknowledge the assistance of several colleagues in the checkered stages of its preparation. Professor Eva G. R. Taylor gave most useful advice and stimulating criticisms of the first general draft. Professor E. G. Bowen also made a thorough reading of the whole and offered comments on the arrangement of each chapter. In its penultimate stage, chapters were read and useful suggestions made by Mr. P. R. Crowe, Mr. E. W. Gilbert, and, above all, by Mr. J. N. L. Baker, who read Part IV and redrafted Chapter 11. Mr. Baker made helpful suggestions and criticisms in a careful reading of the final page proofs. Mr. W. G. East and Miss Phyllis Olsen have also kindly read through the final page proofs. While gratefully acknowledging this help, I am responsible entirely for both the final form and content of the book. I have used English spelling in all quotations from American sources.

ROBERT E. DICKINSON.

October, 1946.

PART I

THE URBAN SETTLEMENT AS REGIONAL CENTRE

CHAPTER 1

THE REGION AS SOCIAL UNIT

1 The Social Unit

The terms region and regionalism have been widely used in recent years with reference to a great variety of problems. All demand, in one form or another, the substitution of new geographical units for existing administrative units, which, as a legacy of the past, are quite unsuited to the requirements and structure of modern society. Regional planning deals primarily with the physical planning of town and countryside, and the term is generally used in reference to an extension of town planning ; indeed, in France it is called Urban Regionalism. It may include the general planning of resources, as in the organization and work of the Tennessee Valley Authority (T.V.A.), and this aspect is becoming increasingly prominent in the planning discussions of other countries. Regionalism is also especially identified with the movement for the re-organization of local government and the devolution of administrative and legislative powers. The outstanding instance of all such aims being put into practice is in the total reconstruction of the economy and social structure of the U.S.S.R.

The term " region " is undoubtedly one of the catchwords of our day among both popular and scientific writers. To the practical man of affairs a region is just an area with certain characteristics (often mere size), in virtue of which it is a suitable unit for some particular purpose of business and administration. To the scientist, and above all to the geographer, a region is an area which is homogeneous in respect of some particular set of associated conditions, whether of the land or of the people, such as industry, farming, the distribution of population, commerce, or the general sphere of influence of a city. The main problems of regional study lie in the selection of suitable criteria for the recognition of such regional homogeneities, their representation, and the elucidation of the forces that condition their formation. Some writers conceive of " natural administrative units ", i.e. units suitable, in virtue of their being social and economic entities, to be used as administrative units. In this sense, a region is envisaged in general terms as a " natural " areal unit, natural in the sense that it is a real, existing unit, arising spontaneously from the very structure of society, in contrast to the " artificial " administrative units, which have been imposed in the distant past and are often ill-adjusted to modern needs.[1] Such a natural unit, it is argued, is the rational one to use as the basis for the organization of modern communities for any particular purpose—be it for planning town and countryside, developing resources, organizing a new system of local government, collecting statistical and census data, or for the regionalization of such public services as health and housing. In all these respects new unit areas are required, differing certainly in character and extent for particular purposes, but better adapted to their purpose than the existing antiquated administrative units. Moreover, assuming that different areas are used for distinct purposes, it is essential that there should be as close geographical co-ordination between them as is practicable.

The essence of this problem has recently been summed up in *The Times*[2] with reference to Britain, and this is typical of the problem in other countries. There is need, states a Special Correspondent, for a new " local unit " of self-government larger and more homogeneous than the parish in the country and the ward in the city, in which there are " enough persons with common interest to form a political community ". Its boundaries should be so defined as to " take account of the natural geographic area, of the historic area, of the economic area for the services, of the financial area ", so that the areas " do not substantially depart from the boundaries of natural community and corporate interest ". The same correspondent goes on to discuss

[1] It should be made perfectly clear that " natural " in this sense is the equivalent of " real " and has no necessary connection whatever with the idea of a unit defined by geographical features of the surface of the land. P. Foncin, a French regionalist, stated many years ago in *Les Pays de France* (1898) : " Notre but précise en ce moment est de rechercher non pas tant de régions naturelles pures, que de provinces administratives qui soient naturelles, ce qui n'est pas en fait la même chose." Geographers have caused confusion by their use of the word " natural ", in the double sense of *physical* in Nature and *real* in human affairs. The term " natural ", as used in this book, refers always to the tendency to the formation of *real* units in human affairs that emerge through areal associations in the structure and organization of society.

[2] *Local Government*, three articles, October 5, 6 and 7, 1944.

the need for intermediate authorities between these local government areas and the State. Health, education, town and country planning—to name but the most important—need areas larger than those of the smaller local government areas. Indeed the country is riddled with such sets of so-called " regions ". But what is required is a comprehensive " unit of planning " as an area of intermediate government larger than the local government unit, and its administration should be in the hands of " a joint or federal authority of which the local government authorities would be the constituent members ".

Such intermediate areas (he continues) exist as *ad hoc* areas governed by *ad hoc* bodies, but this current tendency " breaks into fragments the field of intermediate government ", increases the hold of bureaucracy, and reduces the field of responsibility and interest of the local government authorities and thus of the electorate. " The trend, therefore, to a multiplicity of units must be reversed, and the lessons of the last two centuries learned again."

> To sum up, the future of local government must be seen in association with new forms of intermediate government. Local government cannot itself sustain the role of large-scale scheme-making regional authority without departing from its own proper nature. Local authorities cannot be " upgraded " into huge primary units of regional administration without ceasing thereby to exemplify the self-government principle which gave them birth. The field of intermediate government must rather be integrated by new forms of intermediate authority. It is suggested that these should be federal in composition and, so far as possible, compendious in function.

This is the essence of the problem that faces all the larger countries of Europe and the United States. The concern of this book is not its political aspect, but rather the natural fabric of community-interests upon which the delimitation of new local government and major federal units of intermediate government must be based.

Many of the most vital problems of modern society find their common ground in the basic concept of the Region—what kind of area it shall be, what purpose it shall serve, how it should work. Regionalization as a fact, if not Regionalism as a movement in the French sense, is a fundamental feature of the organization of our national life. Business has long recognized its indispensability. From the standpoint of government, each of the countries of Western civilization is faced with two main problems of internal reconstruction and planning, alike centred in the problem of Regions : first, the need for a new hierarchy of local government areas ; and secondly, the need for new areas for the planning of town and countryside, not as so many isolated fragments, but as parts of a complete nation-wide pattern, comprising a number of organized units. It is upon these two aspects of regionalism that public attention has been concentrated in recent years in all the Western countries. In the words of Lewis Mumford in his *Culture of Cities* (London, 1938) : "The re-animation and re-building of regions as deliberate works of collective art, is the grand task of politics for the coming generation." All these varied problems have a common denominator, namely, the demand for a new unit in place of the existing local government unit, one which shall as far as possible have a social and economic foundation, with the contained settlements as nuclei of life and organization. The same basic idea is found in the notion of the "social unit" indicated by the late Mr. Frank Pick as the essential basis of physical planning. In order to plan for the future, he wrote, it is essential not only to undertake demographic surveys, but also to arrive at "some understanding of the social unit upon which democracy is to be built". For this purpose

> the integration of society demands a special study. . . . A *social unit* must be devised—rather must come to birth—not too large to destroy personal contact and not too small to fail to afford variety and diversity. And the social unit must involve all classes and carry within it no class distinction. How much preliminary thought is needed here, for if the unit is not rightly and naturally conceived, the social structure will never be securely built up. The town, the city, the metropolis itself and finally the region will be aggregates of social units differentiated and combined to fulfil ever higher and broader conceptions of the good life.[1]

This is a timely restatement of the concept of Regionalism which had been elaborated frequently on broad philosophical lines by such scholars as the late Patrick Geddes and, in more recent years, by Lewis Mumford, but which still requires much more attention from the social scientist. It is essential to realize that the community unit is a geographical area with a considerable measure of unity in its services and organization, surrounding and including a focal settlement in which these services are integrated. It is, in other words, an *area of common living*.

[1] Frank Pick, *Britain Must Rebuild*, Kegan Paul, London, 1941.

2. The Hierarchy of Social Units in Theory

The concept of the community group occupying a geographical area of a certain size and enjoying the facilities and amenities that are essential for a healthy social life is prevalent in recent discussions on planning and is a main basis of the *County of London Plan* (1943). Inherent in this concept are a service centre in which the social institutions are clustered, a population adequate to support these central institutions, and a residential area grouped around and within easy access to the centre. The Royal Institute of British Architects in its recent publication *Rebuilding Britain* (1943) suggests a hierarchy of social units at which the planners of town and country should aim as their practical yardsticks—in respect of the size of populations of community groups, the density, type and arrangement of buildings, the general lay-out, and the kind and size of service of institutions to be provided—including under this head not only social or public services but commercial and other facilities as well. A *residential unit* of 1,000 people is suggested as the smallest unit, supporting a small café, a public house, a nursery school, a crêche and a few retail shops for everyday needs. The next grade of unit area would be the *neighbourhood* unit with 5,000 persons and containing five residential units. Its centre would have a few more shops for occasional weekly demands, a restaurant, places of worship, a library, community centre, medical centre and schools. The *borough* unit would contain in its centre all the essential amenities for a fully-fledged town—a theatre, cinema, hospital, specialized shops, public hall and an accessible railway terminus, and it would serve about 40,000 people containing eight neighbourhood units. The *district* unit would be made up of six borough units with, in its centre, specialized services such as technical schools, exhibition and concert halls and special hospitals. It would have about 240,000 persons, the whole forming one urban aggregate. Larger cities, it is stated, may be a combination (as in fact all the big cities are) of several district units, separated from each other by green belts with a central city area for business, finance, entertainment, and administration. An essential feature of this community structure, as suggested in the *County of London Plan*, is that the community area should be compact, the neighbourhood unit, for instance, with its 6,000 to 10,000 inhabitants, being housed on 50 to 100 acres. The size and area of this unit are based on the optimum size of a junior elementary school reached by children without danger of being run over on main roads.[1] Moreover, to ensure the cohesion and unity of the community most of the service institutions should be centrally located in the area, while the main highways, railways, industrial quarters and open spaces should not cut across community areas, but form the physical boundaries between them.

The same approach is advocated by the R.I.B.A. for town and country planning. As in the case of cities, it is suggested that there is a hierarchy of units in the wider countryside,

> beginning as the village grouped around its social services, then the market town which is the focus of several villages and which provides more complicated services, and so on, up through various stages. Much of this structure is actually in being already, but it needs new sorts of buildings, and conscious guidance with national, regional, and local plans if it is to yield the maximum benefits all round.

The ideal hierarchy of community associations, centred in village, town, city or city sub-centre, is not to be thought of as something drawn out of the blue by the planner or the architect. It does really exist in the fabric of our society, and the geographical structure of this society must be thoroughly mastered if we are to discover and rectify its maladjustments and to elaborate principles of planning in accordance with its needs. Consequently we need to know the geographical, i.e. areal, disposition and interrelations of existing neighbourhood relations, for example, the service areas of church, school, and shop. Such relations are complicated and their areas overlap widely, but they are integrated into what may be called socio-economic units through the medium of the central service centre, be it village, town or metropolis. A great deal has been written on these general features in both America and Europe and the main facts are common knowledge. The social sciences must now investigate in selected areas the actual character of the warp and woof of community relations in rural areas, the interrelations of town

[1] Professor C. B. Fawcett, in a recent pamphlet developing this idea, concludes that with a possible range of 20 to 40 pupils per one-year class (rather than on the size of the school as a whole) such a unit should have 1,200 to 2,400 inhabitants, and, if the pupils are to be within a quarter of a mile walk from the school, it should cover an area of 25 acres. Attention should be drawn, however, to the fact that the term "neighbourhood" has a very definite connotation from the standpoint of the American sociologist. It is the smallest social grouping above the family and is comparable, both in fact and theory, to the residential unit of the *County of London Plan*. C. B. Fawcett, *A Residential Unit for Town and Country Planning*, University of London Press, 1944. See also Chapter 3 below.

and countryside, the neighbourhood structure of the great urban agglomeration, and the range of influence of the metropolitan city over the towns and country round it. These are the geographical aspects of the social and economic structure of society, and they form the subject-matter of this book.

3. The "Ideal" Political Unit [1]

There is a dual approach to the concept of the new political region, whatever the purpose for which it is created. In one view, a political region is imposed on society's structure as an arbitrary area with hard and fast boundaries, on either side of which there will be, in consequence, sharp contrasts in social organization and relationships. Alternatively, there is the concept of such a region as something inherent in the structure of society—an area of human associations which we know to exist, although it cannot be defined by exact boundaries. Since boundaries there must be, the need of the future is to establish political regions in such a way as to harmonize as closely as possible with this existing regional fabric of society. And this applies to all grades of "region", for the idea of Regionalism is as applicable to the village and its community area as to the major area in the State with which the term is more commonly identified. It is our aim in these pages to examine some aspects of this "natural" regional fabric, and to emphasize the overwhelming importance of the nucleated settlement, whether village, town or city, either singly or in groups, in effecting the areal integration of the activities and organization of society into geographical community units. We shall also suggest how such units and (more fundamentally) the principles underlying their growth can be utilized in the study of society and in the effective regionalization of local government and administration. In other words, we are concerned with the study of scientific regionalism and its application to current problems.

Definitions of the meaning of "region" as substitutes for the existing political divisions are legion. But all, whether applying to a small section of a town or to a major division in the State, have

[1] The term "political" here implies a geographical unit that is employed for purposes of organizing and administering the political life of the community. These political activities are broadly concerned with matters of public government, and these again fall into two categories, first, self-government on a democratic electoral system, and secondly, administration, normally as a means of devolution from the seat of central government. The functions of local government that fall to the hierarchy of local government areas are varied, but the term "political" is used in preference to "administrative" since it has a wider meaning.

one thing in common. The region is defined as a homogeneous social unit by recognizing the space structure that is inherent in society. The ideal political region, either large or small, has been defined as one which has a maximum number of common interests. As explained by Cole, regionalism is " an attempt to define areas which are at once units of social feeling and, as far as possible, also areas of economic life, and suitable to serve as units for the work of administration." [1] A region as an ideal unit for the purposes of town and country planning and local government has been defined as " an area unified by common economic and social purposes, large enough to permit a reasonable adjustment of necessary activities to subareas and small enough to develop a consciousness of community aims ".[2] From the standpoint of planning on a nation-wide scale in the United States a region has been defined as " a geographical entity as well as an administrative device ", and " a region which may serve economic, social and other needs will coincide closely with actualities of human and physical regionalism, for a region which will function effectively cannot be established solely by edict or enactment (without delimiting factors) ".[3] From the standpoint of the practical planner, states the same authority, " the region should be a convenient device to keep planning problems and functions within manageable dimensions ".[4] The " natural area " as defined by sociologists, economists and, in recent years, by planners, is the social space-grouping that emerges in the city through the operation of unplanned (natural) forces as distinct from the arbitrary existing divisions. All these definitions seek, at various levels, a geographical unit that is a homogeneous social unit.

Professor E. G. R. Taylor writes :

> The concept of a region as a well-integrated whole, characterised as regards all its parts by first-class accessibility to and from the regional capital, and well balanced as regards its resources, economic development, commerce, culture, and occupations is one that so far has not spread outside the ranks of the geographers and the planners.[5]

This conception of the ideal political region contains about six

[1] Administrative convenience is usually the sole criterion accepted by local government officials, and the idea of a social unit is completely foreign to the majority. G. D. H. Cole, *Social Theory*, p. 161.

[2] T. K. Hubbard and H. V. Hubbard, *Our Cities To-day and To-morrow*, Cambridge, Mass., 1929, p. 47.

[3] National Resources Committee, *Regional Factors in National Planning and Development*, Washington, D.C., Dec., 1935, p. 155.

[4] Ibid., p. 157.

[5] E. G. R. Taylor, in *Ground Plan of Britain*, issued by The 1940 Council, 1942.

different basic traits in terms of which, therefore, one might seek its geographical definition. But a homogeneous region cannot be defined, with regard to its general location and its boundaries, on such a composite basis ; some basic interrelated set of criteria must be chosen as a yardstick for the definition of a " well-integrated whole ". In fact, Regions may be defined which are homogeneous units in respect of any one of these sets of facts, and each may have its use for the business concern or for administration or research. Professor Taylor's definition envisages an area which is homogeneous in respect of the group consciousness, organization and interdependence of its people and is distinct from surrounding areas. The key to this summation in the advanced countries of Western Civilization is to be found in the organization and mobility of society. The nucleus of this organization and the focal point of this mobility is the town or city which, as a " capital " or dominant centre, is the most important welding force in the activities and interest of the territory around it.

4. Regionalism and the Region

Much has been written in elaboration of this concept of a Region.[1] The idea has been associated particularly with geographers, and in some recent discussions it is still assumed that the geographer's prime concern is to examine similarities in social structure only in so far as they are attributable to a uniformity in the character of the land and its physiographic features. This limitation is now clearly admitted to be scientifically unsound, for an area of common living can be defined only in terms of the key traits of that common living, that is, in terms of social considerations, not of a particular set of physical factors which condition that pattern of living only in part. This mode of approach to the study of the structure of modern society is akin to recent studies of primitive cultures in relation to their habitat. It throws the emphasis on the study of regional associations, and recognizes that with a marked degree of regionalization there emerges an area of common living, clearly defined in its core, vaguely defined towards its borders, usually where it merges into adjacent regions of similar definition. Thus, " the culture economy of regionalism " takes into account " the whole phenomenon of the new mobility of people, the migrations to

[1] E. A. Gutkind, *Creative Demobilization*, Vol. I, *Principles of National Planning*, The International Library of Sociology and Social Reconstruction, Kegan Paul, London, 1943, pp. 226–39.

and from cities ", and it differs from " the pure geographic area in that it is characterized not so much by boundary lines and actual limits as it is by flexibility of limits, by extension from a centre, and by fringe or border margins which separate one area from another ".[1] There is no need, however, to emphasize unduly this matter of the vagueness of the territorial limits of regions. In practice, as will be shown, there are many occasions when a regional limit is clearly defined, as, for instance, when it corresponds with a political frontier of old standing or with a marked physical divide. Changes in population density and economic structure or the keen competition of two adjacent cities may also give rise to a clearly marked cultural divide or what may be called a steep culture gradient.

The region as a social unit thus demands quite a different approach from that of the earlier geographical determinists. Its character and boundaries must be examined in terms of those social phenomena that are the keys to the social structure. Criteria suggested by Gutkind [2] are the intensity of economic intercourse, as reflected by the interchange of goods as between one district and another, banking and credit relations, communication and accessibility ; and cultural elements—common religious ties, common traditions, the influence, past and present, of a dominant city centre, similarities of habits, standards, knowledge and skills. Tests of regional homogeneity, states an American authority, may be applied " by examining a few factors which would give a clue to the economic similarities and dissimilarities, to political cleavages or cohesion, and to the general cultural likenesses and differences ".[3] In addition to the basic considerations of agriculture and industry, three particular measureable criteria are suggested by the same authority—the volume of movement of vehicles, the more important banking ties, and the service areas of large mail-order houses. Other possible tests are the sphere of influence of educational institutions, as measured by the distribution of students' homes, religious affiliations, long-distance telephone communications, and the circulation of newspapers.[4]

The fullest measure of regional homogeneity, in this sense, is

[1] H. W. Odum and H. E. Moore, *American Regionalism, A Cultural-Historical Approach to National Integration*, Holt, New York, 1938.

[2] Gutkind, op. cit., pp. 234–5.

[3] Ibid., p. 235, quoted from the Report of the Pacific Northwest Regional Planning Commission to the National Resources Committee. See Chapter 12, p. 307, for quotation and further discussion of this view in its relevance to the Pacific Northwest.

[4] Ibid., p. 235.

to be discovered not by considering each of these regional associations *per se*, but through the analysis of the character, intensity, extent and interrelations of them all and the ways in which they are interlocked and separated from each other in space. Clearly this approach demands a central regionalizing principle that lies behind the spatial structure of society and is most efficacious in determining its space associations. This is undoubtedly to be found in the nucleated settlement, whether it be village, town or city.

To sum up, if a region be regarded as a geographical association of human space relationships, then while such a region may be defined from many points of view, as an area with, for example, the same type of farming, the same type of industrial structure, the same type of culture or language, the chief factor in the integration of the life and organization of society into such regional associations remains the settlement centre, be it the village, the town or the great city. The analysis of the functions of these centres and the areas which they serve, provides the framework for the study of the regionalisation of society with its resultant problems and for the readjustment of the administrative divisions to modern needs.

While the rural community finds its needs and services met by the town, the highest seat of culture is the great city. The city has in all ages been the symbol of civilization, and in this era, probably more than in any previous one, the city dominates and leads the life alike of the countryside and of the towns which serve it. This conception of the integration at successive levels of human circulation as the most potent regionalizing principle in the life and organization of society is basic to the whole framework of this book. While it involves, above all, what are commonly regarded as economic circulations, it also embraces social, cultural and political associations viewed in the light of their historical development. The problem of such an approach, involving a composite of factors, is to find a common basis of assessment for regional circulations. A region as defined or conceived on this basis is an area of interrelated activities, kindred interests and common organizations, brought into being through the medium of the routes which bind it to the urban centres.

5. The Regional Capital

The idea of a regional capital or *chef-lieu* was defined by the French advocates of regionalism. Such *chef-lieux* were, in fact, regarded as the capitals of the geographical regions proposed as new administrative units. The great French geographer, Vidal de la Blache, defined the regions which exist in modern France on the basis of their history, culture, and, above all, orientation towards a central capital city, the Regional Capital. He wrote : " It is not the number of its inhabitants, still less the number of its functionaries ; it is not even the type of occupation which constitutes the regional capital. It is a superior element which enters into all aspects of its activity." [1] Such a city is termed by French writers a " natural capital ", which grows to importance without the intervention of high authority, in mere virtue of its favourable geographical position, and the enterprise of its people, as the *chef-lieu* of a large surrounding area, which, in the same sense, may be regarded as a " natural " area, cutting across the administrative divisions, and bearing little relation to them. For this Region the city functions as a Capital, despite its lack of political status. British writers have attempted to divide this country similarly into its " natural administrative units ".

Let us examine further this idea of a regional capital. Professor R. Blanchard writes as follows :

> It applies to a city which owes its importance to its population and its prosperity, to its antiquity and its historical reputation (and generally both causes together), and which is, as it were, the head of the region. In it is established the central (state) authority ; from it are issued orders and decisions ; in it are centred administrative and judicial affairs, and the seats of tribunals and major administrative functions. Thus, it is a political capital. But the capital can have, and should have, other roles. As the head of a region, it must be guarded and defended, for it cannot fall with impunity into the hands of an enemy ; it should, moreover, possess the authority to execute the will of the central State. The capital, then, has a military role. Intellectual influence, moreover, is almost an indispensable function on behalf of the region, whether it be through the medium of newspapers and other publications, or through educational institutions. Finally, the capital has an economic role, especially as it is almost always this economic importance which has brought in its train political pre-eminence. It is a centre of supply of foodstuffs for the surrounding regions ; it is also their market ; it directs their expansion ; and these influences make themselves felt over a more or less wide area, the extent of which depends at once upon the commercial facilities presented by the country and on the degree of economic activity of the city, compared with that of neighbouring cities.[2]

[1] Vidal de la Blache, " Les Régions Françaises ", *Revue de Paris*, December, 1910.
[2] R. Blanchard, *Grenoble : Étude de Géographie Urbaine*, 2nd ed., Grenoble, 1935, pp. 205–6.

In its economic role, he continues : " It is, for a more or less extensive region, the city from whence come directions, in which there is the financial capital, and through which transactions are effected." [1] Its commercial role takes various forms.

> In it are effected the buying and selling for the region which it commands. On the one hand, the capital serves as intermediary between the production of the region and its demands from the exterior ; on the other hand, it sells to the region and supplies it with what it cannot produce, whether the city produces these goods itself, or whether it has them brought from outside the region. Finally, on the routes which cross in the capital there pass in transit both the products going from one part of the region to another, and goods coming from or going to places outside the region.[2]

6. The Metropolitan Concept

This conception of a regional capital is practically identical with that of the city as an economic metropolis, regarded, that is to say, as the headquarters of modern economic organization of society, which has been elaborated by certain American economists.[3]

The highest grade of city which serves as an outstanding centre of human affairs is thus termed the economic metropolis ; the area which is dependent on it, its metropolitan area ; and the type of organization which sustains it, metropolitan economy. Accepting the definition of Gras, metropolitan economy is " the organization of producers and consumers mutually dependent for goods and services wherein their wants are supplied by a system of exchange concentrated in a large city which is the focus of local trade and the centre through which normal economic relations with the outside are established and maintained " ; [4] while a city becomes metropolitan " when most kinds of products of the district concentrate in it for trade as well as transit ; when these products are paid for by wares that radiate from it ; and when the necessary financial transactions involved in this exchange are provided by it." [5] Such a city will have a population considerably larger than that of surrounding towns ; it will be an independent centre of trade, with a large variety of regional industries and a large wholesale business ; it will be a financial centre ; and, finally, a cultural and administrative centre. A

[1] R. Blanchard, op. cit., 1935, p. 221. [2] Ibid., p. 222.

[3] N. S. B. Gras, *An Introduction to Economic History*, Harper, New York, 1922, and R. D. McKenzie, *The Metropolitan Community*, *Recent Social Trends Monographs*, McGraw-Hill, New York, 1933

[4] Gras, op. cit., p. 186. [5] Ibid., p. 294.

really distinctive feature of the structure of the metropolitan city is the central business district—variously named the core, the City, or the downtown district—in which the centralized services are concentrated. This district is characterized by the great congestion and mixture of its buildings, and by the dominance of shops, offices, warehouses and public buildings, often grouped separately, as well as by its steady lateral expansion, its small night population, and by the great daily ebb and flow of workers to and from the surrounding residential districts. The term "*city-bildung*" is used in German to describe this process of concentration of business in the centre of the great city.

The dominance of the economic metropolis is the basic feature of the organization of modern society, since it arises from that geographical specialization of function which is rooted in cheap and rapid transport. Moreover, the great complexity of our modern civilization brings to the city a further variety of functions which it performs for farms, factories and people around it. Metropolitan economy is a universal feature of modern civilization. It *is* modern civilization. In the past, metropolitanism was confined to a few cities. To-day many cities formerly tributary to the older metropolises are becoming increasingly independent of them. Several cities of Britain, such as Manchester, Birmingham, Newcastle and Glasgow, have acquired a large measure of independence of London, as centres of organization, business and opinion. They have become metropolitan in function and structure. Similarly in the United States, the cities of the Middle West and the Far West have passed rapidly through the phases of commerce and industry in serving their regional markets, and are now in large measure independent financial and cultural centres. It is true, however, that the status of such cities is not equal to that of New York or London, each of which not only serves its region, but is the central economic focus of the State, and at a still higher level is an international economic and cultural entrepot. In Germany the case is different, for the historic metropolitan cities have in many cases long been the capitals of independent states, and have developed and maintained the functions of fully-fledged metropolitan cities without relation to Berlin. Examples are Munich, Hanover and Stuttgart. Others, although they have never enjoyed such political status, are outstanding as economic centres, examples being Cologne and Frankfurt. So strong and constant are geographical values that the structure of modern metropolitan

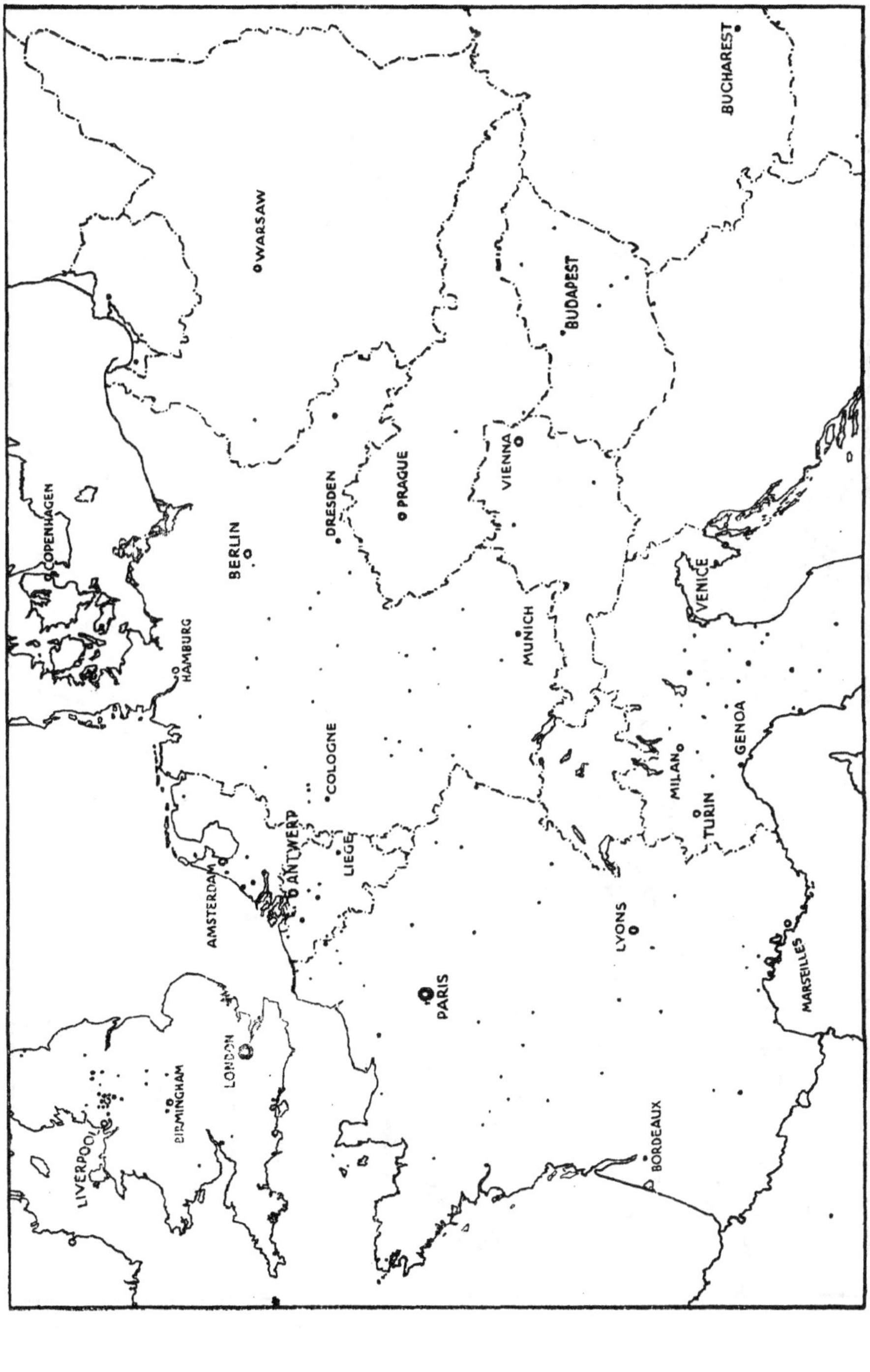

• *Towns with 20,000–50,000 inhabitants* ● *Towns with 50,000–100,000 inhabitants* ○ *Towns with over 100,000 inhabitants*

FIG. 1.—Distribution of Towns in Western and Central Europe, 1830.

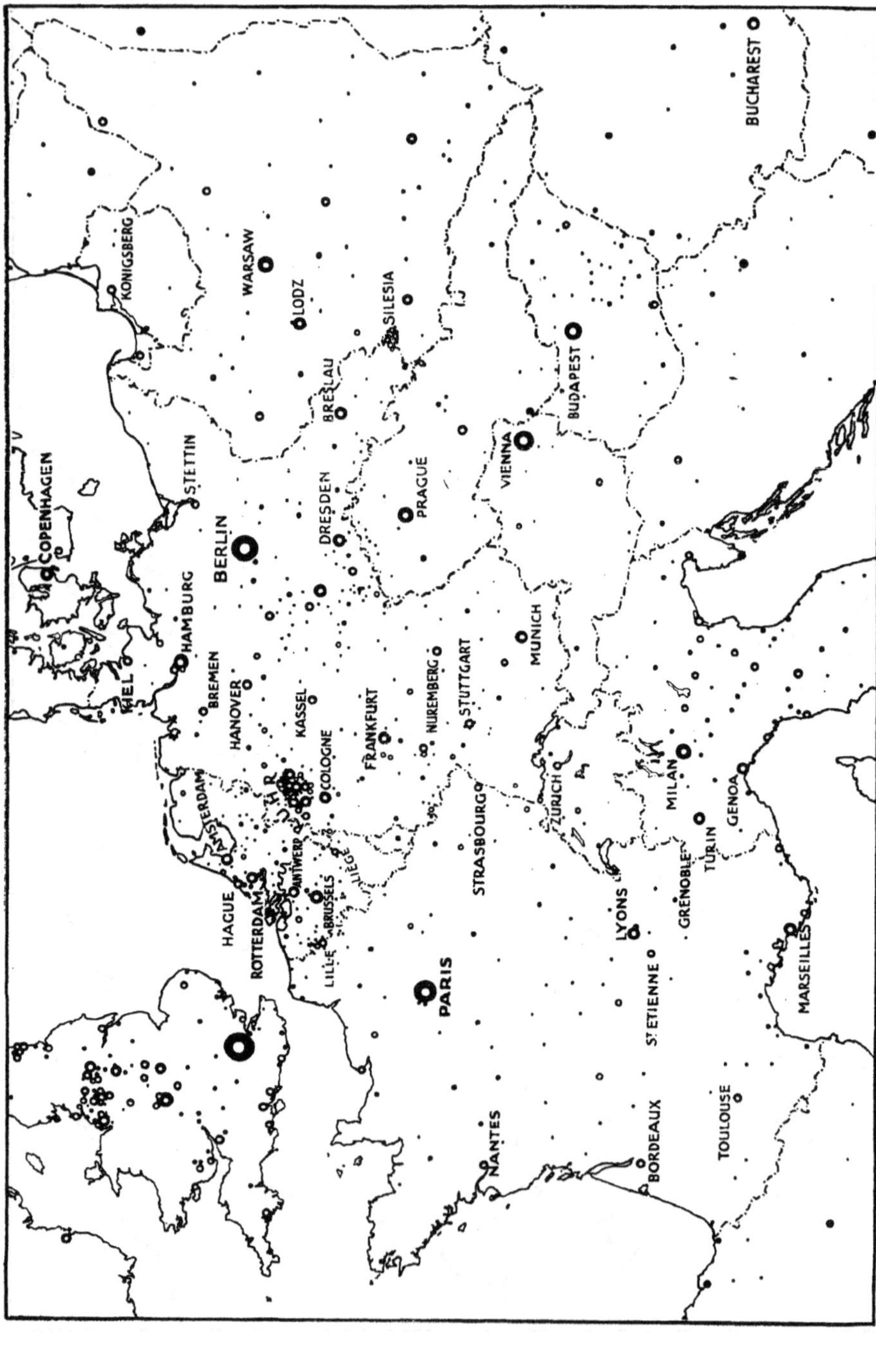

• Towns with 20,000–50,000 inhabitants. ● Towns with 50,000–100,000 inhabitants. ○ Towns with over 100,000 inhabitants

FIG. 2.—Distribution of Towns in Western and Central Europe, 1930.

economy has been superposed upon the pattern of distribution of cities and towns which existed long before the modern era. It is true that the great growth of industrial population has added to their importance as industrial and commercial centres, but, without exception, all the large continental cities that now rank as metropolitan centres enjoyed " capital " status for centuries both as regional centres and as centres for specialized trade and industry.

The structure of the modern city shows two fundamental differences from the city of the past. In the first place, there has been a radical change in the character and complexity of city functions which is commensurate with the change in the structure of civilization itself. Secondly, the absence of rapid transport in the past necessitated the concentration of population in small areas within which all needs could be satisfied and within which all the institutions of town society were concentrated. In a word, centripetal forces determined the structure and spatial distribution of towns. The most fundamental change to-day is the transformation of generalized function in one town into specialization of function by *place*, made possible by cheap mechanized transport ; and the institutions formerly concentrated within the town are now spread over a wide area. " The modern metropolitan community, unlike the pre-motor city, obtains its unity through territorial differentiation of specialized functions rather than through mass participation in centrally located institutions."[1] Centripetal forces still determine the character of both " town " and city, but centrifugal forces have changed the structure of the urban community. The modern city is consequently no longer a compact settlement unit. It is becoming the headquarters of a group of interrelated towns and satellite settlements which yet form one community centred upon the city. This specialization of function, associated with the close interrelations of widely scattered places to form an integrated functional unit with subordinate centres in the towns but with nerve centre in the city, is the essential characteristic of modern society in civilized lands.

Figs. 1 and 2 show the distribution of towns in Western and Central Europe, according to their size and on the basis of their administrative boundaries, for 1830 and 1930. In 1830 the towns were evenly distributed over the land, since they were primarily regional service centres (exceptions were the Low Countries, northern Italy and the English Midlands). In 1930 this same basic pattern of distribution remains, especially in France, but there has been a great growth of population in large cities, and state capitals, and in several great industrial agglomerations, such as the Ruhr and Silesia on the Continent. After A. Welte in *Geopolitik*, 1936.

[1] R. D. McKenzie, *The Metropolitan Community*, 1933, p. 71.

The metropolitan (or city) region thus considered is primarily a functional entity. Geographically it extends as far as the city exerts a dominant influence. It is essentially an extended pattern of local communal life based upon motor transportation. Structurally, this new metropolitan regionalism is *axiate* in form.

The basic elements of its patterns are centres, routes and rims. The metropolitan region represents a constellation of centres, the interrelations of which are characterized by dominance and subordination. Every region is organized around a central city or focal point of dominance in which are located the institutions and services that cater to the region as a whole and integrate it with other regions. The business sub-centres are rarely complete in their institutional or service structure. They depend upon the main centre for the more specialized and integrating functions.[1]

The " city-region " is not to be regarded as a clearly defined geographical unit with sharply defined limits. It is rather a constellation, a cluster of centres round the capital, and the influence of the latter is made evident in its environs by a radiating system of traffic routes, and, further afield, by isolated single strands running to separate towns, each of which in its turn is a local centre of radiating routes through which it, rather than the metropolis, becomes the dominant centre for local affairs. It is in the " suburban " area that the most potent influence is exercised by the metropolitan community, but this influence extends much further over a more vaguely defined " trade area " or hinterland, which, though having little contact with the local institutions and life of the city, provides a wide penumbra to the inner area, and may be included in the concept of the city-region.

7. Cities and Size

The population aggregates employed by statisticians and others are of limited use to designate city character generally, and the status of a city as a regional capital cannot be measured from the number of its population alone. The 100,000 and million figures are now generally used to indicate respectively " large cities " and " super cities " or what have been facetiously called " millionaire cities ". But Clermont Ferrand, for example, like most of the other regional capitals of France, has only about 100,000 inhabitants or less, a size equivalent to that of many one-sided industrial centres in England. The question of size is secondary alike to the density of population and to the degree of concentration of basic industries. There is no doubt

[1] R. D. McKenzie, op. cit., p. 70.

about recognizing the super-city which usually has over a quarter of a million inhabitants and often well over a million. It is normally a political capital, though New York is the great exception, far outstripping in population any other city in the country. Apart from mere size one must look for the head offices of business, banks, stockbrokers, of services, for independent institutions for art and education, and for leadership in ideas. There is a grading in the importance of cities as regional centres of this kind and only a few come into the first category as fully-fledged regional capitals. On the other hand, the larger the city agglomeration and the greater the density of population in its environs the more effective will be its grip on the surrounding area, and the closer the integration of the city and its region.

Among European cities with 100,000 inhabitants (see Fig. 2) there is a clear distinction in the level of leadership as between such small regional capitals and what may be called the "super-city" or true "metropolis", the "primate city", with over 250,000 or usually well over a million people. Small cities though they are, such centres as Grenoble, Nancy, and Dijon, in France, play each the role of a regional capital. Each serves as the leader of the countryside and towns of an extensive tributary region. Each is an historic centre of economy and culture. Some in the past have been political capitals and cultural centres with ancient Universities. To-day, each has a leading Press, a University, and is a pre-eminent centre of wholesaling, retailing, finance and administration. They represent all that a majority of the townsmen and country-folk ever see of city life except for a rare visit to Paris. Yet cities of this small size can scarcely support such activities in very great number or quality, since they serve dominantly rural areas.

In a very different category are the super-cities which, in virtue of their very great size, enjoy all the functions of leadership.[1] Probably the main distinction between one of these super-cities and the smaller cities is that the former is an independent "head office", a heart to a national body. They are financial centres and headquarters of banks, exchanges and big business. In Europe all the political capitals of the present

[1] "All over the world", writes Mark Jefferson, "it is the Law of the Capitals that the largest city shall be supereminent, and not merely in size, but in national influence." This is the "primate city". Indeed, the same writer goes on to declare a law that "a country's leading city is always disproportionately large and exceptionally expressive of national capacity and feeling". Mark Jefferson, "The Law of the Primate City", *Geographical Review*, Vol. XXIX, 1939, pp. 226–32.

States fall into this category. In addition, however, there are many cities of the same order in the provinces, such as Manchester, Cologne, Munich, Lyon, Lille, Milan and at least a dozen in the United States.

Thus, the idea of the regional metropolis is that of leadership over neighbouring tributary towns, and in this respect also the degree of concentration of regional functions is a relative matter. Hence functions of leadership are graded in towns which thus fall into a series, based on their importance as regional centres.

CHAPTER 2

TOWN AND CITY AS REGIONAL CENTRES

It has been suggested in the first chapter that every urban settlement, be it large or small, is in more or less degree a *chef-lieu*, or regional capital. For the very essence of urban character is the function of service for a tributary area. " Cities do not grow up of themselves. Countrysides set them up to do tasks that must be performed in central places." [1] The working and organization of civilized society demand the existence of service centres, and for this reason centripetal forces are fundamental in the localization and structure of the urban community as a seat of industry, commerce, culture and administration. On the other hand, the growth of the city has always been accompanied by expansion, not so much, in the past, of the built-up area—for the houses had of necessity to be clustered as closely as possible around the nucleus—but in commercial and industrial organization. Now, this phenomenon, due to the operation of centrifugal forces, has become of paramount significance with the growth of rapid and cheap transport, so that the built-up area of the city has been able to expand outwards, as well as devolving some of its functions on surrounding towns, which acquire thereby more and more a satellite character. But these centrifugal forces have their complement in the centripetal forces, which are no less significant than in the past. The increasing complexity of modern society demands, as in the past, the concentration of certain functions at nodal points not only in the countryside, but also in the residential districts of the large urban complex, whose centre becomes, in varying degree, the focus for these sub-centres. Centripetal and centrifugal forces, complementary to each other, are thus fundamental to the development, location, functions and physical structure of the urban community in all ages. We are not concerned in this book with the discussion of all the factors, but firstly, with an assessment of the location, structure and growth of the city as a seat of regional integration

[1] Mark Jefferson, " The Distribution of the World's City Folks ", in *Geographical Review*, Vol. XXI, 1931, p. 453.

for the area around it ; and secondly, with the structure of the city's built-up area, when regarded as the result of an ever-active process of functional differentiation.

1. The Town as a Regional Centre

The universally distinctive characteristic of the town arises from the mode of life and activities of its inhabitants. The town differs from the village in the occupations of its people, who are not concerned directly with farming, and who live and work in the settlement, sharing in its life and organization. Rural and urban are not the equivalents of village and town, for the rural community does not necessarily occupy a compact village in the old sense of the term, and a non-rural settlement such as a mining camp is not necessarily a town. True town character implies some measure of community service and organization—what is sometimes called community balance. This is supplied through service and organization, which cater both for the inhabitants of the settlement and for the surrounding countryside. These centralized services must be carried on in central fixed places in order to reach consumers, and they are, and always have been, the very essence of what may be called a town as opposed to an urban settlement in the broader sense. There are considerable areas of brick and mortar to-day which may be called urban, since they are clusters of buildings whose occupants are engaged in non-agricultural occupations, such as mining and factory camps, isolated suburban areas, seaside resorts, but they have no semblance of community organization nor do they provide adequate services—commercial, social or administrative—to meet local needs, many of which must be sought elsewhere. It is this grouping of *centralized services* in a clustered settlement which is the essence of a town, and which, at a higher grade, is the hall-mark of a city. The grouping of these functions in the town is reflected in a certain coherence in the grouping of its buildings—its compactness and orientation around a focal point—be it church, market-place, a particular street, or a city business district. Thus the centralized services are not only concentrated in central places ; they cluster in the centre and form the hub of its life.

The broad principles of the geographical approach to the study of towns have nowhere been more clearly stated than by M. Aurousseau in one of his masterly studies that were published over twenty years ago.[1] As regards the distinction between what is rural and what urban, Aurousseau calls

rural, those sections of the people who are spread over the countryside and are engaged in the production of the primary necessities from the soil, while the dense clusters of folk, who have no immediate interest in the production of the materials for their food and clothing or general comfort, but are engaged in transporting, manufacturing, buying and selling them, or in educating the people, or in managing the affairs of the State, or in merely " living in town ", become the *urban* section.

The power of growth of a town depends on its functional composition.

The primary occupations are those concerned with the functions of the town. The secondary occupations are those concerned with the maintenance of the well-being of the people engaged in those of primary nature. The more primary citizens there are, the more secondary, in a relation something like compound interest. Moreover, there are certain profits especially in the way of amusements, to be made out of both classes. This has to be considered if the limitations of town growth are to be studied.

Since function is the driving force of town life, Aurousseau continues

A town comes into being either at a point having those characteristics of nodality which enable it to discharge that particular function to the best advantage or at a point artificially endowed with nodality. The town will continue to flourish in the discharge of its function until the State finds that it no longer requires the assistance given. The fundamental geographical relationships of towns now become somewhat clearer. In a given State we are able to discern two orders of town, the *active* and the *inactive*. Of active towns there are six classes [administration (the " capital city "), defence, culture, production, communication, recreation]. These terms are used in their widest sense : and, inasmuch as all towns are placed in nodal situations, many are conveniently situated for the discharge of more than one function. There is generally one phase of activity, however, which overshadows the rest.

This approach gives a key to statistical appraisal of town types and for plotting towns over wide areas and studying them in relation to their environs and in relation to each other, as has recently been attempted in an American study.[2]

[1] M. Aurousseau, " The Distribution of Population : A Constructive Problem ", *Geographical Review*, Vol. XI, 1921, pp. 567 et seq.

[2] A recent study of town types in the United States recognizes the following, grouped on a somewhat arbitrary statistical basis : Manufacturing, retail, diversified, wholesale, transport, mining, university centres, resorts and retirement centres. The

Many other studies have been made of the functional classification of towns, and their method is invariably to use the leading industries and services—those upon which the continued existence of the town depends—as a basis. The local and regional services, however, receive scant attention. There is, moreover, a regrettable confusion of criteria in such classifications, oscillating in the same scheme between industry, functional structure, physical or geographical location (break of bulk points and the like), and the actual physical grouping of contiguous or adjacent urban centres.[1] What is needed is a much more careful analysis of the urban community, not only as a seat of specialized industry and service serving a wide market, but also as a seat of industry and service for the " regional " market over and above " local " needs of the urban community itself, and it is not surprising that this consideration has begun to play a more important part in recent studies of the structure of urban communities both in Britain and abroad. The regional functions in sum occupy in many towns, large and small, a considerable cross-section of their working population and of their activity. Such studies will reveal the defects in the structure of the urban community and the nature and degree of its dependence on adjacent and better-equipped cities of a metropolitan kind, and the character of the adjustments needed to give it a more balanced community structure.[2]

605 places with over 10,000 inhabitants each are classed and mapped. The writer adds : " Regional centres (see below) are a clear type of American city ", and of the 41 cities of this type as defined by metropolitan newspaper circulation all except four have important wholesaling functions. In the above classification, 9 are wholesale centres, 19 are diversified, with strong secondary emphasis on wholesaling, 8 are manufacturing cities, and 5 are of other types. Chauncy Harris, " The Functional Classification of American Cities ", *Geographical Review*, Vol. XXXIII, 1943, pp. 86–99. Reference should also be made to the unpublished theses of Dr. Selwood (London) and Dr. K. H. Huggins (Glasgow) on the functional classification of English towns. Selwood reduced the 33 categories of occupations as used in the Census, with some modifications, to the following groups—fishers, land-workers, miners, craftsmen, transport workers, traders, professionals, servants, and clerk store-keepers. For each of the 1,179 administrative units of England and Wales the number in each group was stated as a percentage of the total number of workers and the following types of community were arrived at : mining, craft, transport, service (health), unspecialized or balanced with modifications of these. See *Reports of the British Association for the Advancement of Science*, York (1932), and Cambridge (1938) No fuller published statement of Selwood's work is available.

[1] This is conspicuously the case in The P.E.P. (Political and Economic Planning) *Report on The Location of Industry*, March, 1939, Chap. VI.

[2] The American sociologist, R. D. McKenzie, has submitted a broad classification based primarily on American cities, in which the idea of the town as a service centre is given special weight as follows :—(i) the primary service community is the settlement which effects the first link in the chain between the countryside producing natural products—agricultural products, mineral products, forest products—and the economic metropolis ; (ii) the commercial community, collecting from and dis-

The town in western Europe and North America may be defined as a compact settlement engaged primarily in non agricultural occupations.[1] These occupations fall into two main categories, industry and service. Industry means, in this broad sense, the manipulating or handling of goods, whereas service means the satisfaction of human wants by direct contact with the consumer. The industries are usually classed as primary or basic industries—those which are markedly localized in certain districts and certain places, and are not predicated on the existence of other types of production; and secondary or local industries, which cater for the needs of the working population and for the surrounding towns, such as laundering, baking, public utilities, which are therefore distributed proportionally to the density of population and the local standard of living. The services may be classed as commercial, cultural, professional and administrative. They also fall into two categories, though it is difficult to assess these quantitively. The local services cater chiefly for the townspeople; the centralized services—commercial, administrative and cultural—as well as regional industries—are concentrated in towns in order to serve a widely scattered clientele as well as the townspeople. We need a thorough study of the localization of the centralized services, since they are determinants of the distribution of settlement equal in importance with the localization of industry, and are probably even more vital in the structure of community life. It is implicit in the conception of a regional centre, that all urban settlements exercise the same primary functions, and that the small country town has much the same function as the big city and may be even more important as the centre of local associations for its immediately surrounding district. The big city differs in the variety and quality, rather than in the basic character, of its functions. The local town has direct relations with the capital while the surrounding area has practically all its associations and its organization concentrated in the local town.

tributing to primary service communities, its size depending on the importance of the distributive functions, ranging from a small country town to a great city; (iii) the industrial town which may be highly specialized, but often combines (i) and (ii) with a relative dominance of industry, the latter being of two main types, local and secondary industries, and primary specialized industries; (iv) the places which are not basically endowed with commercial or industrial functions; these include resorts, recreational centres, etc. See McKenzie, "The Ecological Approach to the Study of the Human Community", in *The City*, by Park, R. E., Burgess, E. W., and McKenzie, R. D., Chicago, 1925.

[1] This definition does not pretend to be universal. Many so-called towns of Hungary and South Italy, for example, are primarily agricultural communities, serving as the homes of peasants who have land in the surrounding countryside

The big city has the larger and the directive element in banking and in finance in general, and in wholesaling, education and the Press. Centralization is most marked in the old-established cities and market towns. It is least in the modern specialized communities which have been planted in the midst of the land and have not, as it were, grown out of its needs.[1]

It is difficult to assess from statistics of the Census the importance of these services, but, though inadequate as a complete measure, the numbers engaged in commerce and finance give a ready and sound basis of comparison. In England and Wales there were nearly three million persons engaged in commerce and finance in 1931, out of a total of 18,850,000 occupied and out-of-work, or 15 per cent. The towns with over 50,000 inhabitants fall into several classes. The percentages of workers in the towns employed in commerce and finance range from 10 per cent. to 30 per cent. The great majority of the towns employ over 15 per cent., but a considerable number have only 10 to 15 per cent. This is conspicuously true of the industrial towns of Lancashire, such as Blackburn, Oldham and Bury. High percentages, on the other hand, are found in residential areas on the outskirts of the cities, since these are the homes of many city workers, e.g. Wallasey, 27 per cent., Leyton 25 per cent., and Ilford 29 per cent., the two latter in Greater London. High percentages also are found in the ports and resorts, both coastal and inland, since they have specialized commercial functions, the first as dealers in goods, the second as dealers in services for communities with a relatively high standard of living. The proportion is also fairly high in the great cities. The Administrative County of London has 20 per cent. (City of London 30 per cent.), Manchester 19 per cent., Newcastle 20 per cent., Bristol 19 per cent. Birmingham comes low in the scale with 14·4 per cent., though this low figure is characteristic of its region (Staffs., Warwick, and Worcester with 12 per cent.). Many of the medium-sized country towns with regional and industrial functions have between 15 per cent. and 20 per cent. employed in commerce and finance, e.g. Carlisle, Ipswich, Cambridge and Lincoln. The proportion in the country towns with under 50,000 inhabitants is in general high, e.g. Bury St. Edmunds 29 per cent., Sudbury 20 per cent., Haverhill 19 per cent., Guildford 21 per

[1] Such as mining settlements and residential centres in western countries. In the countries of S.E. Europe and southern Italy the town is often an overgrown village and not in any sense a focus functioning as a *chef-lieu*.

cent., Chichester 21 per cent. These figures give a clear indication of the variations of commerce and finance from town to town, irrespective of size. Even though the great cities have a large population and a great variety of industries, commerce and finance still stand out with a fairly high proportion. Centralized services are also important in the country towns, but they are relatively unimportant in industrial towns and health and holiday resorts. The servicing of the countryside played no part in the origin of a Bournemouth, founded in the 1840's on a heath on the south coast of England as a health resort,[1] of the urban clusters in the coal-mining valleys of South Wales, or of a specialized shipbuilding town like Jarrow on the Tyne. Nevertheless, provided that the environs are settled, for good or ill, the town, in its growth and in catering for its own needs, spreads its net of space relations, through settlement and service.

These comments apply exclusively to the modern town, but they are of fundamental importance in considering the origin and development of the medieval town. A word may be permitted on this complex theme, that must await fuller treatment elsewhere.[2] Until the advent of modern rapid transport, every town had to be a regional focus, because only its surrounding area could absorb the goods of its craftsmen and traders, and, above all, the town had to be fed with both food and men. But it is difficult to unravel the relative importance of local market trade, wholesale long-distance trade, and defence and administration. What is quite clear, as is revealed by numerous individual and regional studies by both continental historians and geographers, is that they varied in importance from town to town, and in groups of towns from region to region. The early medieval town, to which the historian has given most attention, began as a stronghold on a main route, around which clustered itinerant merchants for protection. To the mercantile community, and the local craftsmen who later joined it, were granted rights of self-government, and from this nucleus there had developed, by about 1150, the town that was now known as *civitas*.[3]

[1] E. W. Gilbert, " The Growth of Inland and Seaside Health Resorts in England ", *Scottish Geographical Magazine*, Vol. LV, 1939, pp. 16–35.

[2] See R. E. Dickinson, " The Development and Distribution of the Medieval German Town ", *Geography*, Vol. XXVII, 1942 (two parts), pp. 9–21 and 47–53, and " The Morphology of the Medieval German Town ", *Geog. Rev.*, Vol. XXXV, 1945, pp. 74–97. Both with bibliographies.

[3] The medieval town is defined by Pirenne, in common with most historians, as having three characteristics : its population was engaged in industry and trade rather than in agriculture, it was a community with its own distinct legal status and institu-

It would appear, therefore, that long-distance trade and the mercantile trading colony were decisive in the location and origin of these first towns. But integration with the surrounding country soon became important. The existence of a bishop, count and craftsmen is evidence of this. In the later Middle Ages (after 1200) when, it may be emphasized, probably four-fifths of all the present towns emerged, the town was often primarily a fortress, a seat of administration, a thoroughfare place, or a vintners' village raised to the status of town. Local market trade played a relatively small part in the origin of such towns. They appeared, especially when deliberately founded (as was the case in the German lands) without reference to possibilities of trade, and the endowment of legal status does not make a going concern. But unless such a *civitas* was able to carve out a market area free from competition, from which it could draw supplies and people, and in which its traders and craftsmen could sell their goods, it could not survive as an active urban community. We shall refer to this again when dealing with south Germany in the next chapter.

Let us pass now to a general consideration of the three main functions of a regional centre, the service factor, the administrative and social factor, and the industrial factor.

2. The Service Factor

Human services depend upon direct contact with the consumer.[1] If the service is rendered frequently to individuals of all classes in the community, then it will be located in the midst of its clientele in immediate contact with it. This fact determines primarily the location of those services that must be rendered at the point of consumption, such as doctors, barbers, tobacconists, small general shopkeepers, etc. ; in other words, these are the neighbourhood services that are found in all villages, and are uniformly distributed in urban areas.

This attraction to the consumer market is the ruling force in the location of nearly all types of service industries, but its force is diminished and services tend to be localized at more

tions, and it was a centre of administration and a walled fortress (*Medieval Cities*, Princeton, 1925). This is true of the fully developed town, but the distinction between town and village is by no means so clear cut. The geographical approach, as evidenced by many German studies, reveals that urban settlements differ from region to region according to their environment and historical development. There are towns without markets, villages with walls, and places with the apparatus of government that never became seats of industry or trade.

[1] See H. H. McCarty, "A Functional Analysis of Population Distribution" *Geographical Review*, Vol. XXXII, 1942, pp. 282–93.

widely spaced intervals in accordance with three sets of circumstances. The most efficient size of the establishment, whether a one-man or a large-scale concern, such, for example, as a multiple shop (chain store) or a technical institute, demands a location in a town where it can draw both on the population in the whole town and also on that of its environs. Differences in the intensity of demand for various types of service are also effective in the removal of such services away from the village and smaller town to the larger towns, or from urban neighbourhoods within the big urban areas to the central business district. Thus, the general practitioner may be located in a country town or village, but the brain surgeon is established in a big city. The same fact conditions the location of dealers in high-class goods, such as furniture and jewellery, or specialized retail shops. The ease and speed of transport also tend to reduce the attractiveness of the market for various types of service, and, indeed, every development that permits the service occupations to establish contact with their customers more cheaply and more quickly makes it possible for them to choose locations further from their customers without diminishing the quality of the service. These three conditions account for much of the migration of services from village to town and from the smaller to the larger towns and cities. As a rule, the location of service occupations is consequent upon the location of other types of productive activities, and service occupations in themselves seldom give rise to a settlement. Yet these essential considerations are principal determinants not only of the character and distribution of towns over the land, but also of the character and distribution of such service institutions, together with their various types of segregation, within the urban complex.

The fact that towns of different sizes are fairly equally spaced over the land has often been observed, but not adequately explained. It is clear that certain basic causes must have brought about this distribution, and that the expansion of industry is merely the chief factor in modern urban growth. Before the Industrial Revolution, when the towns were chiefly marketing centres for the handicrafts and farm produce of the surrounding countryside, the centralized services were especially important in determining both the size and spacing of the towns. Then the market town was the seat of services for an area within about one hour's journey by road—a distance of some two to three miles. But in the last hundred years the increasing number and complexity of centralized services, the improvement of roads and vehicles, as well as the great growth of industry and population, have occasioned, as noted above, the concentration of many such services in fewer centres.

There thus emerges a hierarchy of towns, graded according to the degree of concentration of centralized services, which, in considerable measure, is reflected in the size of their population. A German scholar, Walther Christaller, has developed this theme.[1] Working on a theoretical basis, and taking the market town with a service radius of 4 km. ($2\frac{1}{2}$ miles) as the unit area, he has drawn up a scheme of distribution of centralized services which, he shows, is closely borne out by the facts of town size and distribution in south Germany. Theoretically, in respect of the centralized services, a town should serve a circular area (Fig. 3*a*). But towns with the same service status will be

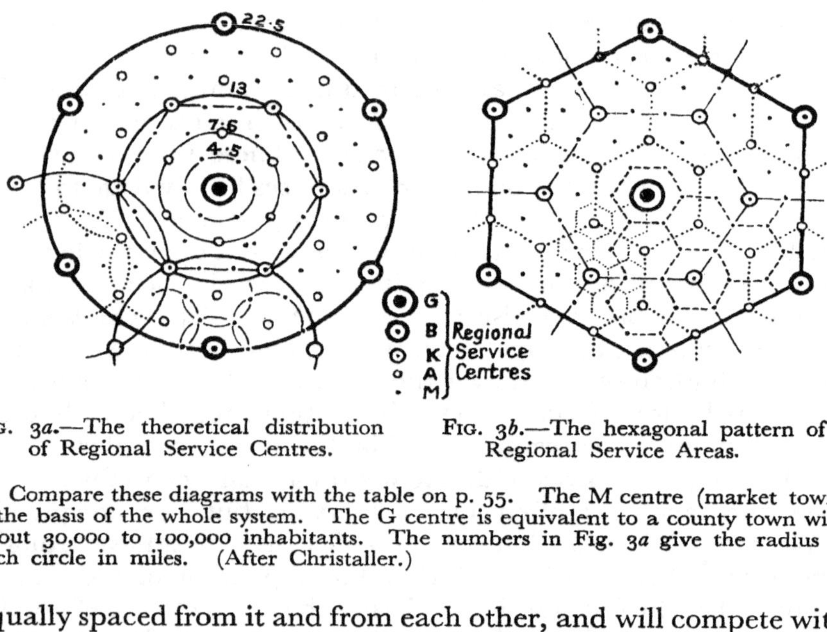

FIG. 3*a*.—The theoretical distribution of Regional Service Centres.

FIG. 3*b*.—The hexagonal pattern of Regional Service Areas.

Compare these diagrams with the table on p. 55. The M centre (market town) is the basis of the whole system. The G centre is equivalent to a county town with about 30,000 to 100,000 inhabitants. The numbers in Fig. 3*a* give the radius of each circle in miles. (After Christaller.)

equally spaced from it and from each other, and will compete with each other in their intersecting border zones, where centres of lower status can supply certain local services more efficiently than the centres of higher order. Thus, one of the latter will be surrounded on the periphery of its service area by six equally spaced

[1] Walther Christaller, *Die Zentralen Orte Suddeutschalands*, Jena, 1933. For a summary, see Christaller, " Rapports Fonctionnels entre les Agglomèrations Urbaines et les Campagnes ", *Comptes Rendus du Congrès International de Géographie*, Amsterdam, 1938, Tome II, *Géog. Humaine*, 1938, pp. 123–38.

centres of a lower order, equally spaced from each other, and from the town in the centre. On this theoretical basis, towns will be equally spaced, in different orders, with hexagonal-shaped market areas (Fig. 3*b*). Since there is a gradation of the services with respect to the extent of the areas they serve, there is a corresponding gradation in the degree of their concentration, which in large measure (and especially before the industrial era) is reflected in the settlement size. But this gradation of service concentration is not gradual, nor does the extent of the composite service area of a town vary proportionally with its size. The concentration proceeds in steps, from which there may be recognized towns of several orders such as are shown on Fig. 3. The smallest seats of centralization are the urban villages and the market towns.[1] In Britain, the smallest urban centres have about 1,000 inhabitants. Such places in East Anglia have specialized retail services and at least one bank. Study of other areas in Britain reveals that this minimum population limit of real urban character holds true throughout the country. Towns of higher orders contain all these services together with those which are more centralized and characteristic of each grade.

The grading and distribution of towns, based on the theoretical distribution of centralized services as elaborated by Christaller, are shown on the table on page 55. This study has special reference to south Germany, and the population figures are averages for south German towns. Christaller assumes that the smallest complete service centre is the country market town situated about 7 km. from its nearest neighbours. Its hexagonal-shaped market area will then have an area of 45 sq. km. with, according to the average density of population in south Germany, a population of 2,700 persons. Centres of the next higher grade will then be spaced at $\sqrt{3} \times 7$ or 12 km. apart with three times the area and population. The distances between each successive grade will increase by $\sqrt{3}$ and the tributary areas and population by three times that of the next lower grade.[2]

Clearly, this fundamental and universal control exercised by services on the functional character, distribution and size of towns is only one determining factor. Even in areas not greatly affected

[1] See Chapter 3, Section 3.

[2] For illustrations of this general discussion, see Chapter 3 on Town and Country Relations. Also see E. Ullman, " A Theory of Location of Cities ", *American Journal of Sociology*, Vol. XLVI, 1941, pp. 853–64, for a summary and criticism of the theory in its relevance to American cities.

by the growth of urban industrial populations, it is modified by such other factors as the relief of the land, the location of routes and river crossings, the distribution of uninhabited land, and the political and economic circumstances of the early medieval development of the towns. Above all, the distribution of modern industry has determined, far more than any other single factor, the actual *size* of the towns if not their basic distribution. It should be noted, on the other hand, that the varying density of population from one area to another does not appreciably affect the spacing of the towns, nor does the size of the enterprise providing the service, since the need for services is universal.

In the nineteenth century came the great concentration of industries in special localities, near to raw materials, or at points where these materials could be cheaply assembled. The overwhelming majority of such industrial centres were simply dovetailed into the existing pre-industrial towns which had come into being in the Middle Ages. In this way the old towns were transformed in functional character and size so that industries became dominant, and the centralized services secondary, in their functional structure. Entirely modern urban communities, such as coal-mining, dormitory, and health-resort communities, which have been planted in the countryside but have no fundamental relations with it, are even more specialized in character. The centralized services, however, together with the predominantly local services, such as distributive trades, building trades, transport services, laundering and confectionery, have increased greatly in the last decades, owing to the rising standard of living of the urban populations, as well as to the general increase in the number and complexity of the centralized services. Specialized industrial towns are deficient in many of such services, which they must perforce draw from a neighbouring and larger city. There are, however, many towns which possess a nice balance of all these functions, because they are especially closely integrated with the surrounding countryside. These towns include the numerous country-market towns with a population from 2,500 to 10,000. Especially characteristic is the county town, which has a variety of modern industries, and is the chief centre for the activities and organization of the satellite market towns in its tributary district. The large cities owe their *raisons d'être* as great urban agglomerations primarily to the concentration of industry, but they have also become, in varying degree, outstanding centres of centralized services, proportional to their importance as the capitals of the economic, social and cultural life of the country and towns around them.

Study of the theoretical distribution of towns according to the service area of Christaller will reveal that this distribution is unsuited to the development of an efficient network of communications, linking the places directly with each other. Commerce demands routes which shall serve best its needs with the minimum cost—with respect to the maintenance of the route and the actual costs of transport. These demands have been met through centuries of trial and error, not by conscious planning. Main military highways and arteries between cities have often persisted

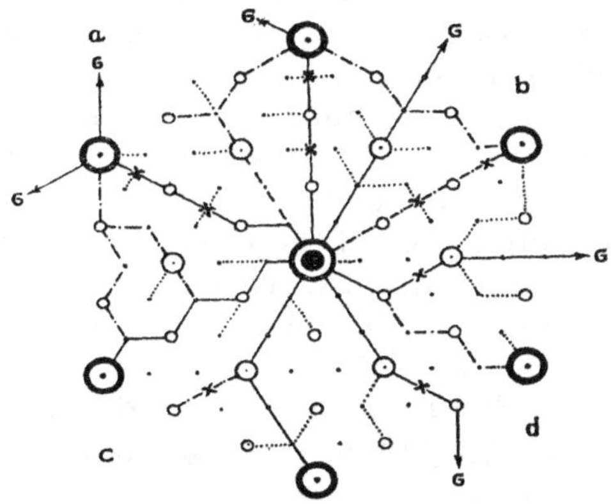

FIG. 4*a*.—Through Trade Routes in the Theoretical System of Regional Service Centres. N.W. sector (*a*)—closely settled area with local traffic dominant; N.E. sector (*b*)—closely settled area with long-distance traffic dominant; S.W. sector (*c*)—thinly settled area with local traffic dominant; S.E. sector (*d*)—thinly settled area, long-distance traffic dominant. (After Christaller.)

since Roman times. Many others emerged in the Middle Ages through linking up short country tracks to form through-routes. From the point of view of commerce, the best theoretical distribution of towns is one which allows the interconnection of the largest number of intermediate places between two large cities along a direct route, leaving the smaller places aside. Moreover, such places should be equally spaced along the route, and the routes should radiate at equal angular intervals like the spokes of a wheel from the central town (Fig. 4*a*). If towns were distributed on a theoretical basis on this principle only, then a considerably larger number of central places of the larger order would be necessary to supply the whole area with the corresponding centralized services. Such a theoretical distribution of towns also reduces the size of the service areas. The latter would lose their theoretical hexagonal shape and be reduced or flattened on the route axes, broader at right angles to them, and larger off the routes. In actual fact, both factors—the market and the route—have operated in determining the distribution, number and size of places in all areas, the relative importance of the two depending on local physical and historical conditions.

Historically, the market factor will dominate in areas where local trade has been more important than through main traffic routes in the development of the towns. The route factor will be more important where through long-distance traffic has played a more important historical role. The service factor will also be dominant in agricultural areas, since a distribution of towns based on this factor offers the minimum number of market centres, whereas only densely populated industrial areas can support the larger number of places required when the route factor is the decisive one. The dominance of the route factor may indeed be dictated by the relief of the land, where routes are confined to single lines through valleys. The basic distinction between the two systems is that one is *areal* in distribution, the other *linear*, and the effectiveness of both in determining the distribution and size of places depends largely upon the geographical and historical conditions in the area.

A third factor which affects the functions and distribution of towns as service centres is politico-social in character, namely, the need for administrative areas with administrative centres, as opposed to the market area. The ideal administrative unit is one which has a capital in its centre with a group of tributary centres of lower order, and a peripheral area of thinly peopled land separating it from its neighbours. The need for defence has been a primary motive in the past in the use of natural barriers as political frontiers, but the divide of trade and cultural associations, which invariably corresponds with a zone of low population density, is still the ideal boundary between adjacent administrative units, be they of the order of States, or provinces or local districts within the State. The farmer will expect to do his marketing, pay his taxes, and use the institutions—such as the courts of justice—in that town which can be most easily reached, not in a town which can only be reached by devious routes and which normally he would not visit. The local administrative area should be determined primarily on this basis of accessibility to a town, and the divide of local trade associations is the best indicator of this. The natural divide, such as a forest or marsh, was often used as a divide in the past, but bears little relation to modern associations. The leading idea in the creation of new units, in the past and no less in the future, is that they should be compact, with a rough equality in area and population, each with its chief administrative centre in its geographical centre, and with its

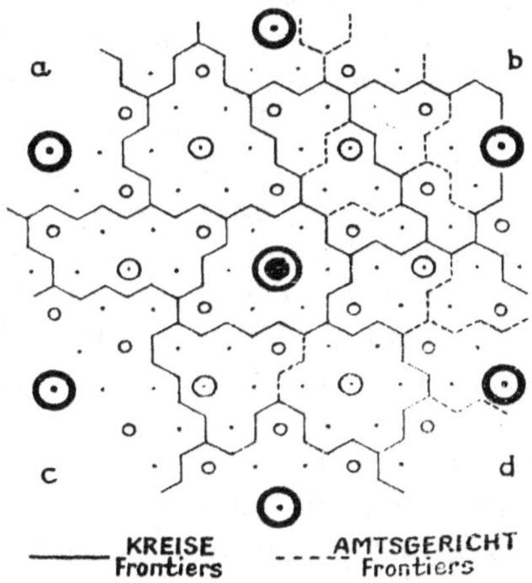

FIG. 4*b*.—Administrative Divisions in the Theoretical System of Regional Service Centres. N.W. sector (*a*)—small *Kreise* without *Amtsgericht* districts ; N.E. sector (*b*)—small *Kreise* with *Amtsgericht* districts ; S.W. sector (*c*)—large *Kreise* without *Amtsgericht* districts ; S.E. sector (*d*)—large *Kreise* with *Amtsgericht* districts. (After Christaller.)

frontiers in thinly populated areas which are divides between local associations. This ideal is reached in part by the considerations of the market factor. But a system based on the latter factor only would always give to the administrative area imposed upon it several places of the next lower order on its borders, which negatives the principle according to which an administrative unit should cut across local associations as little as possible. Given the existing pattern of distribution of towns, new principles will have to be used for defining administrative units (Fig. 4*b*).

3. The Industrial Factor

The relation of the character and importance of industry in the town to the tributary market area served by, and serving it, is a most important aspect of the localization of industry that has received little systematic attention. The role of the regional market in localizing industry in cities is generally underestimated in studies of urbanism, probably for the simple reason that in Britain the big cities are so close together that specialization is possible as between one city and another in respect of what are essentially regional industries, and a city's products are distributed widely in the tributary areas of other cities. On the Continent and in the United States, on the other hand, many factors have contributed to the growth of both industry and commerce in cities in the service of the regional market. Of primary importance is the fact that the historic regional capitals on the Continent and the main commercial centres in the United States, that served initially as collecting and distributing centres, have grown to be great modern cities. Moreover, the great distances between such cities, amounting often to hundreds of miles as compared with tens of miles in Britain, and, in the United States, the system of railway freight rates with " basing points " (see p. 219) at selected towns, have also contributed to the growth of regional functions in the great city. While granting that in one and the same industrial occupation it is impossible to come to a quantitative measure of the relative importance of the nation-wide and international market, the local urban market, and the wider regional market, this is no reason why the last should not be adopted as a main approach to the study of the occupational structure of towns.

The majority of the people of a town are engaged in providing goods and services for other communities, from which they get necessary goods and services from other communities in exchange. These are, according to J. H. Jones, the basic occupations of that community.[1] On the other hand, there are the ubiquitous occupations, both industries and services, that are everywhere proportional to the distribution of population and serve primarily the immediate needs of those engaged in the basic occupations. The basic industries form the foundation of what McCarty has called " the occupational pyramid " of the community.[2]

> The base of the pyramid [in any area] consists of that group of occupations whose presence in the area is not predicated on the existence of other types of production. These basic industries will include most types of agriculture, as well as mining, fishing, lumbering, and those types of manufacturing not tied to sources of materials or to local markets. Occasionally the base will include centralized

[1] J. H. Jones, " Industry and Planning ", in Gutkind, *Creative Demobilization*, Vol. II, *Case Studies in National Planning*, pp. 123–32.
[2] See H. H. McCarty, op. cit.

services institutions such as governmental agencies, insurance companies and universities. In general, however, the basic industries will be concerned with goods rather than services.

The base of the pyramid dictates the pattern of the remainder of the structure. Frequently there must be some types of manufacturing to prepare goods for shipment to distant markets. The basic industries may also require tools and other equipment and supplies, some of which can best be produced locally. On the other hand, the workers in basic industries are not self-sufficing individuals, and the local economic organization must provide them with many types of goods and services including merchandizing establishments, as well as transport facilities, business and personal services, and each of these groups in turn requires workers to care for its needs.

McCarty then goes on to suggest that with the data for basic industries given in a particular area, it should be possible to construct a hypothetical structure of occupations, and to note the difference between the actual and calculated structures. This would offer clues to local economic and social maladjustments or changes in occupational structure that will result from the addition of new basic industries, or from the shrinkage of the basic industries.

The basic industries " export " their products (i.e. send them out from the community), whereas the local services are marketed on the spot. The basic industries fall into three groups. First, there are the industries that are tied down to particular sites by nature—mining, shipbuilding and metal smelting. They are located at the point of production of the coal or mineral, on a tideway, or at a convenient place of assembly. Secondly, there are the industries that have origins hidden in the past—historical accident, early localizing factors that have long since ceased to be operative, the advantages of a going concern, local skilled or cheap labour, or the presence of related industries or services. Typical in this category, in Britain, the Continent and the United States, are the textile industries. These two groups make up practically all the immobile industries that are highly localized in fixed localities. Thirdly, there are the relatively mobile or " foot-loose " industries, many of which are of relatively recent development. These are normally concerned with the production of highly fabricated and fairly light products in which transport costs and the cost of raw materials (that are also small in bulk) are small items in the total cost of production. They produce primarily consumers' goods. These industries have been located at places with good communications, with ready access to a market, near existing sources of labour supply, or near the manufacturers' place of residence. These mobile industries are in part highly localized, as in the case of the automobile and electrical engineering industries, but are normally in the big cities. But there is a vast range of new consumers' industries that are placed in all big cities so as to assure effective cover of the regional market.

The local services are considered by J. H. Jones [1] to occupy from one-fifth of the total occupied population in a highly industrial area to about two-fifths in others. Jones draws a broad distinction between local services and basic industries. The former include " local industries and services, such as transport, electricity, and gas production, retail distribution, municipal services, hairdressing, tailoring and dressmaking, domestic service, law, education, accountancy and religion ". Most of these are essentially local, but some serve a wider clientele. And what is the meaning of local? It cannot mean the administrative area of the city. Some of the above and many other services of commerce, finance and industry are of a local character but, as the turnover of any one firm would show, their clientele is more widely distributed geographically. The proprietor of a newspaper, a retail establishment, a wholesale concern or an insurance firm could (if willing to do so) tell us exactly what proportion of his turnover was effected within the city boundaries, in the surrounding towns and in the wider countryside. He could prepare a detailed analysis of the geographical distribution of his customers from which it would be apparent that many of the local and even of the basic industries have markets covering both " local " and wider " regional " needs, as well as serving the outside world, with the products of its basic industries and services.[2]

[1] J. H. Jones on " Industry and Planning ", in Gutkind's *Creative Demobilization*, Vol. II, p. 125.

[2] The industrial structure of Merseyside, as studied by W. Smith, reveals four main groups of related industries.

(1) Industries and services involved in the working of the port of Liverpool—the only large manufacturing industry is shipbuilding and repairing.

(2) Manufacturing industries working up raw materials imported through the Port, e.g. oil, seed-crushing, sugar-refining and matches, accounting for one-quarter or over of the total in the industry in England and Wales.

(3) Manufacturing industries ancillary to the above, e.g. packing-cases, sack-making.

(4) Services characteristic of a great conurbation, ministering to the industries and to the population. These include the distributive trades, national and local government service, professional service, entertainment, hotel, restaurant and laundry service, gas, water and electricity supply, building and contracting, printing, publishing and bookbinding. These accounted for just half of the insured employed on Merseyside in 1939. This figure is high since some belong to (1) and (2) above,

It will be clear that, in fact, *regional* industry and service may be so important in a city as to constitute basic industries ; that is, the basis of the occupational pyramid may, in fact, be the regional services, rather than specialized industry as is generally assumed. Its functions as a regional capital are then its main *raison d'être.* Industries that have a predominantly regional orientation may be considered from a twofold viewpoint. They may be concerned with the processing of raw materials drawn from the region—lumber, livestock (slaughtering and meat-packing), agricultural products (sugar-beet factories, brewing, flour-milling, canning, etc.), and finishing processes to manufactured products (e.g. dyeing and finishing in the textile industries). On the other hand, both consumers' and producers' goods (certainly wholesaled by import from other centres) may be manufactured for distribution throughout the tributary area—agricultural machinery, fertilizers, hardware, machinery, and the great range of consumers' goods that has grown so rapidly in the last twenty years. This development can be seen very clearly in the case of American cities, which have passed through the three phases of serving first as collecting and distributing centres, drawing manufactured supplies from the East in the mid-nineteenth century ; then developing industry, at first by processing local raw materials and manufacturing certain regional supplies as the consuming market grew ; and finally, in the last fifty years, emerging as fully fledged financial and manufacturing centres independent of the older big cities in the East. Among the universal industries of all big cities are printing, publishing and bookbinding, light engineering, clothing, public utility services, and the building trades. In spite of certain industries of recent development dominating the industrial structure of such cities (e.g. rubber at Clermont Ferrand), it will invariably be found that commerce, administration and public service in aggregate exceed the industrial occupations, which are of a diversified character.

A final feature should be noted with regard to the basic industries in many cities. One can frequently trace the tendency for the factories in the city to close down and move to its outskirts, a process that has been accelerated in the last twenty years,

but it is quite impossible to separate transport services, for example, into the various categories.

(5) A fifth group includes those engaged in the facilities for export which the Port provides. These occupations again overlap with the above and cannot be isolated. See W. Smith, *The Distribution of Population and the Location of Industry on Merseyside*, University of Liverpool Press, 1942.

and has become one of the most significant traits in the development of urban economy to which we shall return later. It must again be emphasized that the big cities of Britain, with the possible exception of Edinburgh, assumed the aspect of industrial agglomerations by the middle of the nineteenth century; they are still industrial agglomerations and are emerging gradually as true metropolitan cities with diversified industry and commerce, whereas on the Continent the regional capital emerged primarily as a seat of regional as well as specialized industry in the last decades of the nineteenth century in virtue of its already being a focus of railways and roads and a historic capital: a going concern (often a political capital) with a great momentum. In the United States, the cities originated in the middle nineteenth century as commercial centres on rivers and then attracted the rail, becoming the foci of converging routes. It was not until the last decades of the century that the big cities of the Middle West developed as industrial centres serving primarily their regional markets, and only since 1900 have they reached the final stage of their development as seats of diversified industry and as independent financial centres.

4. The Administrative and Social Factor

The compact settlement, be it village or town, does not function in a vacuum; it utilizes, in more or less degree, the district around it. It is upon this basic fact that the existing system of administrative areas in all European countries was based in the Middle Ages.

Administrative units in western and central Europe, with remote historical origins, took definite shape in the Middle Ages and gradually became units of group feeling and cultural units at different levels. Social groups emerged from the tribal organization in the Dark Ages, each tribal group occupying a specific territory with its centre in an open, easily cultivated area, and its borders formed by wide zones of uninhabited forest and marsh or mountains. These border zones were not narrowed down until the Middle Ages when the process of forest clearance and the drainage of marsh commenced.

The word "city" is derived from the Latin *civitas*. Its Greek equivalent is *polis*. In ancient Greece and Rome these words embodied the concept of the city-state, that is, of the state as a small unit with a central focus of its life and activities, the whole being not too large in area or population to prevent effective government through the assembled body of its citizens. Gallic tribes and the areas they occupied were called *civitates* by the Romans and the Roman civil divisions were based upon them, the Roman town centre being usually placed on the site of, or near, the preceding Gallic hill-top *oppidum*. The ecclesiastical dioceses were later adjusted to the same framework with the same centres, the latter being called *civitates* in virtue of their being seats of bishops. But, in general, the word *civitas* acquired a much wider meaning, and after about 1150, when the medieval concept of the town was fully developed, *civitas*, the name given to the town, implied a settlement with industry and commerce, a law of its own, and walled fortifications (see p. 27). The French version, *cité*, was used throughout the Middle Ages as applying more or less strictly to the seat of the bishopric. During the Middle Ages, with the growth and expansion of towns throughout western and central Europe, the town became a centre of defence, administration, industry and commerce—although these functions were not always combined in the very small towns that cover most of the countryside. In spite of the splitting up of territory during the era of feudalism and after this period, the town was made *de jure* or became *de facto* the centre for its surrounding local territory. New administrative districts were formed, beginning far back in the Carolingian era, and were remoulded in various countries about 1800, with the towns as their centres. The organization of the church demanded a hierarchy of ecclesiastical divisions with central Bishoprics and Archbishoprics. Defence in the early Middle Ages called for the organization of defined territories (though we seldom know their extent) around centres which served as places of refuge and administration. In the later Middle Ages feudal lords in the German lands in particular established new fortified towns to organize and defend their tattered territories as small compact geographical units. The weekly market was an accompaniment of the medieval town and has remained so until this day, though its function, aspect, and organization have changed. There were laws forbidding the holding of markets within a competitive radius of one already existing—a radius of about four to six miles—although other considerations often outweighed this, for the basic factor of distance to market and the competition of neighbours eventually sorted out the active centres, while those that were superfluous declined.

In the Middle Ages the town as the centre of civilization was a most important factor in moulding areas of group feeling and organization as well as in the formation of administrative units. The towns of the early Middle Ages had their origins as centres of secular and ecclesiastical administration, and as such the dioceses had their central bishops' seats, and the counties had their central *burgs*—places of defence and refuge in which the count was installed as the representative of the king or emperor in order to maintain law and order. The establishment of such burgs, each with a tributary area, accounted for the emergence of the administrative units over large areas of Europe. In England we may instance the foundation of the counties of the Midlands which were established as the Danelagh was reconquered by Ethelred and Athelstan of Wessex.

The town also emerged as the dominant centre for the trade and administration of justice for a surrounding district. In Germany in particular as the new castle towns were established in the Middle Ages by independent lords and dukes, they established administrative centres for small districts, and selected for this purpose an existing town or founded a castle which became the nucleus of a town. The association of the countryside with the local market and administrative centre in the past, no less than in the present, is reflected, for instance, in France by the frequent occurrence of a *pays* or province with a name taken from that of the local capital town, e.g. Mâconnais, Laonnais, Anjou, Touraine, Poitou.

The seventeenth and eighteenth centuries were the heyday of the provincial political and cultural capital. In Germany each of numerous small states had its own capital, with its own court, and all the apparatus of sovereign government. In France, the traditional capital of each province was the seat of provincial administration, the seat of the nobility, of government, of the *Parlement,* all of which, however, were abolished in 1789, when their place was taken by the *Départements,* and the capitals robbed of their administrative functions. In England, the county town became the recognized centre of social life and movement for each county.

The country gentry with their wives and daughters came to regard a visit to the county-town and indulgence in a round of balls and feasts, visits and functions, in the same light as a season in London is regarded at the present date.[1]

[1] Edwin A. Pratt, *History of Inland Transport and Communications,* 1912, p. 94.

The county town was (the countryman's) Metropolis. He was attracted there by business and pleasure, by assizes, quarter sessions, elections, musters of militia, festivals and races. There were the halls in which the judges opened the King's Commission twice a year. There were the markets at which the corn, cattle, the wool and the hops of the surrounding country were exposed to sale. There were the great fairs to which the merchants came from London and where the rural dealer laid in his cutlery and muslin. There were the shops at which the best families of the neighbourhood bought grocery and millinery.[1]

Many important changes were made at the end of the eighteenth century and in the early nineteenth century in the administrative units of western Europe. These changes were made in the heyday of road transport, just before the development of modern factory industry and the advent of the railway. The most important change was the creation of the *Départements* in France in place of the historical provinces. Each *Département* was so defined as to have a central city chosen as its administrative centre accessible to all parts of the *Département* in a day's journey. The historical province was abolished from the administrative system, although the *Département* often shows a close relation to the earlier province, usually by a simple division of one province to form two or three departments. Each *Département* was later divided into *Cantons* and *Arrondissements* defined on the same principle of the accessibility of a central town. In Germany, similar changes took place, partly under the Prussian kings, partly through the reforms of the Napoleonic regime in the southern states. We find, in Prussia to-day, a fivefold hierarchy of administrative units, which came into being at the beginning of the nineteenth century, the *Provinz*, which normally was the direct successor of the old historical unit, the *Regierungsbezirk*, the *Kreis*, and the *Amtgerichtsort*, the successor of the type of unit established in the later Middle Ages as centred on a castle town, while this in turn contained several parishes or *Gemeinde*. The main features of this system are repeated in the other states, although the names of the districts vary. The States of the Reich, referred to as *Länder*, are equivalent to the provinces of Prussia in this hierarchy. It may be noted, however, that many of these states are very small, though they enjoyed for centuries the full powers and status of independent sovereign states. In consequence, the small capital city is a very distinctive feature of the towns of Germany.

[1] Ibid

In Britain, at the opening of the nineteenth century, there were only two types of administrative district—the parish and the county. In 1834 new districts were established for the administration of the Poor Law, and these " Unions " were usually defined so as to correspond with the market area of a central town where the Poor Law Guardians met and the Poor Law Institution was located. In the 'seventies, with the establishment of Rural and Urban Sanitary Districts, the Rural Districts normally corresponded with the Poor Law Union districts, and though the latter have now been abolished, the Rural Districts remain with areas in many ways too small for modern conditions of organization, function and transport. The counties also remain as the historic units without any substantial change, but they too are so small that for many purposes of administration they are grouped to form larger units or new areas are used that fit more closely to modern needs.[1]

Thus, there are three main factors which determine in varying degree the distribution of towns when considered as regional centres, namely, marketing, long-distance traffic, and local administration. The relative importance of each of these in the history of town development varies with geographical and historical conditions. It is important to note that Christaller was seeking the laws that govern the size, distribution and number of service centres. He equates service centre with town. This interpretation is quite inadequate for a study of modern towns, since it does not give enough weight to industry and residence, that is, to the localization of industry and the territorial expansion of the city, as well as to the growth in modern times of specialized residential centres as places of residence, recreation and retirement.

[1] It is of interest to note the comparative size of these administrative units. In Prussia the *Landkreis* and its equivalent in Saxony are remarkably equal in area. Each is compact in shape with a central town and has an average diameter of 20 to 30 km. This is comparable with the French *canton*. The English Rural District, based on the earlier Poor Law Union, has an average diameter of about 15 km., that is, it is of the same order. The *Kreis* in Bavaria is the equivalent of the Prussian *Regierungsbezirk* and is roughly equal to a French *Département*, with a radius of about 60 to 80 km. from its capital. Each of these has about 15 to 20 divisions, each about 20 km. in diameter. By way of comparison, Mayenne, a French *Département*, has three *Arrondissements* and 27 *Cantons*, the diameter of the last being about 20 km. The Bavarian *Kreis* and the French *Département* and the German *Regierungsbezirk* are roughly equal in area to a medium-sized English midland county, such as Leicester or Warwick, which have a diameter of about 50 to 60 km. There is no work that deals with the character and development of politico-geographical divisions and the principles underlying their definition. The reader may be referred to G. Montague Harris, *Local Government in Many Lands: A Comparative Study*, London, 1933.

5. The Functional Classification of Towns

It is because the functions of towns can be approached from so many angles that there is such inconsistency in the basis of classification. Once it is recognized that the town does not function within its arbitrary administrative limits, but serves and is served by a surrounding area, then the functional structure of the town is revealed more clearly as involving a seat of specialized industry, the base of its functional pyramid, with a system of local services built upon it. The assessment of the functions of regional service has figured largely in recent comparative town studies, and several of these must be briefly mentioned.

We have already described in some detail the theoretical system of Christaller and its application in south Germany. Christaller was seeking principles of centralization, and while of great value in establishing a new mode of approach in the geographical study of towns, his system is so theoretical as to be somewhat unreal in detailed application. The method of quantitative analysis is to take, for example, telephone connections. This may give a measure of " centrality ", but states nothing of the actual composite functional structure of the individual town, while the problems of unemployment and regional planning have directed attention in recent years to precisely this appraisal. The difficulties of unravelling the census data in all countries are great and more research is needed. The alternative is to assess empirically the services that are definitely centralized, and grade towns according to their aggregate of such selected key services. On both bases town types may be recognized.

For a broad picture of the functions of British towns with over 50,000 inhabitants the reader is referred to the *Ground Plan of Britain*, prepared for the 1940 Council. Here on the map of " Wheels of Industry " each town is shown by a circle proportional to its population and each circle is divided into three segments showing the proportion of workers engaged in (1) Extractive Industry—fishers, farmers and miners; (2) Manufacturing Industry—factory and workshop hands; (3) Service Industry—transport workers, builders, government servants, professional men, shopkeepers, personal servants and others.

In different parts of the country now one now the other of these three categories predominates among employed persons, and dictates the local view point. The balanced community is far less vulnerable in times of economic depression than the occupationally specialized community, but perhaps it is equally important that such a community is also well-balanced as regards talent, opportunities and thought.[1]

The proportion engaged in services, in this broad sense, is normally between 50 per cent. and 75 per cent. of the total employed. Only in very specialized *industrial* communities does it fall below this figure, to about a third. The higher proportion is reached in specialized service centres, such as ports (e.g. Liverpool, Hull, Southampton), and, above all, in London, in each of which over 20 per cent. of all the employed are engaged in commerce (including shopping) alone. In the same category are the coastal resorts, like those on the south coast, or inland resorts, like Bath or Harrogate, in which the high proportion is due not to the handling and distribution of goods or business and finance, but to shopping and personal service, to the resident population—the holiday-makers, the elderly retired, the invalids and the residents working in distant cities. The "towns" on the outskirts of the cities, containing a large proportion of "black-coat workers" (i.e. service occupations), fall into the same category. Another type of town in the category is found in such places as Exeter, Norwich, York, and Gloucester, which are primarily service centres, with a small but significant proportion in manufacturing—25 per cent. to 50 per cent. Many of the smaller towns fall into this category. They are clearly concerned not only with service to the town community but also, in greater degree than in the preceding types, to the surrounding countryside. The great cities which we recognize as the provincial capitals, like Birmingham and Manchester, have a proportion of about half and half as between service and industry, although in some cases, as in Edinburgh and Newcastle, the proportion in service is much higher. The smaller specialized manufacturing towns, as in Lancashire and the West Riding, Nottinghamshire and Leicestershire, have up to three-quarters of the workers in manufacturing and extractive industries.

Every town must, of course, be analysed as an individual

[1] In addition to showing this threefold division of occupation in each town, there are shown those industries which absorb 20 per cent. or more of the employed population (mining, pottery, chemicals, metals, textiles, clothing, commerce (shopping), administration (armed forces), personal service (hotel and domestic). This predominance of one group of occupations gives the key to the particular character of the town. *Ground Plan of Britain*, Wheels of Industry, Map 10, p. 24. This scheme seems to follow closely the quantitative method devised by Dr. Selwood.

entity in order to assess its real character and to know how to handle its particular problems and to know how and in what degree its industries can be diversified. But the foregoing remarks give a general guide to the functional composition of a town in respect of industry and service.

The typically urban institutions are as follows :

Banks.
Specialized retail shops.
- General shopkeepers.
- Special retailers—draper, chemist, grocer, butcher, footwear.
- Department stores.
- Chain stores, e.g. Woolworths.

Offices—commercial, administrative and professional.
- Solicitors.
- Doctors.
- Dentists.
- Insurance firms.
- Auctioneers.
- Other aspects of commerce and finance.
- Government Departments—Inland Revenue, Post Office, Food Office, etc.

Social institutions.
- Elementary Schools (several in a town).
- Secondary School (one in a small town).
- Cinemas.
- Newspaper.
- Places of Worship of various denominations.
- Assembly Hall.
- Hospital.

Each of these functions is graded in itself, ranging, for instance, from the specialist retailer in the small country town to the great department stores in the city, or from the local office of a Government department to a regional head office in a central city. They also tend to occur in groups at different levels, so that grades of urban settlement may be recognized. This approach was used by the present writer in a study of the small urban centres of East Anglia that will be summarized in the next chapter, and has more recently been extended by A. E. Smailes to cover all the urban settlements in England and Wales.[1]

It is pointed out by Smailes that a group of three or four banks in one centre is the most reliable indicator of its significance as a shopping and business centre, and that with this economic

[1] A. E. Smailes, " The Urban Hierarchy in England and Wales ", *Geography*, Vol. XXIX, 1944, pp. 41–51. Also Chapter 3, below, for the idea in its application to East Anglia. Reference should also be made to our treatment of English cities in Chapter 10.

status there are normally associated a secondary school, a cinema, a weekly newspaper, and a hospital, that express the cultural and social functions of the centre. These five service institutions tend to hang together as a *trait complex*, and they are the key criteria of what may be called a *fully-fledged town*. Other typical features are the professions, branch insurance offices, specialized retail establishments, including multiple shops such as Woolworths in particular, and administrative offices, such as the local Head Post Office, an Employment Exchange and an Inland Revenue office. Many places that are well equipped in some respects as towns are deficient in others. They lack one or more of the five minimum key institutions. These Smailes calls *sub-towns*. Thus, there are many shopping and entertainment centres that lack secondary schools and hospitals ; they are usually near larger centres with a full range of services. A few possess the social services but have inferior shopping facilities ; these are usually in thinly peopled rural districts beyond the range of influence of fully equipped towns. Most numerous are places with three out of the five key institutions that are small service centres, some growing and active, others decadent and inactive. Finally, he recognizes what we have called *urban villages*, that are intermediate between the rural village and the town proper.

The city is characterized by the greater variety and quality of its services, and it integrates the activities of a number of towns around it. Certain *major cities*, at the head of the hierarchy of urban settlements, have the following distinctive features : they have the regional offices of Government Departments and of private organizations that operate on a national scale, a Stock Exchange, a branch of the Bank of England, the greatest concentration of personnel engaged in wholesale trades, a daily morning newspaper, a University or College, a Medical School, and a large general hospital. Apart from London, outstanding cities, in this sense, are Birmingham, Bristol, Cardiff, Leeds, Liverpool, Manchester, Newcastle, and Nottingham, to which may be added Norwich, Southampton, Plymouth, Hull, Sheffield, Leicester and Bradford. Smaller *cities* with a wide range of services but with a more limited sphere of influence bring the total of cities to fewer than forty in all. Between these cities and the towns are what may be called *major towns* or *minor cities*. These form a large and heterogeneous group of about 100, and include most of the specialized urban centres that are less important as regional centres and are, indeed, in many ways dependent on the latter. These places are grouped into three classes : industrial, county, and resort towns.

A recent German study of the functional structure of towns is available [1] for Germany, a study presented with a view to determining the basis of the existing community structure of towns so as to estimate in what measure towns are resistant to economic depression and in what measure they are able to offer opportunities for permanent employment and so fix the demand for new houses—an aim which, in essentials, is similar to that of the P.E.P. report on the location of industry.[2] The approach is from the standpoint of the town as (1) a seat of specialized occupations, producing goods both for an extensive market and for the local service area ; (2) the trade with the service area (described as its " hinterland ") ; (3) and the food supplies drawn from the hinterland. The *living possibilities* (*Lebensmöglichkeiten*), or, in the more limited sense, *occupational possibilities* (*Erwerbsmöglichkeiten*) are assessed on this threefold basis with the following tentative classification.

The small country town (*Landstadt*) is normally situated in an agricultural area. It was until the nineteenth century the seat of the markets and handicrafts serving the surrounding rural parishes. If situated on a main route, it was also a post-station for through coach traffic. It was normally in the Middle Ages the seat of a religious house, of an *Amtmann*, a count or other great landlord. Some such towns lived almost entirely by agriculture, and were distinct from the *Gemeinde* (parish) only by their wall and their special law. Many of these townlets, usually with less than 2,000 inhabitants, have been unable to take advantage of the modern growth of industry, commerce and administration. Many which have no railway have sunk to the level of villages ; they bear the name of towns, but have none of their special functions. Young people, in the absence of economic and social opportunity, leave such towns. Thus, about 40 per cent. of all the towns of Brandenburg had more inhabitants sixty years ago than they have to-day. Such a town draws about half of its income from the farm products of its own *Gemeinde*, and 20 per cent. from

[1] Gerhard Isenberg. *Erwerbsmöglichkeiten und Krisenfestigkeit als Voraussetzung für die Siedlungstätigkeit. Dargestellt an der Hand der Untersuchung von 10 Mittelstädten im Land Sachsen.* Based on work carried out in 1933–5 in the *Seminar für Städtebau und Siedlungswesen* at the Technical High School, Dresden (Mimeographed).

[2] P.E.P. (Political and Economic Planning), *Report on the Location of Industry*, London, 1939. Chapter VI, Towns and Industry.

trade with its hinterland—a total of 70 per cent. About one-third of its occupants are engaged in the local services (*Näherer-werbstätigen*).

The *Kreisstadt* is the small administrative centre. It is named from the *Kreis*, the local government district in Prussia of which it is usually the centre. If the administrative area is small, as in south Germany, the town often has under 2,000 inhabitants, but if the area is on the large side, with more than, say, 50,000 inhabitants, the town may have from 2,000 to 5,000 people. Other service functions accrue to the town in virtue of its being a centre of administration. Those in pleasant surroundings attract retired people or others of independent means. There is a little industry. The backbone of the population is its official and professional classes—*Landrat*, *Finanzamt*, *Arbeitsamt*, *Versorgungsamt*, the personnel of the small garrison, the railway station, the post office, the pensioned officials and retired residents. Special types of *Kreisstadt* arise when one of these classes is dominant—officials, teachers, or retired persons—or when a town is situated at a railway junction. Such towns, though small, are well balanced in their social and economic structure and in the sources from which they draw their income. These are calculated as : agriculture, 15 per cent. ; hinterland trade, 22 per cent. ; public funds, 25 per cent. ; unearned income, 20 per cent. ; and industry, 16 per cent.

A third type has the same basic features as the *Kreisstadt*, but differs from it in the much greater importance of industry—due to the development of old-established industries or the introduction of new ones. But the economic basis is still sufficiently wide for the town community not to be seriously hit in time of economic depression, for the unemployed, though only temporarily, can be given some other kind of work. The type occurs as small and medium-sized towns, the latter only when there is a group of officials and a garrison. It normally has 10,000 to 40,000 inhabitants. Functions of administration and commerce in relation to the hinterland are not adequate to account for a population of this size, and industry is much more important. In the Reich all towns of 20,000 to 50,000 inhabitants are found to average 52 per cent. in industry and handicrafts. Such a town is capable of providing through its local services all the normal requirements of a fully-fledged urban community.

In the industrial town the occupants depend dominantly upon industry, and upon industrial occupations depend in turn the other occupational groups and sources of income. The residential function is of small importance. The industries of the hinterland are identical with those of the industrial town, and here too live workers who work in the town. Industry makes up over four-fifths of the town's income, and this clearly has repercussions on the remaining occupations.

The last type is the metropolitan city, the great centre of business, industry and culture, serving an extensive area around it, and having nation-wide and international connections. These cities have as a rule about 100,000 inhabitants as a minimum. They are many-sided in their activities and offer a varied range of opportunities for employment.

CHAPTER 3

TOWN–COUNTRY RELATIONS

Every nucleated settlement, whether hamlet, village, town or city, is, in varying degree, a centre of services and organization for a surrounding area. It is, indeed, common knowledge that the village is normally the centre for many of the activities of its parish, that the town is the centre for an area within a radius of about five to ten miles, and the great city for a still wider area which it serves in its capacity as a king among towns and a regional centre of economic and social organization. The location of these nucleated settlements has been fixed since the time of their origin in the Middle Ages, but, in the last hundred years, industries have caused some to grow as great excrescences upon the countryside without any organic relation to it, and services have tended to concentrate in fewer centres, since in that way they can be more efficiently rendered. Thus, fundamental changes in the social structure of the village and in the interrelations of town and village have appeared. The village community has been affected in many ways by the growth and concentration of services in neighbouring villages or small towns. Rural cultures have been profoundly affected by the impact of urban ways of life. In the vicinity of all urban agglomerations there appear changes in the conditions and ways of rural living which upset the balance of rural society. So deeply associated are town and village in their interrelations that in effect no clear-cut distinction can be drawn between the urban and rural ways of life. Towns are able to compete with each other as service centres, and to specialize on particular industries to the virtual exclusion of others. They are thus more closely bound up with neighbouring industrial towns and particularly with the metropolitan city, to which they look for their deficiencies of materials and services. The metropolis becomes ever more metropolitan, not only as the focus for its nearby " town-area " but also for the many towns that lie over a much wider area. In this chapter we shall attempt to examine some aspects of these trends, with particular reference to the village community and the smaller towns in rural areas and to the relations between the village and the town. We shall illustrate from Germany, the United States and England.

1. Germany. Classification and Distribution of Settlements as Service Centres

General consideration has already been given to the significance of the centralized services—the urban functions that cater primarily for the surrounding service or market area—in urban development. This general conception has been applied in some detail to the structure of both the urban and rural communities of to-day in south Germany by Christaller, whose work was noted in the previous chapter, and the main results may be summarized as follows.

The centralized services, as listed by Christaller, are concerned with administration, culture, health, social service, organization of economic and social life, trade, finance, service industries, the organization of the labour market and traffic. He lists the specific services and institutions in each category, grouping them, according to grade, into lower, middle and upper. This empirical method of assessment cannot give an adequate composite measure of the centrality or nodality of a place, and Christaller seeking, like other investigators, a statistical approach, took the number of telephones as a basis, on the assumption that this is the best measure of general relationships between one place and another.[1]

The centrality (Zz) of a place (that is, the services it performs over and above the local needs of its inhabitants) he measured by the formula $Zz = Tz - Ez \cdot \frac{Tg}{Eg}$, where Tz is the number of telephones and Ez the number of inhabitants in the place, Tg and Eg the number of telephones and inhabitants respectively in the area (g = *Gebiet*) served by it. $Ez \cdot \frac{Tg}{Eg}$ is what the importance of the centre *ought* to be in proportion to its population, Tz is the *actual* importance, and the difference between them is the centrality. The fraction $\frac{Tg}{Eg}$ is the telephone density of the whole area.

From this assessment there emerges a gradation of urban centres as seats of centralized services as follows. The distribution of these centres is shown on Fig. 5 (p. 56). The *market centre* (M) has a normal distance of 7 to 9 km. from its neighbours,

[1] See Christaller, op. cit., p. 146.

and a service radius of 4 to 5 km. It has centralized functions of the lowest grade—registrar's office (*Standesamt*), police station, doctor, dentist, veterinary surgeon (in some centres), a small hotel, a local branch of a district bank, craftsmen, repair shops, breweries and mills, and almost invariably a head post office and telephone and railway station. The *Amtsort* (A) corresponds with the lowest administrative centre, which usually serves three market towns and their parishes. It has a police court, library, elementary school, museum, chemist, veterinary surgeon, bank, cinema, a local newspaper, local trade associations, specialized shops, and almost always it lies on a railway. The *Kreis town* (K) is so named from the frequency with which it occurs as the capital of the *Kreis* in Prussia and Hesse, and of the corresponding *Amtsbezirk* in Baden, the *Oberämter* in Württemberg, and *Bezirksämter* in Bavaria.[1] It includes many functions that are centred in towns of this administrative category and is a well-defined type. The *Bezirk* centre (B) is not so clearly identified administratively, but economically it is very important, and there is a tendency to group *Kreis* districts for administrative purposes into units with centres of a higher order (see below). In Germany, the distinction between the *Kreis* and the *Bezirk* centres is popularly expressed in the terms *Städtchen* or townlet and *Stadt* or town. The latter has all the main characters of a fully developed town. It has acquired many functions that it can carry out more effectively than the *Kreisstadt* by serving a larger area that includes several *Kreise*. Thus, in addition to all the functions of the *Kreisstadt*, it also has, for instance, a district labour office, an institute of higher education, specialist doctors, several cinemas, specialist shops and dealers, warehouses, daily papers, several district banks and post offices. A corresponding centre in Britain would be a country market town such as Bury St. Edmunds. The *Bezirk* area corresponds in size and grade with the French department. The G or *Gau* centre (G) is named after the old German social unit called the *Gau*, comparable with the French small province. The *Mittelstadt* of the statistician includes up to 70,000 inhabitants, and corresponds administratively with the centres of the *Landeskommissariatsbezirk* of the province of Hesse. The P or *Provinzstadt* (P) corresponds in Prussia with the *Regierungsbezirk* centre and in Bavaria with the groups of *Kreise*. The L or *Landstadt* (L), with about 500,000 inhabitants, is very

[1] The *Kreis* came into existence as an administrative unit in the early nineteenth century.

prominent in Germany, Italy and Spain. In southern Germany the *Landstadt* centres are Munich, Frankfurt, Stuttgart and Nuremberg-Fürth. Similar cities are Strasbourg and Nancy in France and Zurich in Switzerland. Strasbourg, if serving its geographically contiguous districts on the German side of the Rhine, would have a much higher status, whereas, in fact, the centralized services of the upper Rhineland are divided between several major centres that are peripheral to it—Strasbourg, Stuttgart, Mannheim, and Basel. The R or *Reichstadt* is a capital city, but between it and the L centre, Christaller suggests an intermediate type of the order of size of Hamburg, Cologne, Düsseldorf, Essen, Munich. He calls these R.T. or *Reichsteile* centres, since each dominates and serves a large part of the Reich. In France, such centres are Bordeaux, Lyon, Marseille, and in Italy, Milan and Naples.

THE STATUS AND DISTRIBUTION OF TOWNS AS BASED ON THE THEORETICAL DISTRIBUTION OF CENTRALIZED SERVICES (after W. Christaller)

(*Die Zentralen Orte in Suddeutschland*, 1933, p. 72)

This table shows the theoretical distribution of service centres in a predominantly rural area such as south Germany, and the figures of average population are broad estimates for this area. The system is regarded as a network of service centres falling within the service area of the *Landstadt*.

	Grades of Town		*Approximate Population*	*Distance Apart (Miles)*	*Service Area (Sq. Miles)*
I.	*Marktort*	M	1,000	4·5	18
II.	*Amtsort*	A	2,000	7·5	54
III.	*Kreisstadt*	K	4,000	13	160
IV.	*Bezirksstadt*	B	10,000	22·5	480
V.	*Gaustadt*	G	30,000	39	1,500
VI.	*Provinzstadt*	P	100,000	67·5	4,500
VII.	*Landstadt*	L	500,000	116	13,500

It will be noticed on Fig. 5 that if *all* the centres be taken into consideration, they are spaced at an average distance apart of 7 to 9 km.—a distance that gives a local service area with a radius of about one journey-hour. This is a basic determinant of the distribution of villages and small country towns that is found all over the Old World in closely settled areas. This cell unit is the market area (M). The uneven distribution of the historic towns is evened out by this map of service centres which shows

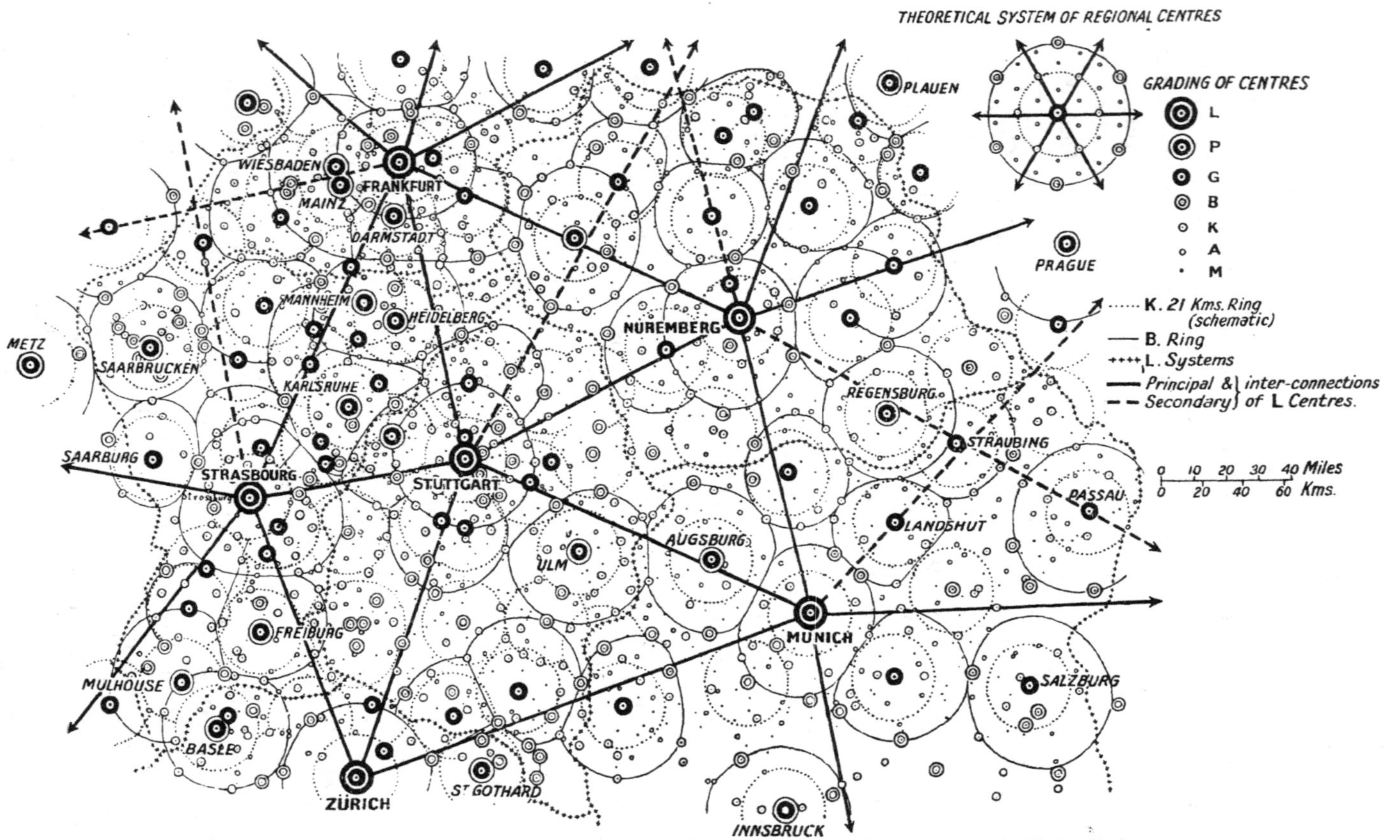

FIG. 5.—The Distribution of Towns as Service Centres in south Germany (after Christaller).

Service Centres are graded from the *Landeshauptstadt* to the *Marktstadt* (see text).
Services Areas are shown as follows: 1. K centres are given a schematic radius of 21 km. 2. B centres are given actual radius of influence (schematically it is 26 km.). 3. Boundaries of the L systems. 4 and 5. Principal and secondary interconnections of L centres.

clearly, for instance, the even spacing of market centres in the Bavarian plateau south of the Danube.

Christaller claims that the laws of the distribution of modern centralized services account for the spacing, size and functions of urban centres, modified occasionally by the factors of long-distance trade and administration. It is, however, quite clear that he underrates the importance of modern industry as an urbanizing factor. A great deal might be written at this stage on the significance of industry in the growth of the medieval towns of southern Germany, but the important facts may be briefly stated.[1] Reference should be made to the distribution of towns in 1830 and 1930 in Figs. 1 and 2 (pp. 15–16). The vast majority of the present towns and market settlements were in existence by the end of the Middle Ages. The basic *spacing* of urban settlements arose therefore at this time. The few towns founded subsequently were court towns and places for religious and political refugees. But these introduce quite minor alterations in the medieval distribution that persisted with little alteration a hundred years ago. Seventy towns or *civitates* (as defined on p. 41), were in existence in south Germany in 1200 and about 500 by the end of the Middle Ages. Possibilities of urban growth were afforded by overland or long-distance trade, local market trade and defence and administration. Let us consider each of these. Merchant traffic was confined to the public highway (*via publica* or *regia*) for the regulation of traffic by the Emperor. In considering the importance of long-distance trade in the origin and development of the medieval town, these routes alone come into account. There were very few main routes in south Germany in 1200, the main network developing in the later Middle Ages with the growth, in particular, of trans-Alpine traffic. There were many main routes of early origin in northern Germany, especially on the northern border of the central uplands, and here most of the early medieval towns began as resting-places (*Rastorte*), mainly at river-crossings on these routes. They were used as strongholds or bishops' seats, then attracted small settlements of merchants, who ultimately obtained rights of complete self-government. Nearly all the main roads in southern Germany, on the other hand, developed after 1200 and, according to Gradmann, the

[1] For further reference to this subject, see the articles on "The Development and Distribution of the Medieval German Town", by the writer in *Geography*, Vol. XXVII, 1942, pp. 9–21 and pp. 47–53. See also the paragraph on pp. 27–8 bearing on the origin of towns.

authority on this area,[1] many of the towns lie *off* the main roads, and long-distance trade cannot have played a decisive part in their growth, whereas in the north the great majority lie *on* the routes. Local market trade played a main part in the development of the town, as is proved not only by their position in relation to the main routes, but also by the fact that the majority were granted the right of holding a weekly market at the time of their foundation. But, against this argument, many were founded without reference to the possibilities of local trade, and, for various reasons, were never able to integrate the trade of their surroundings. The two factors here concerned are long-distance trade and defence and administration.

Numerous towns developed after 1200 on the routes which were then used for through traffic on a big scale with the development of trans-Alpine trade. They are spaced at regular intervals of 25 to 30 km., as, for example, on the roads radiating from Nuremberg and (outside our area) on the skein of routes that ran from north to south through the lowland corridor of Hesse across the central uplands. These are comparable in origin and function to the coaching towns that developed in the seventeenth and eighteenth centuries.

One of the chief considerations in the foundation of towns in south Germany in the later Middle Ages was defence and administration. Germany was then plunged into feudal chaos and local territorial lords (and bishops), in defending their territories, established towns or walled existing settlements, and in organizing their lands on a territorial basis, they erected towns and castles adjacent to them for the administration of fixed districts tributary to them. In this way, the *Ämter* districts came into being. Towns appeared, then, adjacent to castles, monasteries, and villages, and markets were raised to town status, to serve as political centres, and the grant of a weekly market often made a tardy appearance. Similarly, rival lords built or acquired towns in rich areas such as the vine-growing districts of Alsace, and along the Rhine gorge where, in addition, tolls could be levied on traffic using road and river. In these areas of extreme territorial disintegration, such as Alsace and the Neckar basin (in Württemberg), many towns were established on naturally defended isolated hill sites or on river meanders, very close to each other (in different political territories) and off the trade routes. On the other hand, in large areas that were under the control of one political authority, as in

[1] R. Gradmann, *Süd-Deutschland*, Vol. I, Stuttgart, 1931, pp. 158–69.

Bavaria and the Tyrol, there was less demand for fortresses or castles, so that towns are fewer and the unwalled market much more characteristic. Elsewhere in south Germany the unwalled market settlement is rare. Thus we find that many of these places, through competition with near neighbours, became superfluous. Though enjoying the features of a medieval town, with planned layout, wall, and the machinery of self-government with a town council or *Rat*, they were unable to function as seats of trade and industry, and to-day have a few hundred inhabitants and have dwindled to agricultural villages. Such are the *Zwergstädte* or dwarf towns named by Gradmann, of which Dinkelsbuhl is a famous example.

During the mercantilist era in the seventeenth and eighteenth centuries crafts were fostered in town and countryside and to an increasing degree the towns became centres of economic organization. This was especially true of the textile industries that developed during this period. Numerous small towns have become seats of specialized industry and seats for the district organization of domestic industries carried on in the countryside. But in general very few urban centres in south Germany owe their existence to the appearance of industry, and industry has accrued to the towns in such a way as to be proportional to their size a hundred years ago. The great majority of all the towns in south Germany, as in other countries, have under 10,000 inhabitants. Those with over this figure, which may be safely taken as a broad indicator of real urban character, are fewer in number, but they account for most of the urban population and are the chief seats of industry, commerce and administration. Towns increased normally in the nineteenth century in proportion to their size and their size was largely proportional to their nodality as route centres. The great bulk of the increase, however, was due to the growth of industry and secondarily to the rendering of services. The towns that are predominantly industrial—taking 50 per cent. of all employed persons as a crude criterion—are markedly concentrated in the south-west, clustered in the Saar coalfield, around Mannheim — Ludwigshafen, around Frankfurt — Mainz — Wiesbaden, and in the larger area of the Neckar basin in Württemberg with its focus in Stuttgart, and a group of small towns in the extreme south-west between the Danube and Lake Constance. Munich stands out as a separate great industrial focus, while there is a sprinkling of small industrial towns in north-eastern Bavaria, Nuremberg-

Fürth being the chief centre. The distribution and size of the specialized industrial centres are determined by very different factors from those governing the mainly service centres, which, when shown with the remaining towns with over 10,000 inhabitants and under 50 per cent. engaged in industry, are evenly spread over the face of the land.[1]

We have left for final consideration the functions and distribution of the small, auxiliary centres that serve the countryside.[2] These are small nuclei which have several centralized services in them, but do not reach the status of an urban centre such as the market (M) town. They are called by Christaller *Hilfszentralen* and may be best described, in translation, as auxiliary centres. They are described below (in Section 3, p. 78) as " urban villages ". First, there are places that were formerly towns or market centres that have lost functions, and even urban character, with the concentration of their functions in other places. Such has been the fate of towns on isolated hill sites and within river meanders, sited originally for reasons of defence, but to-day lying aloof from the main roads without adequate local connection with their surroundings. Some lie so near bigger towns as to have lost functions to the latter, or even to have been completely absorbed by their expansion. Others have declined through the natural poverty of their surroundings or through changes in political frontiers. All these centres are in the *inactive* order and have been degraded from the status of small towns.

Secondly, some places are definitely in the *active* order. Though small, with an incomplete range of urban services, they fill a useful purpose in servicing their surroundings. They include the small service centres that have often grown around a railway

[1] See H. Bobek, " Über Einige Funktionelle Stadttypen und Ihre Beziehungen zum Lande ", in *Comptes Rendues du Congrès International de Géographie, Amsterdam 1938, Tome Deuxième, Géographie Humaine*, pp. 88–102 with two maps. Bobek reckons that in Germany only about a fifth of the population, at the maximum a third, is specifically engaged in servicing the market area of the town. The proportions are probably highest in the big cities—Stuttgart, Frankfurt, Munich and Karlsruhe having 40·7, 41·2, 43·3 and 46·7 per cent. of their population respectively engaged in trade, commerce, administration, public service and the professions.

[2] In Germany, isolated farmsteads are characteristic in certain areas, notably in the north-west and in the upland and mountain areas of the south. Villages fall into the following classes : (1) *small villages*, with 70 to 200 inhabitants ; (2) *normal villages* (also called the School villages), with 200 to 600 inhabitants ; (3) *church villages*, with 600 to 1,600 inhabitants ; (4) the *market villages* with 1,600 to 4,500 inhabitants, and the *urban villages* with over 4,500 inhabitants. The last three classes are especially characteristic of the rich, open arable lands in the Rhine Valley, the middle Elbe basin (between Magdeburg and Leipzig), the Erzgebirge (where they are mainly industrial settlements), and the land on the northern border of the central uplands around Hanover. See W. Christaller, *Die Ländliche Siedlungsweise im Deutschen Reich und ihre Beziehungen zur Gemeindeorganisation*, Berlin, 1937.

station, or in areas where there is a rising standard of living, or where new industrial establishments and settlements are planted sporadically in an agricultural area. Especially do they appear in tourist areas (*Fremdenverkehr*)—a type that is found, for instance, in England at numerous points along the coasts where there are small residential settlements.

Thirdly, there are *permanent* as opposed to *decadent* and *emergent* auxiliary centres. These are permanent centres in the pattern of settlement in an area and best serve the special needs of its communities, whose social and economic conditions are relatively stable. Such are the places in mountainous or other thinly peopled areas that have too thin a veneer of settlement to support towns, however small, at frequent intervals ; but must nevertheless be supplied with civilized needs and have outlets for their products if they have any for sale.

These auxiliary centres usually lie on the borders of the service areas of the towns where, at points furthest removed from the latter, they are able to supply more effectively some of the local needs. They are also found in isolated and thinly peopled areas, and on the borders of the big cities.

The service area of the market town of the M grade with a service radius of 4 to 5 km. is the unit settlement area in southern Germany, and indeed occurs throughout the country, in spite of its great variety of countryside and types of settlement distribution. The arrangement of approximately equal-sized districts focused on small seats of centralized services is excellently illustrated in Westphalia. Church districts (*Kirchspiele*) all have approximately the same area and shape, and, if defined arbitrarily by lines of equidistance from neighbouring church centres, they are roughly hexagonal in shape. These church centres often have markets and a distinct law (*Wigboldrecht*) comparable with that of the market town (*Marktrecht*). The church district is also the political *Gemeinde*, and this in turn is divided into smaller groups of neighbouring farms of ancient origin, known as *Bauernschaften.* The road net that links the centralized places appears as a series of isosceles triangles with the settlements in their corners. This particular area, as for example in the neighbourhood of Münster, has a dispersed rural settlement, but the same characteristics are found in areas with an entirely different settlement distribution. In the Paderborn area, for example, the rural settlement is clustered in large villages with very little dispersed settlement. These villages are spaced about 3·5 km. apart, but the larger urban centres are about 6·7 to 9·4 km. apart. Again, throughout *Niedersachsen* or Lower Saxony, the centre of the *Kirchspiel* is also the market centre and its area is often clearly geographically defined by moor, heath, or woods. In the areas of completely dispersed settlement in the reclaimed marshland and the drained moors along the North Sea coast larger places appear as service centres, accommodating chemist, doctor, bank and telephone and post office, and serving a district with about 2,000 inhabitants. In Brandenburg, where are to be found characteristics of settlement that are fairly typical of the eastern lands of German medieval colonization, there are regularly spaced small villages with 300 to 500 inhabitants. Here there was no system of "mother churches" serving a large church area within which smaller districts evolved, served by daughter churches. There is also lacking that very close network of market towns such as developed under medieval conditions in many parts of western or old Germany. Thus, in origin, the historical development of settlement and of the administrative districts did not allow for the growth of the central places of the lowest order noted above, so that selected villages have gradually acquired such functions and risen above their neighbours in more recent times. These market units have regularly 2,500 to 3,000 inhabitants as in western Germany. Such places have a church, they are road centres, and have a larger population than the surrounding villages. The same development is apparent in Pomerania and East Prussia.[1]

2. United States: Town-Country Relations

Work on the areal or spatial aspects of social organization has been done mainly in the United States, and we must turn to the works of American sociologists to understand the scope and method of this type of investigation. In the United States, local government units, such as townships, villages and counties, which were created in the pioneer days, in the era of neighbourhood economy, were of necessity small in area and population. With the increasing complexity and multiplying functions of government, hundreds of special districts have been created, each separately administered—for fire protection, library service, lighting, irrigation, and so on. The citizen is also burdened

[1] W. Christaller, op. cit.

with a bewildering number and variety of elections each year. Thus, the existing units are inadequate to cope with modern needs and there is lack of uniformity among the many local government boundaries.

> . . . It is being increasingly recognized [write Brunner and Kolb] that this archaic system is breaking down under the social and economic changes—such as the changes in the mobility and characteristics of population, the increasing interdependence of country, to town and city, and the greater emphasis upon education, health, and welfare activities. Local units of government no longer conform to the social and economic relations and functioning of rural society. As a result, discussions and studies looking toward improvement have been initiated and reorganizations have been attempted here and there. The belief that the persistence of these tens of thousands of such local units is a factor in mounting taxation has further stimulated this movement.

Thus, in such discussions, there rise the questions :

> What criteria should be used to delimit the boundaries of the new units ? How may social considerations be given their proper weight along with those of efficiency, economy, or administrative convenience ? For it must be emphasized that considerations for the reorganization of local government will need to go far beyond the immediate interests of economy or efficiency.[1]

The solution to this vast problem lies not only in the reorganization of the system of local government, but also in carrying out research and working out principles for the redefinition of local government areas so that they shall conform to the optimum demands of social and economic needs, of economy, and for efficiency of government.

The first social study of this type to be attempted in rural areas was that of Dr. C. G. Galpin on *The Social Anatomy of an Agricultural Community*, published in 1915.[2] The area examined was Walworth County in the State of Wisconsin, in which twelve service centres were selected for an exhaustive analysis. This study revealed the significance of the emergence of a rural community, larger than that of the small country neighbourhood. This rural community is made up of the scattered farms and the small town or large village which serves them. Galpin called it the " rurban community ", since it combines the simplest traits and functions of rural and urban life. He showed by means of

[1] E. de S. Brunner and J. H. Kolb, *Rural Social Trends, Recent Social Trends Monographs*, McGraw-Hill, New York, 1933, pp. 286 and 287.

[2] Research Bulletin 34, Agricultural Experiment Station of the University of Wisconsin, 1915.

maps (probably the most important innovation of his study), based on detailed farm to farm questionnaires, that surrounding each town there was an area or " zone of land " in which the farm-houses depended on the town centres for certain regular and fundamental services. These services were : general trade, banking, newspaper, milk marketing (the county studied is a commercialized dairy farming area), school, church and library. Farm families and village or small-town families used the same centre and together form the larger " rurban " community. This community area is not static. The rural society of America is still a rapidly evolving community, adjusting its areal organization to new forces, and, in order fully to understand its ecological structure, the centres and their community areas must be studied periodically. Literally hundreds of studies on these lines have been undertaken in recent years, notably by the Agricultural Experiment Stations under the direction and stimulus of Dr. Galpin. These have been summarized, together with their independent researches, by Brunner and Kolb in their monograph on *Rural Social Trends*.

The smallest social unit is termed the *neighbourhood*. This is usually defined as the first grouping beyond the family which has social significance and which is conscious of some local unity. An alternative title is " rural primary group " or simply the " rural community ". Numerous studies have revealed the general tendency for the identity of the neighbourhood to be lost owing to the transference of its functions to the village centre, which serves as the first integrating service centre, and to the town. Consequently, the neighbourhood only functions actively in the isolated areas, and a new rural community grouping is emerging centred on selected, larger, village centres. After the definition of Dwight Sanderson in his *Rural Community* : " A rural community consists of the social interaction of the people and their institutions in a local area in which they live on dispersed farmsteads and in a hamlet or village which forms the centre of their common activities " ; or again : " In its geographical aspect, a community is the local area tributary to the centre of the common activities of its people, which centre is normally a village." [1]

In Otsego county in New York State [2] there were found 150 neighbourhoods, with an average of 14 houses apiece and an

[1] Dwight Sanderson, *The Rural Community : The Natural History of a Sociological Group*, Ginn, New York, 1932, pp. 481 and 484.

[2] See footnote 4, p. 66.

area of 2¼ square miles. They were grouped as follows according to the main basis of their unity: (i) a hamlet or group of houses, usually associated with some institution or business; (ii) a group of houses tributary to an institution such as a school, church or community building; (iii) a business neighbourhood, tributary to a store, mill, cheese factory, railway station, or factory; (iv) an ethnic neighbourhood distinct in view of its common alien descent; (v) a kinship neighbourhood, that is,

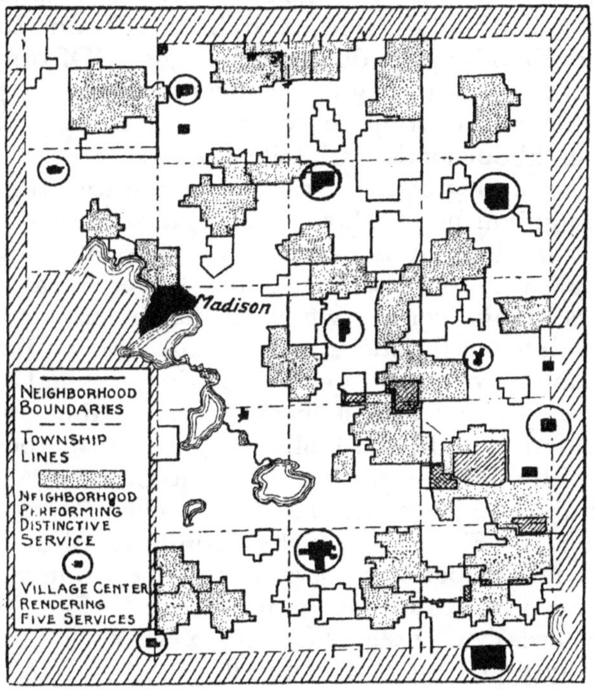

FIG. 6.—East Dane County, Wisconsin, showing country neighbourhoods and village service centres in 1920 (after J. H. Kolb).

a family group; (vi) a topographic neighbourhood, grouped in a valley, on a hill, near a lake, etc.; (vii) a village neighbourhood, centred on a village from which it derives its name. The hamlet persists, the business neighbourhood is "practically gone", the kinship neighbourhood is "passing", the topographic neighbourhood may foster neighbourliness but usually only gives the name to the group. "In general, the rural neighbourhood in Otsego county is ceasing to function as a social unit except where its life is centred on some local institution." A half of the neighbourhoods had the same name as the school district, and the neighbourhood and the school district were identical. Frequently, though to a much less degree, they correspond with the church parishes or the small trade area of the hamlet—the shop or the creamery or the post office. The neighbourhood is in no sense self-sufficing. It depends upon larger service centres both " rurban " and urban, for many of its needs.

In Dane County, in the State of Wisconsin [1] (Figs. 6, 7*a* and 7*b*), the institution such as a school, church or club, is the strongest binding factor, so that the neighbourhood boundaries often correspond with school districts, church parishes and trade areas.[2] 121 neighbourhoods were identified in 1921. Of these, 95 were considered " active " or " going concerns ", carrying on group activities and with a group consciousness. In 1931 the number of active neighbourhoods had fallen to 70.

The active and growing " rural community area " is larger than the neighbourhood, though since it is endowed with the same basic functions, it is in effect not possible clearly to distinguish the two. Actually, it is the larger rural community area and its village centre, which is displacing the neighbourhood, absorbing the functions of the latter and adding to them.

Two examples of town–country relations in the United States are shown on Figs. 7 and 8 from works by the same authors, Dwight Sanderson and J. H. Kolb. Fig. 7*a* shows the changes in the extent of the trade areas in East Dane County, Wisconsin, for the period 1921 to 1931, as well as the classification of service centres in 1921. Fig. 7*b* shows the boundaries of the trade, high schools and newspaper areas of eastern Dane County, Wisconsin, with the town of Madison in its centre.[3] Fig. 8 shows the community areas of Otsego County, New York State.[4]

In the study of Otsego County it was found that the rural community area is smaller than the township.[4] The average distance between village centres was 5·1 miles. The population of the community centre averaged 380, and that of the surrounding community area 416, so that the total population was about 800. Some of these communities (a third of the total of 43) are of a higher order (comparable in size and function with Kolb's Grade I centres discussed below), furnishing all the necessary

[1] J. H. Kolb, " Trends of Country Neighborhoods : A Re-study of Rural Primary Groups, 1921–31 ", Agricultural Experiment Station of the University of Wisconsin, Research Bulletin 120, November, 1933.

[2] J. H. Kolb, op. cit., p. 18. [3] Ibid.

[4] Dwight Sanderson and Warren S. Thompson, " The Social Areas of Otsego County ", Agricultural Experiment Station of Cornell University, Research Bulletin 422, 1923.

local services. The three services common to the larger villages are hardware shops, high school and banking, together with occasional professional service, undertakers and cinema. These

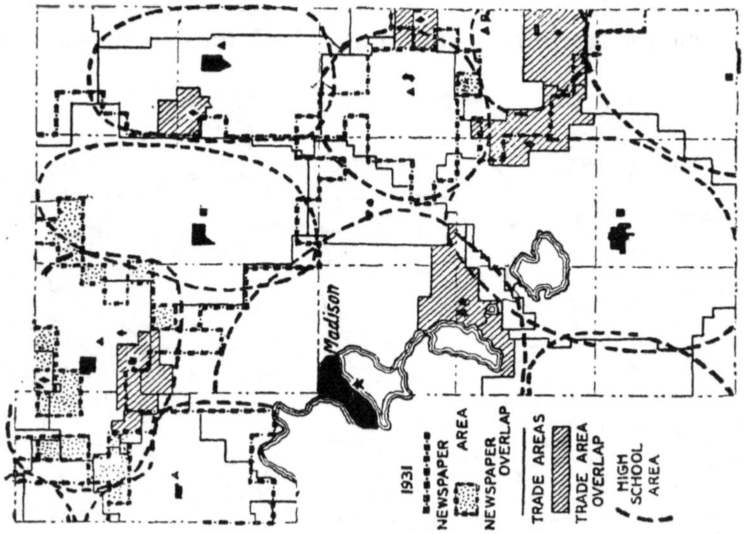

FIG. 7*b*.—Trade Areas, High School Areas, and Weekly Newspaper Areas in East Dane County, Wis., for 1931 (after J. H. Kolb).

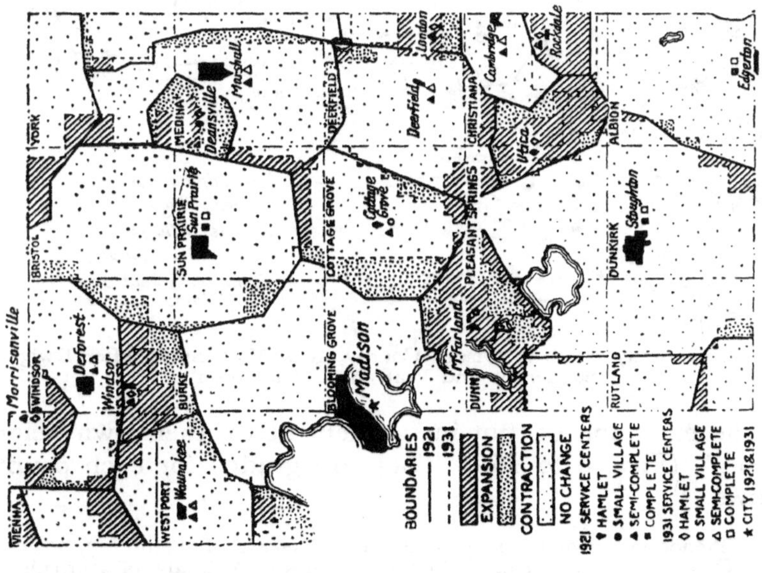

FIG. 7*a*.—Changes in Trade Areas from 1921 to 1931 for East Dane County, Wis. (after J. H. Kolb).

larger community areas include two or three of the smaller ones, and their centres are on railways.

These service centres and the extent of their service areas have received noteworthy attention for Walworth County in

Wisconsin, as noted above, by Galpin, and, after a ten-year interval, by J. H. Kolb ; as also, by the latter author, for Dane County, again in the State of Wisconsin. In the latter study, Kolb describes these service centres as " the centres beyond the neighbourhoods to which farm families look for essential services,

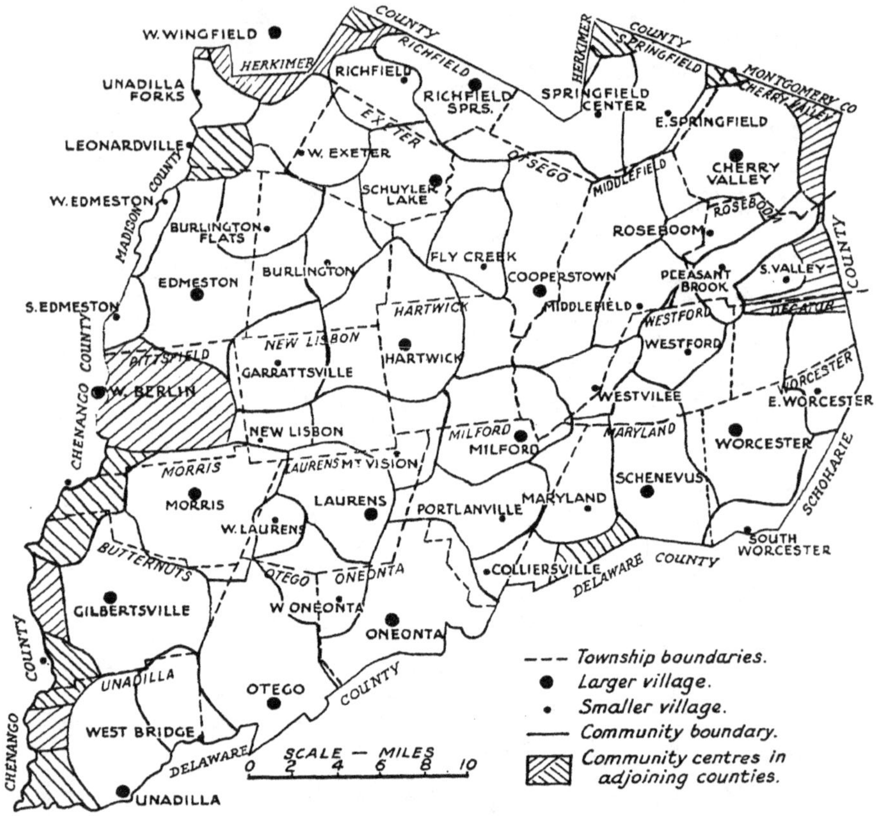

FIG. 8.—Community Areas in Otsego County, New York State (after Dwight Sanderson and Warren S. Thompson).

not to the exclusion of, but in supplement to, their own neighbourhood " (Figs. 9 to 16). They are graded for each county as follows :[1]

[1] See " Trends in Town–Country Relations ", Research Bulletin 117, 1933, on Walworth County, Wis. ; " Rural Primary Groups : A Study of Agricultural Neighborhoods ", Research Bulletin 51, 1921 ; and " Trends of Country Neighborhoods : A Restudy of Rural Primary Groups, 1921–1931 ", Research Bulletin 120, 1933, both studies of Dane County, Wis. ; all published by the Agricultural Experiment Station of the University of Wisconsin. Also J. H. Kolb and D. G. Marshall, " Neighborhood-Community Relationships in Rural Society ", Research Bulletin 154, 1944.

SERVICE CENTRES IN THREE COUNTIES IN WISCONSIN, U.S.A.

	Major Types	Kinds of Centre	Number	Population 1930
		East and West Dane Counties		
I.	Rural or Country Centre [1]	(a) Neighbourhood or Hamlet Centre	20	25–200
		(b) Small Village with incomplete service	4	250–300
II.	Rurban or Town-Country Community Centre	(a) Village or small Town with semi-complete service	13	400–1,000
		(b) Town or small City centres with complete service	6	1,000–5,000
III.	Urban and City Centres	(a) City Centre (Madison, State capital)	1	58,000
		Walworth County		
I.	Rural or Country Centre [1]	(a) Neighbourhood Centre	13	10–200
		(b) Small Village Centre	5	300–400
II.	Rurban or Town-Country Community Centre	(a) Village or small Town	10	500–1,000
		(b) Town or small City	6	2,500–4,000
III.	Urban and City Centres	(a) City Centre	5	17,000–67,000
		(b) Metropolitan Centres (Milwaukee, Chicago)	2	578,000 and 3,376,500

The component services of these various centres are education (the high school), trade (the trade area), religion, social intercourse, commerce and transport. The composite service areas of towns and village centres have an average radius of 4·3 miles, and the distinctive neighbourhoods continue to be found in the open country between, and well removed from, the village and town service areas. The neighbourhood, which is based on a geographical locality, is widening the sphere of its contacts through the medium of locality contacts with the town or country community, and of its interest associations through such activities as breeders' associations, county health committees, and voluntary associations.

[1] "Country centres" include the neighbourhood or cross-roads centre and the hamlet centre. The country centre is the meeting-place for farmers' families—creamery, general store, school and church.

DISTRIBUTION OF SERVICES ACCORDING TO TYPES OF CENTRE
based on Kolb's Study of Walworth County, Wisconsin, for 1929.
(Wisconsin Research Bulletin, No. 117, 1933, p. 28.)

Farm Family	*Rural* or Country Centres	School Church Social Centre Grocery Store Filling Station
	Rurban or Town-Country Centres	High School Library Banking Grocery Store Dry Goods Store Ready-to-Wear Store Hardware Store Furniture Store Farm Machinery Depot Recreation Movies Milk Plant
	Urban or City Centres	Recreation Special Medical and Hospital Care Ready-to-Wear (men's, women's, children's) Store Dry Goods Store Furniture Store
	Mail Order	Dry Goods Ready-to-Wear Tyres Hardware

The technique of the investigation in these studies is shown on Figs. 9 to 16, which are taken (with the author's permission) from Kolb's study of the character of the relations of town and country in Walworth County, Wisconsin, and the changes in these relations since 1913, when the same area was first studied by Galpin in his pioneer work. The basic data were obtained from circulars sent throughout the county to individual farm families. Fig. 9 shows the position of the county and the grading of its service centres. The following maps show the changes in specific service areas—library service areas (Fig. 10), milk marketing areas (Fig. 11), high school areas (Fig. 12), grocery areas (Fig. 13), church areas (Fig. 14), dry goods areas (Fig. 15), and banking areas (Fig. 16). It is upon these services and service areas that Kolb bases his classification of service centres given above, into the three major classes of country centres, town-country community centres and urban centres.

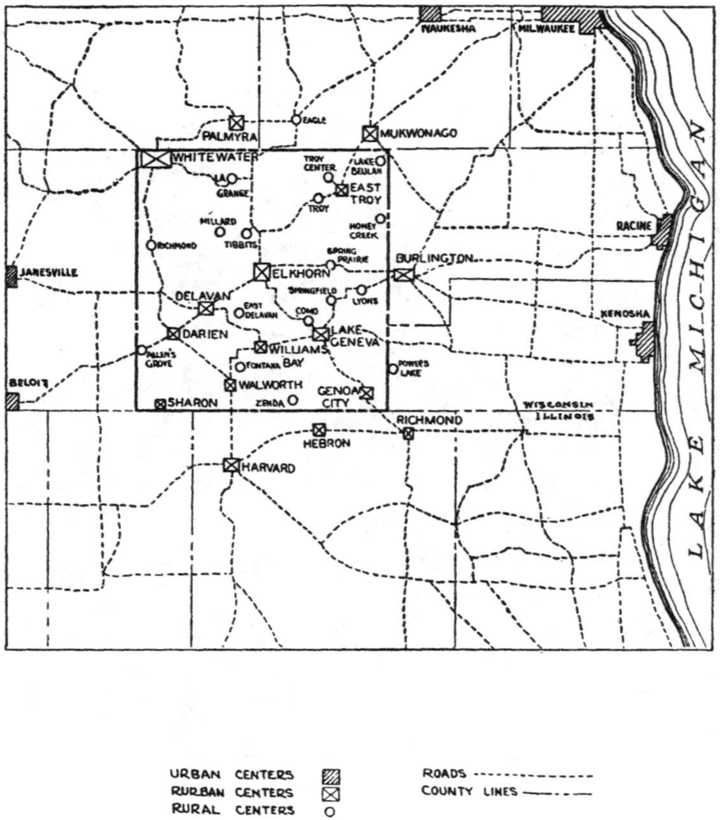

Fig. 9.—Service Centres in and about Walworth County, Wisconsin.

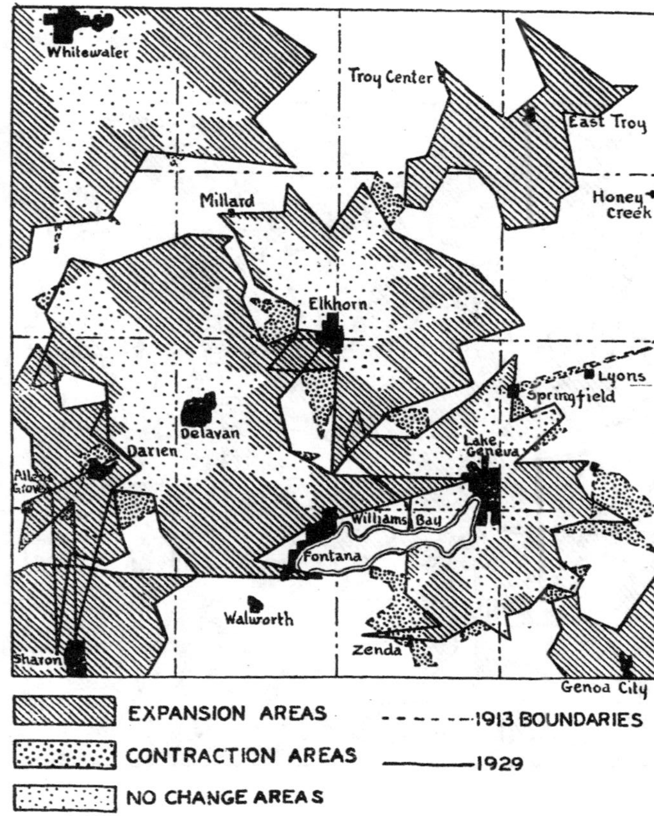

Fig. 10.—Walworth County, Wis. Changes in the Library Service Areas. The same key applies to Figs. 11 to 16.

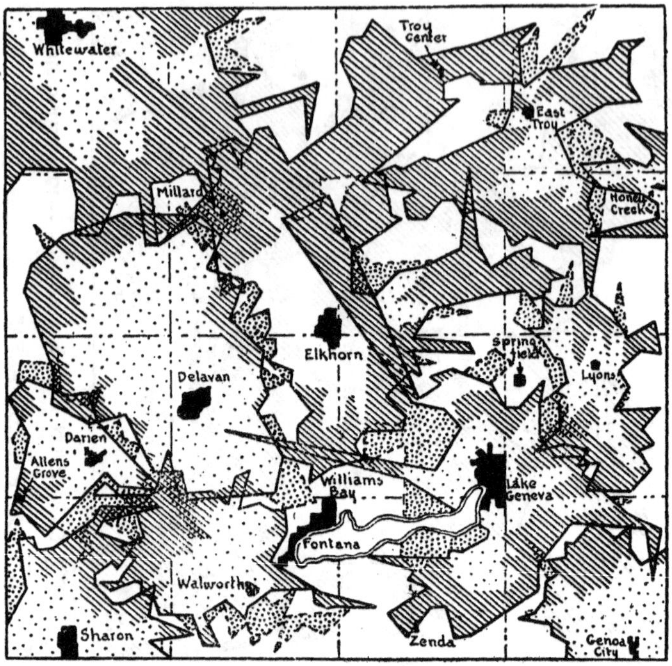

FIG. 11.—Walworth County, Wis. Changes in the Milk Marketing Service Areas.

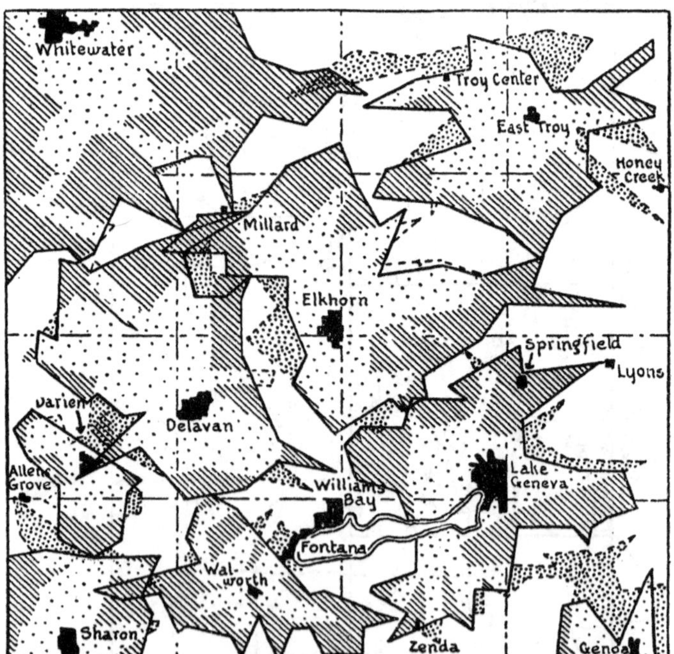

FIG. 12.—Walworth County, Wis. Changes in the High School Service Areas.

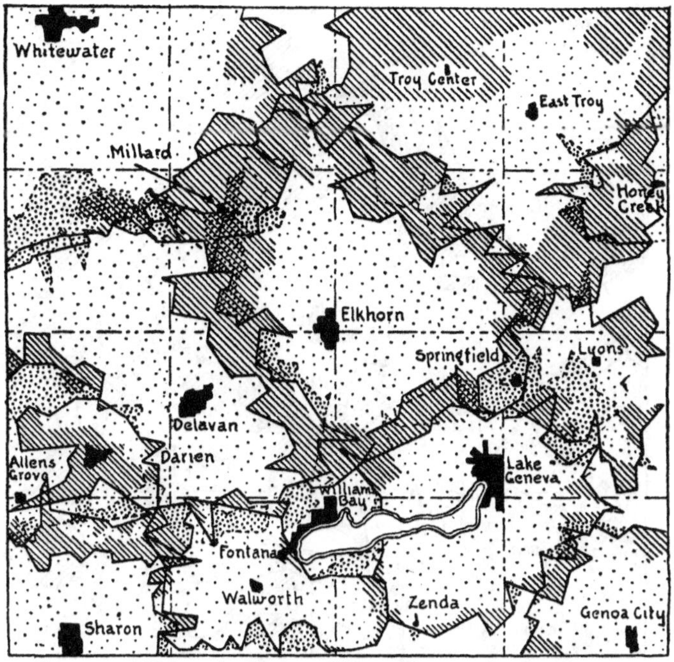

FIG. 13.—Walworth County, Wis. Changes in the Grocery Service Areas.

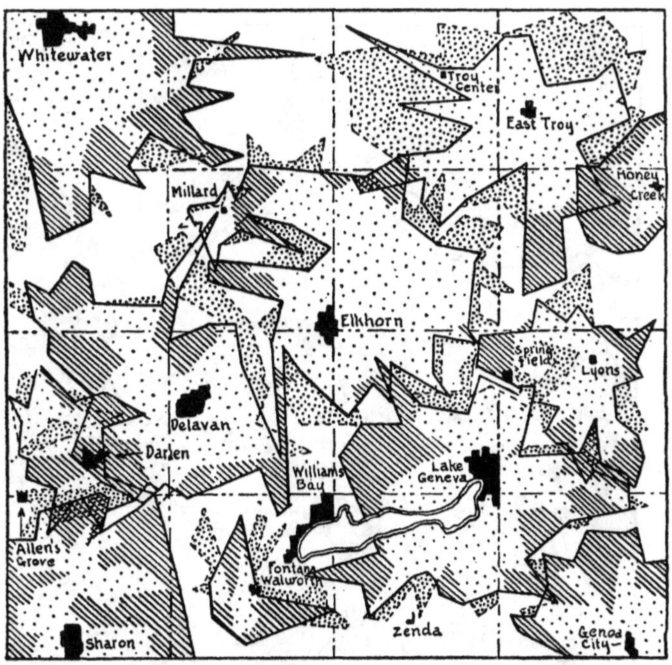

FIG. 14.—Walworth County, Wis. Changes in the Church Service Areas.

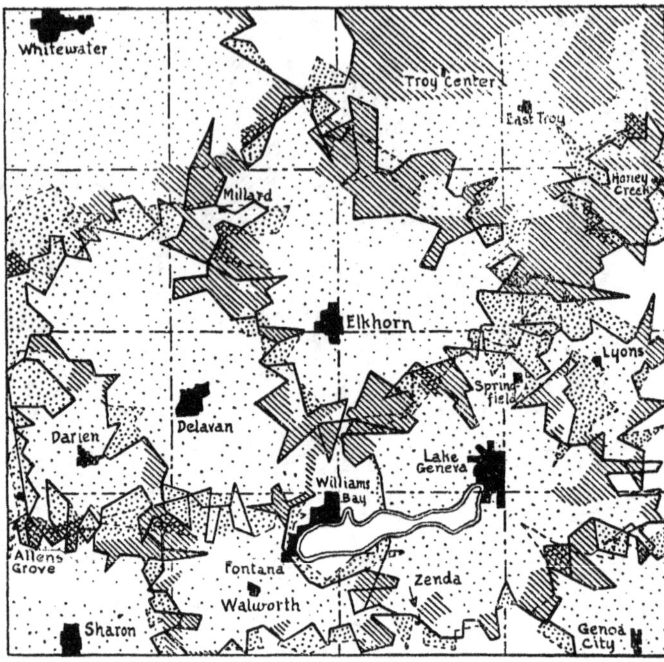

FIG. 15.—Walworth County, Wis. Changes in the Dry Goods Service Areas.

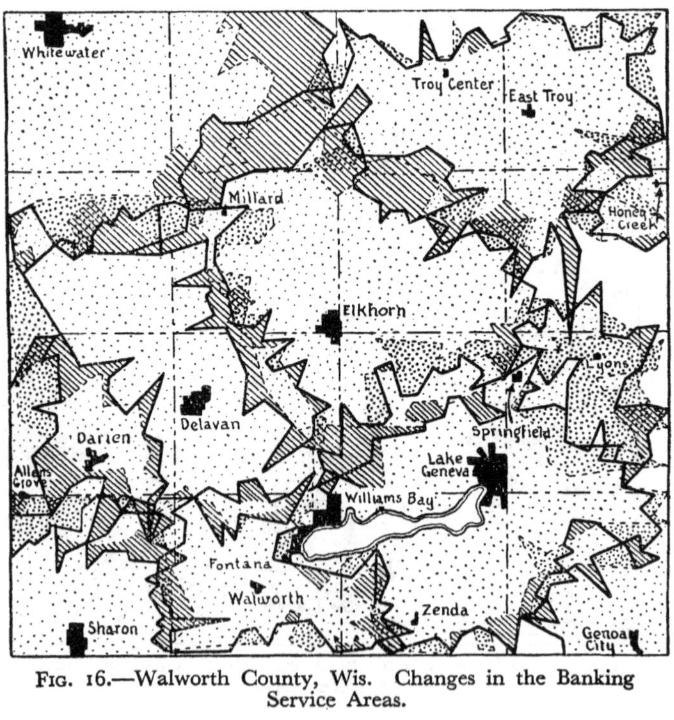

FIG. 16.—Walworth County, Wis. Changes in the Banking Service Areas.

The composite community areas of 140 village centres and the changes in areas from 1924 to 1930 have been studied in various parts of the United States for purposes of the monograph on *Rural Social Trends*. "In every region, and by every method of study, came the report that the high school was the most important single factor gauging village country relations and areas"[1] and "there is a growing tendency for such services as trade, education, religion and recreation to be organized about the village as a centre."[2] The average size of the service areas of centres in different parts of the United States is given below from the same study. It will be noted that the larger the village, the larger its tributary area in each region, and the sparser the rural population density, the larger the tributary area.

SIZE IN SQUARE MILES OF THE AVERAGE VILLAGE COMMUNITY AREA BY REGIONS (1930)

From Brunner and Kolb, *Rural Social Trends*, p. 95.

	All Villages	*Small*	*Medium*	*Large*
Mid-Atlantic	50	43	46	87
South	108	77	111	146
Mid-West	114	96	113	148
Far West	251	121	365	223

Small Villages	250–1,000 inhabitants
Medium Villages	1,000–1,750 ,,
Large Villages	1,750–2,500 ,,
All Villages	250–2,500 ,,

This whole position in the United States has been summed up in the following words[3]:

> From the standpoint of merchandizing, three types of trade centres are developing in rural areas. First, there is the primary service centre, a small town offering goods that are well standardized and frequently demanded. These towns are usually under 1,000 in population. Secondly, there is the shopping centre, a town which, in addition to convenience goods, offers goods in specialty stores. Such places may vary from 1,000 to 5,000 in population. Finally, there is the terminal centre, which is large enough to offer the most specialized kinds of services. These centres are usually the larger cities in a State or other area. A process of integration of service centres is taking place to secure maximum efficiency of the services to the donors of the service and its recipients. The chain store illustrates this point. There is specialization as between shop and shop and the shops of different towns, all being dependent upon the terminal trade centre.

[1] E. de S. Brunner and J. H. Kolb, op. cit., p. 97. [2] Ibid., p. 92.

[3] C. R. Hoffer, "Services of Rural Trade Centers", in *Social Forces*, Vol. X, No. 1, October, 1931, pp. 66–71.

The same writer goes on to point out that the country marketing services, such as livestock pens, are dying out in many country places and becoming concentrated in larger centres, facilitated in large measure by the development of "trucking", i.e. road haulage by lorry. Only in a few cases, such as grain elevators and creameries, does the original distribution of marketing agencies remain. Banks in villages failed during the Great Depression. A bank can exist on an independent basis in a small town with 500 people, but requires support from larger banking centres in times of financial stress. In professional services, the shift to the larger centres is less marked. Research studies in the States show that approximately 1,000 people can support a doctor. It is estimated that 10,000 people are required to maintain a well-equipped hospital. In social welfare, the county is becoming the unit of support and administration, and there is a vital need for the co-operation of town and country groups. In Michigan, over 50 per cent. of the towns with above 500 in population have cinemas, whereas only a fourth of the towns smaller than this have them, and even in the bigger towns the terminal or metropolitan centre is visited for choice. Newspapers require a town population of at least 1,000. Probably 1,000 people is the requirement to support a church. Many social services will stay in the country, for the country population depends on simple organizations and face-to-face contacts, whereas specialized organizations and indirect relationships are centred in the towns. Country and towns, however, are becoming increasingly interdependent, and the provision to the countryside of the services existing in the towns is one of the most effective ways of increasing the general standard of rural living.

3. England : Town–Country Relations in East Anglia

It is important to realize that throughout most of the United States the township is an arbitrary geographical unit defined as a rectangle six miles square. With the settlement of the Middle West in the latter half of the nineteenth century, neighbourhood groups had to organize themselves so as to provide the essential social services of all civilized communities—church, school, and meeting-places and a general store. The neighbourhood unit is thus comparable in area, population, and function with the parishes of the original thirteen states and with the parish in England ; and the tendency for the neighbourhoods to be integrated towards nearby towns and large villages is precisely analogous to the integration of activities in larger centres and to the ever closer service bond between town and country.

The " rurban " community in Britain is the small town with less than about 5,000 people and the villages around and dependent upon it. This service area is most clearly defined by the market area of the town, since where the farmer buys and sells his farm goods, he and his wife will carry out much of their other business and make most of their social contacts. The integration of services in this country in towns of different sizes has been going on for over a hundred years, and the structure of the centre and the extent and potency of its service area are always in a state of change. This process has been much quicker in the United States owing to the universality of the automobile, and the changes in the last decades are far more obvious and easy to assess. Little work of this kind, however, has been attempted in Britain.

The distributions and functions of the urban settlements of East Anglia (the counties of Norfolk, Suffolk and Essex, excluding Greater London) have been the subject of a special study by the present writer on the same lines as the American studies quoted above. This may be taken as an example of typical conditions and trends of town–country relations in eastern Britain.[1]

East Anglia is an area of low relief and is one of the chief grain-producing areas of Britain. About 60 per cent. of the farmed area of the three counties was devoted to arable farming before the 1939 War, the remainder being under permanent grass and heathland. The whole area is characterized by the relative absence of industry, though formerly its woollen and worsted manufactures were important, and all three counties are overwhelmingly agrarian in character.

On the basis of the 1831 Census supplemented by early directories, which portray the conditions prevailing prior to the changes wrought by railways and the general material progress of the nineteenth century, the nucleated settlements of East Anglia have been divided into three functional groups,[2] according

[1] R. E. Dickinson, " The Distribution of Functions of the Smaller Urban Settlements of East Anglia ", *Geography*, Vol. XVII, 1932, pp. 19–31 ; " The Markets and Market Areas of East Anglia ", *Economic Geography*, Vol. X, 1934, pp. 172–82 ; " The Markets and Market Areas of Bury St. Edmunds ", *Sociological Review*, Vol. XXII, 1929, pp. 292–308. A. W. Ashby, " The Effects of Urban Growth on the Countryside ", *Sociological Review*, Vol. XXXI, 1939, pp. 345–69.

[2] Compare this classification with one suggested for the Craven district of Yorkshire, based on the Poll Tax Returns by A. Raistrick, " A Fourteenth Century Regional Survey ", *Sociological Review*, Vol. XXI, 1929, pp. 38–49.

to the nature of the non-agricultural operations pursued, the number of persons engaged in them, and the proportion of non-agricultural to agricultural workers.[1]

Rural villages are mainly engaged in agriculture, with a small proportion of their inhabitants supplying non-agricultural services, mainly to meet the needs of the parish which each serves. A large variety of the latter services is found as a rule only in those parishes which have over three hundred inhabitants.

Urban villages have higher functional status. They have a larger proportion of non-agricultural workers, engaged in retail trades and handicrafts—plumbers, saddlers, cabinet-makers and specialized retail vendors (drapers, grocers and chemists as distinct from the village shopkeeper). They are found owing to the necessary grouping of these more specialized services in fewer centres, and are usually located at local road junctions, while, when grouped with the towns (see below), they are fairly evenly distributed at intervals of two to four miles. Most of these villages have a population of seven hundred to a thousand, but since 1850 they have lost population by emigration in sympathy with the depopulation of the rural areas. Usually the decrease is least, and urban functions have been best maintained in those villages which lie on railways, and greatest (over a third) at a distance from the railways in the midst of the areas of heavy boulder clay soil of east-central Suffolk and northwest Essex.

The *towns* are distinguished by the greater variety and number of those services which are also found in the urban villages, and by their distinct commercial, professional, financial and cultural elements (schools, solicitors, banks, etc.). The number of non-agricultural workers always exceeds that of the agricultural, but the percentage of the former to total workers (including industrial) does not always exceed 50 per cent. ; a minimum of 33 per cent. is a more accurate criterion of distinction between urban village and town.

The number of persons employed in non-agricultural pursuits is a truer criterion for defining urban status than the ratio method, which is vitiated by the varying size of the parishes. On Fig. 17 all places with over a hundred males engaged in retail trades,

[1] From 1801 to 1841 the Census Reports give full details of occupations for *every parish*, but subsequently for only a few of the larger towns, and in recent reports for urban and rural districts, the many small urban settlements being included in the latter. Moreover, there have been considerable changes in the classification of occupations. For these reasons, while detailed statistical investigation of urban settlements is possible for the early nineteenth century, comparison on the same basis with present conditions is impossible.

handicrafts and commerce, are classed as towns. This distinction is based upon the occupations of the smallest settlements, in compact parishes (thereby reducing the agricultural element to a minimum), which possess typical urban functions. The majority of these places were market places in the early nine-

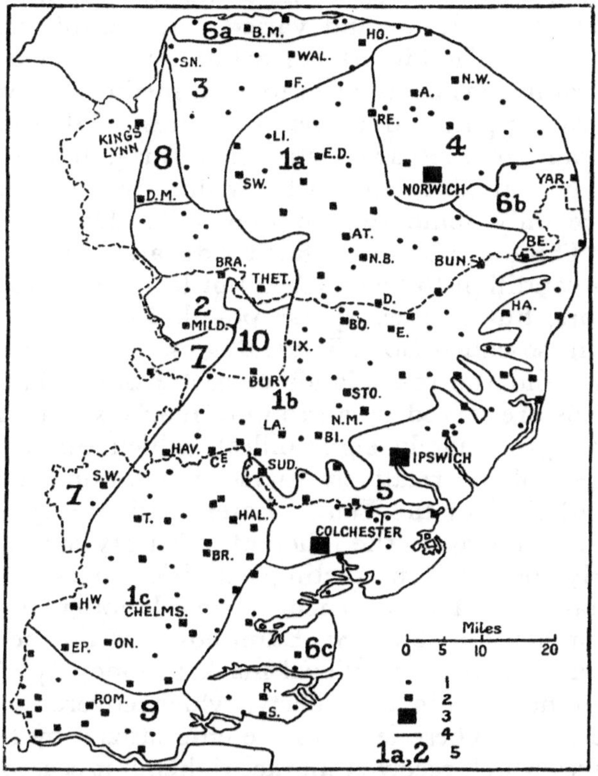

Fig. 17.—The Urban Settlements and Physical Regions of East Anglia in 1831. The symbols show 1. 50–100 persons engaged in retail trades, handicrafts and commerce. 2. Over 100 persons so employed. 3. County towns. 4. Boundaries of the physical regions, as follows: 1. *a*, *b*, *c*, High Anglia. 2. Breckland. 3. Good Sands. 4. N.E. Norfolk Loams. 5. East Suffolk Sandlings. 6*a*. North Norfolk Marshland. 6*b*. Broads. 6*c*. East Essex Marshland. 7. Chalk Downland. 8. Greensand Region. 9. London Clay Region. 10. Light Loam Region of West Suffolk.

teenth century, and, in addition to a number which subsequently declined, they are shown on Saxton's county maps of 1570–6, from which are to be obtained the earliest complete records of old market towns. Of the remainder, some are handicraft centres, with recently introduced manufacturing industries, e.g. Long Melford and Glemsford; a few are merely coaching (or "thoroughfare") towns; and several are places with a variety of retail trades and handicrafts.

The *maximum* range of influence of the medieval market was about six miles; indeed, it is still illegal, in accordance with an old law, to establish a market within six and two-thirds miles of an existing legal market. The actual market area, however, rarely reached this limit, and varied considerably according to local conditions. In Fig. 18 an arbitrary market area of only four miles radius has been given to each medieval town (to avoid excessive overlapping), but even with this small radius, a belt of overlapping market areas stretched through the richer and more closely settled inland belt, within which, therefore, all places were only some two to four miles from one or more markets. Three areas lie at a farther distance from their nearest markets, a condition to be ascribed not to the absence of suitable siting factors, but to their sparse population till the nineteenth century—the so-called Good Sands Region of north-west Norfolk,[1] the Broads, and the marshlands of east Essex. The Breckland and the Fens are served by a group of markets spaced at greater intervals (approximately eight miles), which originally catered for the needs of the relatively sparse population, and marketed the fish and fowl of the Fens. Later, with the reclamation of the latter, and owing to the inherent infertility of the Breckland, these partly lost their marketing activities to larger and more distant markets. The absence of overlapping market areas around Norwich and Bury St. Edmunds will also be noted, due in the latter case to the medieval market monopoly of the abbey of St. Edmunds within its Liberty, which covered a large part of west Suffolk. These features are emphasized by the distribution of places which received market rights, but ceased to exercise them before Saxton's day. Their concentration in the centre of the region, and absence in the east of Norfolk and Essex, and in the west, around Bury and in the Fens and Breckland, is very marked.

The markets in the early nineteenth century (Fig. 19) have been divided into three groups—large, small, and disused or declining. The decline of so many markets before the advent of the railway is due to three main causes:

(1) The development of road transport after 1780, a result of which, as shown by ample evidence, was the con-

[1] Several places were granted market rights in this region, in the early Middle Ages, but their markets had ceased to exist before Saxton's day.

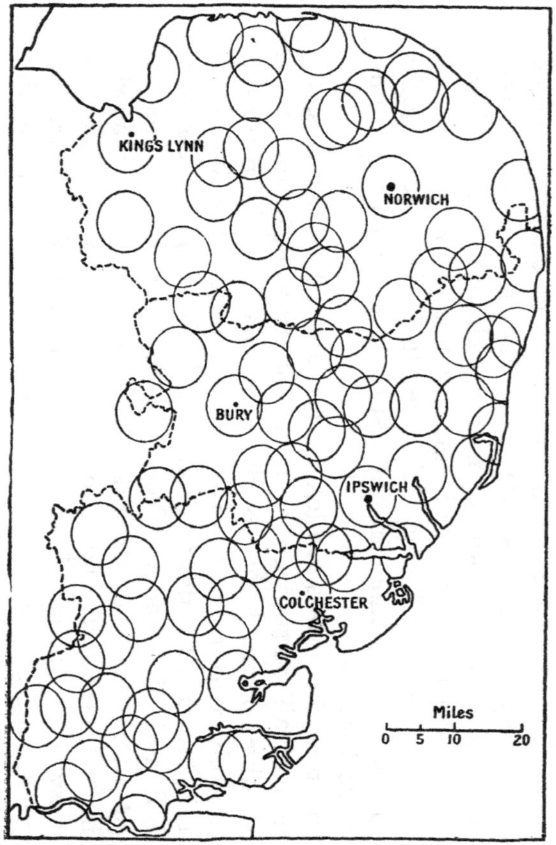

FIG. 18.—Markets in East Anglia in the Sixteenth Century (after Saxton).

Each market town is given an arbitrary market area of four miles radius.

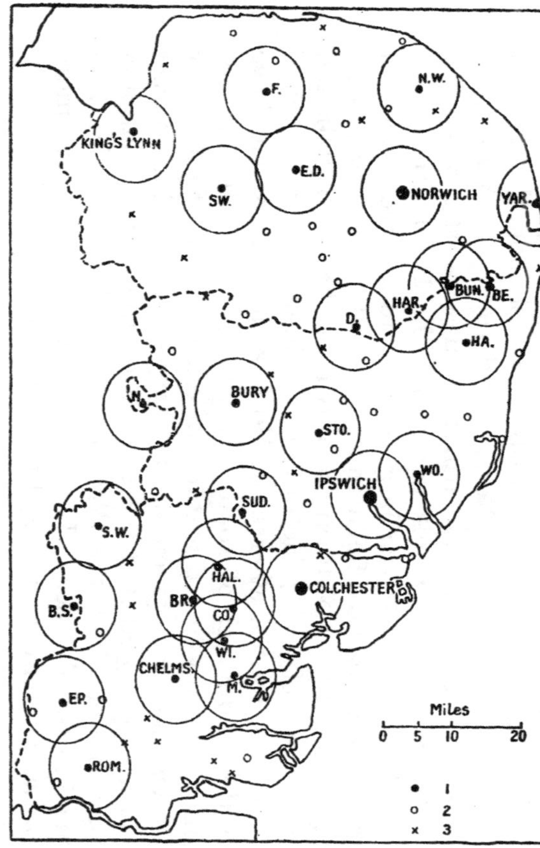

FIG. 19.—Markets in East Anglia in the Early Nineteenth Century.

Each large market has an arbitrary six miles market radius. 1. Large Markets. 2. Small Markets. 3. Disused or declining Markets.

centration of marketing on fewer towns, due to their increased marketing radius.

(2) The decline of old-established industries, mainly the worsted and hemp industries, which notably affected the towns of south Norfolk and the Stour Valley.

(3) The lack of productive tributary areas as shown by the towns of the Suffolk coast and the Breckland.

It will be noted that the large market towns are spaced at intervals of from ten to fifteen miles. In all cases they possess the greatest nodality. Natural nodality is marked in Suffolk and Essex, where relief and penetration of tide-water are the chief determinants of routes, whereas in Norfolk the chief towns have become the main foci of roads the course of which was not determined by pronounced features of relief. Each of the larger market towns has been given an arbitrary six-mile radius market area. It will be noted that there are two groups of overlapping market areas, one along the Waveney from Diss to Beccles, and the other in central Essex. It is in these areas that, with more rapid railway communications, further selection would be likely to occur.

In the nineteenth century, there were advances in material progress which greatly augmented the functions of urban centres as foci of social, economic, administrative and cultural integration ; the advent of the railway, and latterly the motor-bus, occasioned the concentration of many such functions in fewer centres, although to meet new local demands a few new market towns have come into being. The result is that the majority of the towns have declined in population since 1851, while only a few have increased.

In most of the towns, which to-day are urban districts, the proportion of all persons engaged in commercial, financial and administrative occupations to those engaged in industries is approximately two to one. In the remainder, the industrial element is larger (Thetford, Sudbury, Braintree) or dominant (Haverhill and Halstead). Towns in the last two groups all increased after 1851. In the first group, however, the majority (including such relatively large towns as Swaffham, Eye, Bungay and Halesworth) decreased because of the lack of industries to offset the decline of the handicrafts, and because of the decline of their commercial functions ; while the remainder grew at their expense through the augmentation of their commercial functions, e.g. Dereham and North Walsham.

This conclusion is substantiated by a comparison of the commercial functions of the chief towns in 1831 and 1931, based on the Census Reports, which reveals a marked decrease in the numbers employed in commerce and administration in many old-established towns which were flourishing market centres the beginning of the nineteenth century. The same feature is apparent in the development of livestock markets, many of which

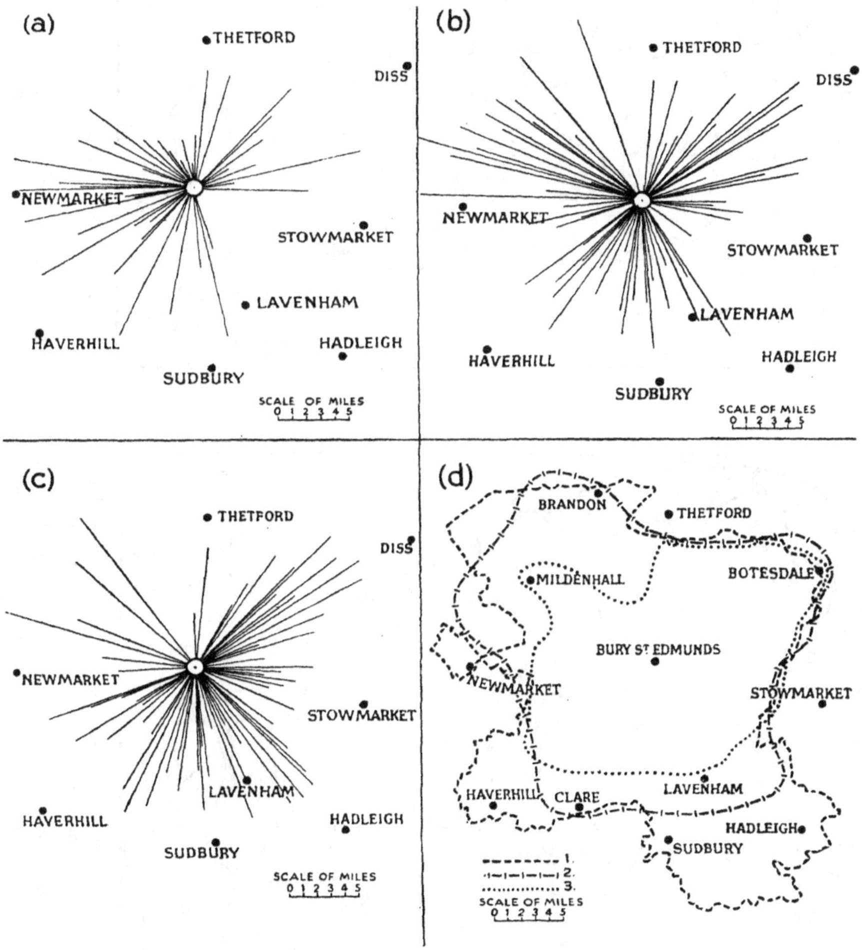

FIG. 20.—The Market Areas of Bury St. Edmunds.

The central circle represents Bury St. Edmunds. The lines are drawn from Bury to the village centres of the parishes from which stock was sent in 1929. (*a*) Sheep area, (*b*) pig area, (*c*) cattle area. The composite market areas are shown on Fig. (*d*) in which 1. is the West Suffolk county boundary, 2. the composite livestock area and 3. the egg and poultry area.

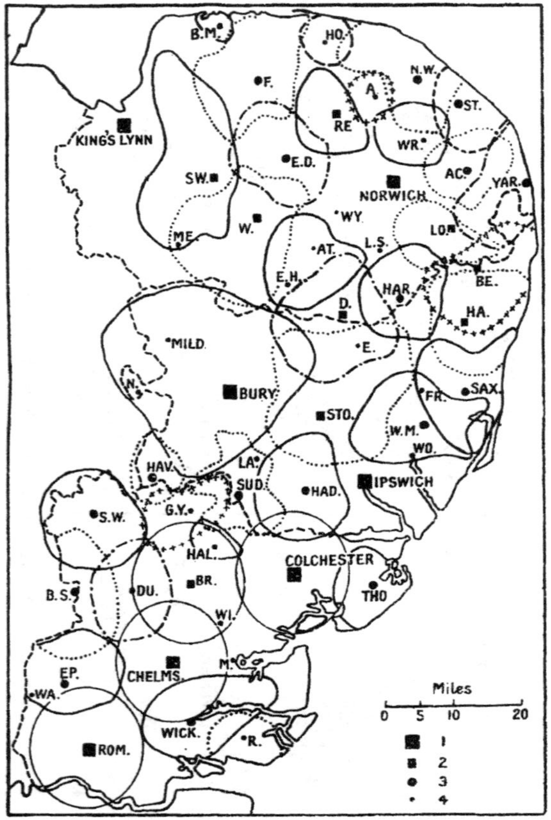

FIG. 20*a*.—Market areas in East Anglia in 1931, based on information obtained from auctioneers.

In several cases where the actual area could not be obtained, circles with an 8-mile radius have been drawn, within which most of the local marketing takes place. In order not to complicate the map, the market areas of the chief markets are not directly shown, but their main areas may be inferred. The line of plus signs encloses the area served by the seasonal sheep sales at Swaffham.

Symbols show annual livestock sales: 1. Over about 50,000. 2. 20,000–50,000. 3. 10,000–20,000. 4. Under 10,000.

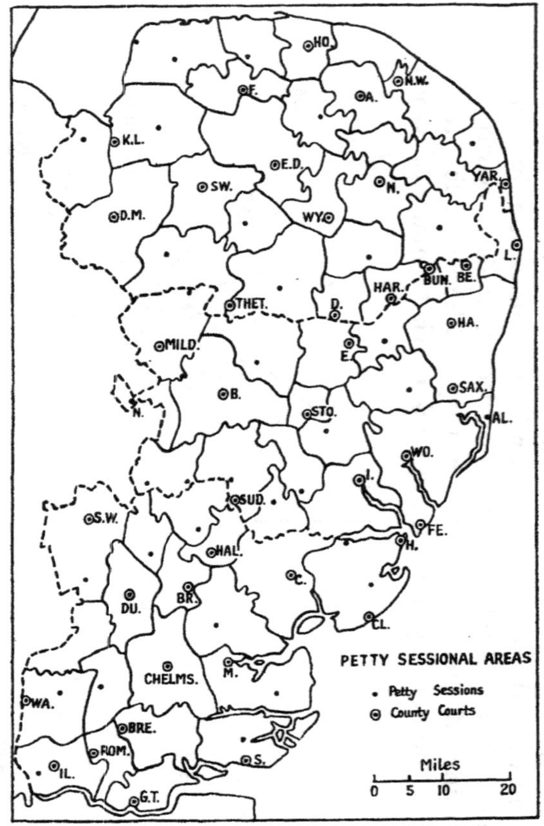

FIG. 21.—Petty Sessional Areas in East Anglia.

are to-day disused or extremely small. It is repeated in the case of corn sales. These were effected, and still are to a small extent, in public houses in many small towns, but the main sales, according to the returns made to the Ministry of Agriculture, are effected in a few of the smaller towns. The great increase of business done at Norwich, Ipswich, Lynn, Bury, Colchester and Chelmsford, as a result of rapid communications, indicates that the process of concentration is not yet complete.

Among the towns with disused markets shown in Fig. 19, Bungay is one of the group of towns along the Waveney which at the beginning of the nineteenth century had overlapping market areas, and Coggeshall, Witham and Maldon are members of the Essex group. But new markets in small towns, each with an increasing population, have grown up in the east of Norfolk and Essex. These markets deal with fat stock, marketed over short distances, and supply, particularly in Essex, the coastal resorts near to them.

There are certain services, the specialized retail services, which are common to *all* urban settlements, but difficulty has been found in assessing their distribution and the urban status of settlements in which they are located. For, as stated above, accurate statistical treatment of existing urban settlements is impossible because of lack of census data. A functional analysis was attempted, the data being obtained from recent directories. The places were listed in which each of the specialized commercial services were located. The services considered were all the specialized retail services (grocers, chemists, outfitters, etc.) and banks. These lists suggest that the presence or absence of one or more banks is the best general indicator of the importance of the retail services ; and that, in drawing a dividing-line between town and village, of the many places with only one bank (often subordinate to a branch in one of the large towns), those which have the *full* quota of retail services are towns, and the remainder urban villages. The latter form about 50 per cent. of the urban villages as defined for 1831. The establishment of banks in such places has simply been a process of selection of as few centres as possible to serve the whole area adequately. It is thus no accident that nearly all these places are situated on railways. The absence of banks in a large area of west Norfolk, the region of Bury St. Edmunds and north-west Essex, is to be related to the absence of large villages in these areas, and the concentration of services in central or peripheral towns.

All places with two or more banks are undoubtedly urban, and those with more than three include most of the principal market centres, though Downham, Wymondham, Bungay, Halstead and Witham are exceptions.

The administrative, social and cultural services show, in some measure, the same features of concentration. The only administrative functions at the beginning of the nineteenth century were *legal* in character, and most of the small market towns were and still are centres for Petty Sessions, the areas being based upon the Hundreds (Fig. 21). In many cases, this is the only administrative function which such towns possess. At the larger towns were held the Quarter Sessions and County Courts. In 1834 the Poor Law Unions were established, with the chief towns as their centres, and their areas roughly defined by the extent of the market areas (eight to ten miles) of the towns. The Rural Districts were subsequently based upon these areas (Fig. 22). As the medieval market towns were usually at four-mile intervals, small towns are peripheral to the unions and rural districts. Similar concentration is characteristic of the distribution of the new social services—cinemas, newspaper office, secondary schools and hospitals.

From the foregoing it will be obvious that, while all towns have certain common functions and characteristics, their " urban status " varies greatly according to the number and variety of functions with which they are endowed. An attempt will now be made to grade the towns of to-day according to function and correlate functional status with population. The scheme lays no claim to detailed accuracy ; it is merely an attempt to correlate and co-ordinate the functional data summarized above.

First, all the towns have a population over a thousand. This seems to be the critical dividing-line as far as can be gathered from comparative study of functions, between towns and urban villages. The smallest towns, with one bank, the essential retail services and sometimes petty sessional meetings, exceed this limit. Secondly, the majority of these places are distinctly urban in that they are big enough to support, or their size necessitates, the provision of public services—water, gas and electricity and main drainage.

With these features as basic criteria, the towns may be graded, according to their functional status, as follows (see Fig. 23) :

I. The first group are the towns in which all the services are concentrated. These may be described, from this point of view, as fully-fledged. They have general livestock markets with aggregate sales exceeding 10,000 head per annum, three or more banks, a cinema, newspaper, secondary school and usually some local industries. Moreover, they frequently have subsidiary concerns (banks, co-operative societies) in neighbouring smaller towns. They are all the culminating points of local urban integration, and the majority record increases in their population and commercial functions since the middle of last century (e.g. East Dereham, Beccles, Braintree). Their population is usually about 5,000 or above.

II. The second group are towns with most of the services characteristic of Group I, but they differ from the latter in that they have small or decayed markets. They have a population of 1,500 to 4,000, but the majority have lost population and declined in commercial functions since 1851 (e.g. Swaffham, Bungay, Halstead). The towns in both these groups had large markets a hundred years ago, and the decline in the status of Group II, as a result of railway transport, is clearly shown by a comparison of Figs. 19 and 23.

III. A third group includes settlements with good markets (livestock sales exceeding 10,000 head per annum), but with none of the other services, i.e. no cinema, newspaper, etc., and only one or two banks. The distinctive members of this type are in the east of Norfolk and Essex. They usually have 1,000–2,000 inhabitants and all record an increase of population since 1851, at which time a number were only urban villages. These are in effect centres of very low urban status, and owe it mainly to their good markets, which, in these cases, are in no real sense indicative of town character.

IV. The lowest grade comprises a number of small towns with a population of 1,000–2,000, which have decreased greatly since 1851. Though they have banks and the specialized retail services, they have none of the social services noted above, no industries, and they have lost (recently or several centuries ago), or never had, markets, e.g. Thaxted, Debenham, Wickham Market.

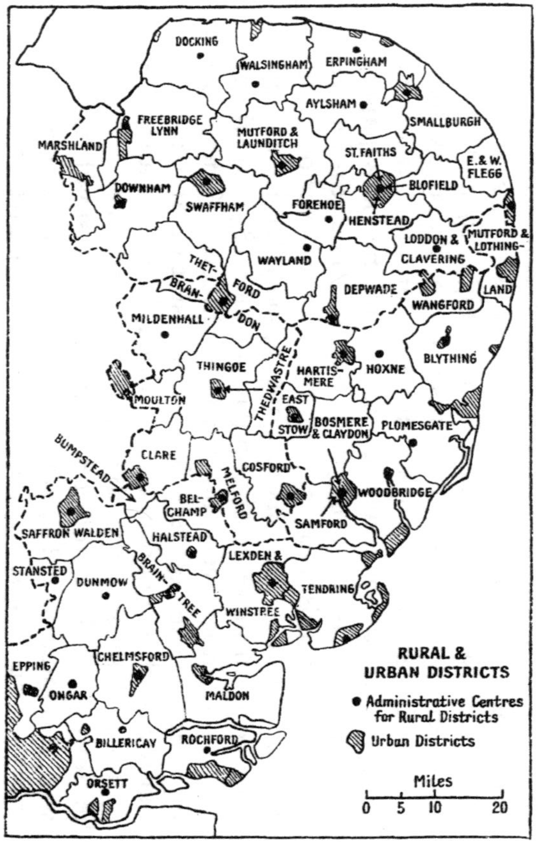

FIG. 22.—Rural and Urban Districts in East Anglia.

Urban Districts include Municipal and County Boroughs. Boundaries are as in January 1927, and subsequent changes are not shown. Aldeburgh U.D. should be added.

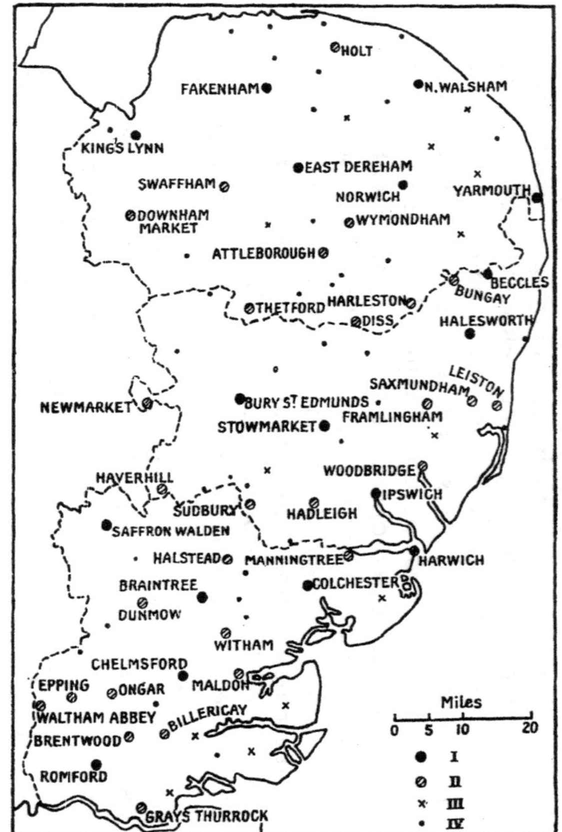

FIG. 23.—Urban Settlements in East Anglia.

Four grades of smaller urban Settlements are recognized and described in the text. (Add Lowestoft.)

The distribution of these four grades of towns is shown on Fig. 23. It will be seen that the fully-fledged towns of to-day are fewer than the " large market towns " of a hundred years ago. (Compare Figs. 23 and 19.) Those of the latter, which have lost some of their commercial functions as a result of railway development, are located in the two belts of early nineteenth century overlapping market areas, viz. Harleston and Bungay, Halstead, Witham, Coggeshall and Maldon, and Sudbury. Apart from these, Swaffham is an excellent example of a town, first increasing its marketing activities at the expense of its neighbours (e.g. Downham) and later itself succumbing to the competition of larger markets. To-day, though it has large seasonal sheep sales, it has no weekly market, and has lost much of its commercial importance.

Grouped round the Grade I towns are smaller towns of varying urban status at about four-mile intervals. The latter serve areas of more limited extent with a radius of some four to eight miles. To complete the picture, the urban villages have retail and other services to serve groups of easily-accessible villages within a distance of two to four miles.[1]

While the actual distribution of these towns dates back to their origin some six hundred or more years ago, the development of transport and of marketing organization, and the growth of new social services, have occasioned the concentration of functions in a few, and this involves considerable extension of the areas which the latter serve. This maladjustment of present economic, social and administrative services to the distribution of towns, which is in the main a medieval legacy, is reflected in three fundamental and general problems of rural activities and organization.

(1) Administrative areas, which are based upon the parishes and hundreds, established many centuries past, and the Poor Law Union areas (now abolished), established before the advent of modern transport, bear no relation to the existing areas of economic orientation. Our present system of administrative areas needs drastic

[1] Compare this with Smailes' scheme (summarized above on pp. 47–8) and with his map, op. cit., p. 43. According to Smailes' map, Norwich ranks as a *major city*, Ipswich as a *city*, Kings Lynn, Great Yarmouth, Lowestoft, Bury St. Edmunds, Colchester, Chelmsford and Southend as *major towns* (or *minor* cities). His *towns* correspond generally with our Grade I and II settlements, and his *sub-towns* with our Grade IV settlements. He rightly excludes markets as criteria of urban status so that our Grade III settlements are omitted from his scheme.

revision, and in its more rational reorganization the distinctive functions of, and the areas served by, urban settlements should be adopted as basic criteria.[1]

(2) In social surveys of rural areas, prior to the formulation of any idealistic scheme of reorganization, the distribution and character of all existing social services should be thoroughly investigated as a first step. This should be followed by a systematic survey, as far as this is possible from the data available, of the distribution of and areas served by hospitals, schools, cinemas, libraries, retail firms, wholesalers, etc. To determine the nature of the necessary reorganization of these services, i.e. to determine whether certain services are redundant or inadequate, selected criteria would have to be adopted as a basis. Little work has been attempted on these lines in this country, but the question has received considerable attention in rural areas of the U.S.A.[2]

(3) The study of markets and the extent of the areas served by them reveals the fact that many are redundant—and this in East Anglia, where, as compared with other parts of England, there are relatively few. Many markets have sales under five thousand head of stock each year. They draw few buyers and sellers, and the farmer would get better terms by travelling farther to bigger markets. In fact, for livestock and corn sales and general retail trade this is an almost universal tendency in East Anglia, and, with the exception of areas elsewhere, where small markets sometimes do serve a useful purpose, it is strongly encouraged by the Ministry of Agriculture as a healthy development in modern marketing organization.

The "market district" in Germany, as defined by Christaller, has a radius of about three miles and a population of 1,600 to

[1] See H. J. E. Peake, "The Regrouping of Rural Population", *Geog. Teacher*, Vol. IX, 1917, pp. 71–7, and "Geographical Aspects of Administrative Areas", *Geography*, Vol. XV, 1930, pp. 531–46.

[2] See, for example, J. H. Kolb, "Service Institutions for Town and Country", Agric. Expt. Station, Univ. of Wis. Research Bulletin 66, 1925. The ideal distribution of services (hospitals, libraries, schools) he determines from the minimum number of service units (bed, book, pupil) necessary for the successful organization of each service, and this in turn depends on the total population of each town together with that of the area which surrounds it.

2,700 inhabitants.[1] Its centre is the small market town (M) with 1,000 inhabitants. The corresponding feature in East Anglia is the small town with between 1,000 and 2,000 inhabitants, or even the urban village with 750 to 1,000 inhabitants. This nucleus, together with the half-dozen or more villages served by it, form a district with a total population of about 2,000 to 3,000 in East Anglia, and this figure may be taken as typical of other rural areas in Britain with the same density of population. It seems that such a district might serve as a suitable unit area in a new administrative system.

NOTE

Since this chapter was written there has appeared the symposium of the Agricultural Economics Research Institute, Oxford, entitled *Country Planning: A Study of Rural Problems* (1944), published by the Oxford University Press, under the direction of Dr. C. S. Orwin. This is an intensive survey of a particular rural area containing twelve villages covering 24 square miles. (Three parishes have just over 1,000 inhabitants each and the remaining twelve have 150 to 350 inhabitants.) A main conclusion of this survey is the inadequacy of the small village to function as an active seat of rural activity.

The village of a few hundred people cannot survive as a healthy organism. It cannot maintain any of the social services ; it must send its senior, and sometimes all its children away for their schooling ; it must share the services of a district nurse ; it cannot bear the overhead costs of water supplies, sewerage, or electric light ; it has few shopping facilities ; it cannot support the usual recreational organizations, cricket and football clubs, Women's Institutes, Young Farmers' Clubs, Guides and Scouts, and so on, solely because there are not enough men, women and children of the various age-groups to run them ; it cannot give a living or a life to a resident parson or Free Church minister " (p. 274).

Orwin, in applying Fawcett's proposal for a rural residential unit of 1,200 to 2,400 inhabitants, points out that the three larger villages in this survey area " approximate to the lower of these figures, and they confirm in many ways that this popu-

[1] " We find throughout the Reich that the smallest market district is centred on a place of the lowest grade (Grade I) which is a commercial centre and a seat of administrative and professional services." The service areas specifically referred to, in addition to those of a commercial kind, are police, telephone and postal districts, professional organizations, and medical services. See Christaller, *Die Ländliche Siedlungsweise im Deutschen Reich und ihre Beziehungen zur Gemeindeorganisation*, Berlin, 1937.

lation can support a vigorous community life " (p. 275). They are identical with the urban villages of East Anglia. The same writer favours the controlled development of industry in rural areas. " Dismissing the idea of new industrial towns built round the factories as being anti-social and artificial, the alternative would be to repopulate all the little villages within a certain radius of the factory " instead of establishing " housing-estates " in the towns near the factories (p. 278). " Villages thus enlarged by the influx of industry should find themselves emancipated from most of the disabilities from which small rural communities . . . are suffering to-day " (p. 281).

PART II

THE STRUCTURE OF THE CITY

CHAPTER 4

THE STRUCTURE OF THE CITY PROPER

In the previous chapters we have examined the structure of both urban and rural settlements as service centres for the areas around them. In this section we shall carry this concept of the region as a social unit a stage further by examining from the same point of view the geographical structure of the large urban complex. This has two aspects, for the city is not only a single entity in space, but there are also regional groupings within it that differ greatly from each other.

This conception and its technique are of fundamental importance in the understanding of the structure of the city and planning for its future development.

The great city, in spite of its abnormal growth and the heterogeneous character of its build and population, is a unit of social life and organization, as defined in the last chapter, and is in itself an aggregate of such smaller homogeneous units, each having distinctive characteristics and playing a special role in the life of the city as a whole, and all finding their common nerve-centre in the central business district, the site of the original town and the nucleus from which expansion has taken place. In the small town with under about 10,000 people, differentiation by districts does not appear, but there is the same ordered arrangement of types of human occupance in relation to road, river and rail, grouped around a centre—usually a market-place or a High (or Main) Street—in which are clustered shops, offices and public buildings.

Now, in general, the administrative and local government districts of the large city bear little relation to these regions that have emerged through the development of uncontrolled processes, and, moreover, divisions used for census purposes, such as the borough and the ward, are too large and too diverse to reveal the distinctive and essential characteristics of such regions. In other words, we need a more rational division of the city into areas for local government and census purposes in order to understand more fully its structure and the processes that determine it. This is necessary not only for rebuilding derelict or obsolescent urban areas, but also for building new towns in the post-war world on sound principles.

This field of study—the actual structure of the city as a geographical unit—has received much attention in the last twenty years or more from geographers and sociologists both on the Continent and in the United States. We shall have to forgo detailed presentation of the geography of particular cities, but, for examples of such studies, the interested reader may be referred to the excellent articles in the American *Geographical Review* (New York) on both European and American cities. It is on the basis of such studies, geographical and ecological, that we present the following survey of the structure of the city as a whole and then of the smaller homogeneous units or regions that make up its component parts.[1] We have drawn freely on American studies, both individual and comparative, the latter, in particular, providing useful data that are much more difficult to arrive at for European cities. Specific points have been drawn, particularly from the case of Chicago, a city that has been a laboratory for the researches of social scientists at the University of Chicago, and which is well known to the writer.[2] European cities that are quoted as examples are Paris, Berlin and Stockholm.

1. The Process of Urban Growth

The general process of urban growth that underlies the location, growth, and structure of cities has been summed up as follows.[3]

(i) *Concentration.* This is the tendency for people to cluster in cities as near as possible to each other, to their work, and to the amenities of city life. It is essentially centripetal in

[1] The fact that little space is given to detailed appraisal of British work and that no British city is studied in detail in no sense means an underrating or neglect of such work or of the obvious importance of applying techniques discussed here to the study of our own towns and cities. This subject is now before the public eye, and many valuable research publications have appeared in recent years and these are accessible to the interested reader. Our aim is to stimulate work on new lines and to put the structure and problems of the great city in a clearer perspective by closer consideration of foreign researches, with which we are probably not all so familiar.

[2] T. V. Smith and L. D. White, *Chicago : An Experiment in Social Science Research*, University of Chicago Press, 1929.

[3] See R. D. McKenzie, " The Ecological Approach to the Study of the Urban Community ", in *The City*, by R. E. Park, E. W. Burgess and R. D. McKenzie, Chicago, 1925 ; and " The Scope of Human Ecology ", in *The Urban Community*, edited by E. W. Burgess, Chicago, 1926. See also N. P. Gist and L. A. Halbert, *Urban Society*, second edition, New York, 1941, Chapter 6, " The Social Ecology of the City ".

character. The " friction of space " has been described [1] as the essence of the process of urban growth and of differential segregation. It is also the prime cause of concentration in cities at all, since the services demanded can most efficiently be carried out in clustered communities at suitable locations—whether for purposes of industry, commerce, or administration—and where the work is, there the workers must live, with daily access to their place of work. The nucleus of this urban growth is the old town, that to-day is normally the hub of the city, from which the built-up area has spread steadily outwards both concentrically and radially, or by the growth of physically separate centres that gradually merge with the main urban area. The phenomenon of urban concentration is one of the outstanding features of our modern civilization, and has been frequently studied, so that we need pursue it no further.

(ii) *Centralization* denotes the " distributive pattern of population and institutions in the area of (urban) concentration, and the process whereby the patterns appear ".[2] It refers to " the drawing together of institutions and activities, i.e. the assembling of people to work rather than to reside in a given area ".[3] This process of centralization is exactly analogous with the clustering of the centralized services in urban centres in their relations to the countryside. It manifests itself in the emergence of a central business district in the heart of the city and of commercial subcentres around it, in the concentration of factories in distinct areas, and, in consequence, in the daily rhythm of movement of workers between homes and workplaces. The essential feature, therefore, is the separation of workplace and residence, and the segregation of workplaces into distinct districts according to function. By a process of competition, establishments seek out, and segregate themselves in, that area in which their optimum conditions are to be found and in virtue of which they are normally able to exclude others.

It was with reference to Stockholm, that the late Sten de Geer, Professor of Geography in the University of Stockholm, wrote :

That geographical laws determine the best position and distribution of each particular phenomenon is rarely a matter of conscious appreciation. Yet it is such laws that determine the areal growth

[1] R. M. Haig and R. C. McCrea, " Major Economic Factors in Metropolitan Growth and Arrangement ", in *Regional Survey of New York and Environs*, Vol. I, New York, 1927, pp. 38–9.

[2] Gist and Halbert, op. cit., p. 148.

[3] S. A. Queen and L. F. Thomas, *The City, A Study of Urbanism in the United States*, New York, 1939, p. 262.

and outer shape of the city as well as the distribution of function between its different parts and so differentiation in interior structure.[1]

In regard to the segregation of functions in different sectors of the urban area, the most outstanding feature, common to every city, is the law that its business and services concentrate in its centre, the " 100 per cent. locations ", from all parts of the urban area. This central district is very small—often facing on to a market-place or a single street in the case of the small town —but it becomes larger and more compact in larger cities, until not only is there a general concentration of business and services —shops, hotels, offices, restaurants, public buildings, etc.—but the different services are also segregated in districts within it. Owing to the great demand for space in this centre, land values are always here the highest in the city, and the great demand for space has a twofold effect on its structure. First, there is the trend of vertical expansion, that is, the construction of multi-storeyed buildings, which reaches its extreme in the skyscraper. Second, there is the trend of horizontal expansion, that is, the expansion of the business area to the adjoining streets and districts. This horizontal expansion usually takes the form of the conversion of residential properties, often quite unsuited for the purpose, to the use of business. Ultimately such properties are destroyed and make way for modern structures which adequately serve the new function, be it a modern hotel, office or civic building. This extension may be caused by private enterprise or by municipal slum clearance. The effect of the existence and lateral expansion of the city centre, with its high land values, is to raise values on the land around it. Since the city expands normally from its centre outwards, buildings become progressively older towards the centre. Thus, the combination of high land values and obsolescent buildings, ripe for demolition, accounts for the dingy-looking " zone of deterioration " that surrounds the central business centre of almost every city. In the American city in particular, where " real estate " is fluid, old buildings are let at high rentals and the sites (plus buildings) eventually sold at high profits. The expansion of the centre—especially of its high-class shopping and hotel quarter—is normally

[1] Sten de Geer, " Greater Stockholm, a Geographical Interpretation ", *Geog. Rev.*, Vol. XIII, 1923, pp. 487–500. This statement of urban growth is clearly substantiated in the thorough study of Stockholm recently undertaken by Professor H. W. Son Aklmann and his colleagues, entitled *Stockholms Inre Differentiering, Meddelande fran Geografiska Institutet vid Stockholms Högskola*, No. 20, 1934. See a summary of this investigation in " Stockholm : Its Structure and Development ", by W. William-Olsson, *Geog. Rev.*, Vol. XXX, 1940, pp. 420–38.

in the direction of the high-class residential district, as in the shift towards Kensington in London, and northward in Manhattan to Central Park in New York City.

(iii) *Deconcentration.* This term refers to the tendency for people and institutions to shift out from the existing urban area to the open land on its outskirts. It is the result of centrifugal, as opposed to centripetal forces, and has developed particularly during the last fifty years, and especially in the inter-war period. It arises from the availability of cheap transport services to the city centre and to all parts of the urban complex, and from the congestion in the central business district, and from the obsolescence of the old, often condemned, living areas clustered around it. This is a most remarkable and fundamental feature in the development of the modern city during the past thirty or forty years.

(iv) *Decentralization* is to be distinguished from deconcentration. The latter implies simply the expansion of the brick and mortar of the urban complex. Decentralization, on the other hand, implies the shedding of certain of the city's activities—such as industry or commerce and administration (the dispersal of the latter has been particularly important during the recent war)—to a distinct and separate town that itself functions as an independent local and regional centre. This process has been going on for a considerable period both around the big cities and over the wider countryside. Through the development of existing small towns and the establishment of " garden cities " and " satellites " (in the full sense of these terms), planned decentralization is advocated as an alternative policy for the future reconstruction of urban life, as opposed to excessive concentration in most urban complexes.

(v) *Residential Segregation.* This refers primarily to the concentration of residents into districts, similar to the district concentration of distinct economic uses. It implies, according to the sociologist, " the clustering in space of persons or institutions ". Individuals tend to gravitate not only to areas in which they can compete for a livelihood more efficiently, but to areas populated by others of similar race, interests, culture or economic status.[1] The principle of residential segregation is, in a measure, inherent in building practice, for houses are built in groups or in large estates, usually nowadays of exactly the same type, so that they automatically cater for people of the same economic and social level. Thus, areas of residential segregation are normally clearly defined by the type of housing, and the

[1] Gist and Halbert, op. cit., p. 175.

geographical extent of that type may be blocked out as a district. This does not mean, of course, that houses are necessarily tenanted in the way originally intended, for with time they become old-fashioned or dilapidated and their status may completely change preparatory to demolition. It is in this latter special type of segregation that the sociologist in particular is interested.

(vi) *Invasion and Succession* are terms used to indicate this process of change in buildings, their use and their occupants. It is referred to by American geographers as " sequent occupance ". The process is one of the displacement of one dominant type of land use or population group by another. The history of a housing district in the inner zone of London or of houses near the business district of a town illustrates the process. They are let off in floors and single rooms as residences, offices or workshops. The building appears dilapidated and the social and economic status of such a district is changed, usually for the worse. This process is usually most marked—this can certainly be considered as a basic law—on the margins of the city centre and along the great highways radiating from it, and has its root causes in the growth of land values in districts of old property. This phenomenon of deterioration, in various stages and degrees, covers a large part of the inner and middle zones of our cities and constitutes one of their greatest problems.

Once a dominant use becomes established within an area, competition is less ruthless between its units, and an invasion of a different use is for a time obstructed.

2. The Urban Plan

Centrifugal and centripetal forces, the differentiation of functions within the urban area, and the invasion and succession of uses outwards from the centre, are evident at all stages of the development of the Western city. While centripetal forces were dominant in the past, especially when the wall separated the town from the countryside, centrifugal forces drove certain activities outside the town walls or to the edge of the built-up area, whether it were inside or outside the walls. The town wall and the land next to it acquired distinct features of build and use that are clearly traceable in the plans of the modern city, as may be seen, for instance, in the case of Berlin (Fig. 26). The shift outwards from the outworn houses of the centre did not commence on a significant scale until the slum clearances of the nineteenth century on the Continent, but the beginning of the factory era was marked by the appearance of factories on the outskirts and the growth of houses next to them. The wealthier people settled in their new town houses in the seventeenth and eighteenth centuries (often moving in from the countryside) in the quieter quarters of the town, usually on its outskirts inside the walls or on pleasant sites—high land or a high river bank. In London the wealthy began to shift outwards to Chelsea and Islington so as to avoid the noise and smells of the city. As for the so-called " ribbon development ", this is the most natural form of housing extension—the building of houses with gardens along a main road. It has been a primary characteristic of town growth in all ages, and in the Middle Ages was a mode of suburban expansion along the roads from the gates in the town walls. The earliest maps extant all show ample evidence of this. But the dominant trend throughout all the ages until the present was centripetal, for by choice or necessity people and institutions sought a place within the city walls.

During the nineteenth century the forces of urban growth were still dominantly centripetal, since through the lack of ubiquitous, cheap, and rapid transport, the concentration of buildings on the smallest and most compact possible area was essential. In the last fifty years centrifugal forces have been of increasing importance, involving the spread of urban land uses outwards from the compact urban centre as well as the location of new enterprises on the fringes. The railway permitted the growth of separate urban centres outside the city, both as residential and industrial centres. Since the advent of the automobile and the lorry, there has been linear expansion from the compact urban area along the roads. Long tentacles extend into the countryside, especially in Britain where the phenomenon is known as ribbon development, and the interstitial land gets filled up usually at a slower rate—especially where unfavourable topographic conditions prevail. The whole process takes place outwards from the fringe of the compact urban area. The railway, offering the quickest means of transport, attracts separate settlements like beads on a chain that have spread outwards from the central city. The waterway, if navigable, has long been selected for industrial sites, but the importance of a valley floor as a zone of urban extension depends entirely on topographic and historical conditions. Trends vary from one city to another, depending on the historical development and the site of the individual city, and of the urban settlements in its near environs. But we may note some broad contrasts between the American, British and continental city.

The American city invariably has a grid plan—sometimes conforming to one grid, as in Chicago, sometimes made up of a mosaic of grids badly fitted together through the piecemeal development of separate real-estate operations, as in San Francisco and Los Angeles. But over this grid plan one can trace often a radial arrangement of main highways which can sometimes be traced back to the convergence of highways upon a centre which preceded the grid plan, while even the original nucleus of settlement is often traceable from its irregular and crooked streets and irregularly shaped building blocks. The railway pattern above all has a radial arrangement cutting right across the grid of streets and built-up blocks. Expansion has proceeded often by frontal expansion of whole districts as well as along the highways, where, owing to lack of public control, all kinds of undesirable enterprises have been localized on approaches to the city.

In the British city the star-like formation is more clearly developed for several reasons. First, the central section, the early nineteenth-century town nucleus, had no walls or other fortifications to impede expansion. Secondly, the town was almost invariably of " natural ", that is unplanned, growth, based on the framework of old highways, some older than the town itself—such as the Roman roads—which were used as the skeleton of the town, about which the new rectangular or herring-bone lay-out of houses was built in the nineteenth century. The result is that there is no clearly defined break in the stages of expansion of the urban area of a city and, since 1900, its tentacles extend along the routes from the centre, which has an irregular congested nucleus that is invariably grouped around a market-place.

In the continental city, we find, on the other hand, great contrasts. In the first place, the nucleus of the town, its modern core, sometimes has an irregular plan with narrow, winding streets and irregular blocks ; but it often has, in part at any rate, a rectangular or circular plan and a more open lay-out, dating either from its medieval foundation or from extension and rebuilding of fortifications effected during the seventeenth and eighteenth centuries. Second, the continental town was almost always enclosed by walls right through the Middle Ages down to the nineteenth century. In the growing cities these were eventually replaced by boulevards, in the manner set by the *grands boulevards* of Louis XIV in Paris, and new and more extensive fortifications were built farther out, beyond which lay a wide zone in which permanent building was forbidden. These fortifications set clear limits to urban expansion, and beyond them growth continued along the highways. In Germany such ribbon extension is far less marked than in Britain, owing to the greater degree of public control over building rights. The net result is that the continental city shows frequently a double feature, first, one or two circular roads in the built-up area forming wide traffic thoroughfares, and secondly, a radial arrangement of main roads, giving to the whole plan a radial-concentric pattern.

The broad topographic features of the city also depend on the character of its surrounding settlements. In a dominant rural area where the central city far outweighs any competitors, surrounding villages and a few small towns become absorbed by it, and new urban centres bud off from it, though these usually have old nuclei. Such has been the case with the big capital cities such as London, Paris and Berlin. In other cases, in industrial areas with a long history and a close distribution of industrial centres, depending perhaps upon the use of water power before the Steam Age, several adjacent towns have expanded and merged. One usually dominated the rest, as Birmingham in the Black Country, and Manchester in south-east Lancashire. Separate towns, serving as local market centres as well as seats of industry—and also as local centres for cottage industries in their environs—may expand and then merge through ribbon expansion along roads and valleys. This has been the case typically in south-east Lancashire, and the same sort of trend is to be found, for instance, around Chemnitz in Saxony and at Saint-Étienne and Lille in France.

This urban expansion results in the spread of the built-up area beyond the administrative limits of the town, so that one urban area embraces several contiguous administrative units. For such an urban unit, or rather for the contiguous administrative districts that make up the urban unit, the name " conurbation " was coined by Geddes.[1] One of the greatest problems of local government and regional planning is to give to these " natural " urban units the *de jure* unity which is needed for their efficient administration, organization, and development, in

[1] P. Geddes, *Cities in Evolution*, London, 1915. For further discussion of the " conurbation " see p. 168.

respect of housing, roads, public utility and other services, as well as local government and taxation.

Since its appearance in the middle of the nineteenth century, the railway has become a vital factor in the life of the city as well as in the nature and problems of its topographic expansion.[1] Three primary facts appear almost invariably in the development of the railway net of every big city. The first railways sought terminals as near as possible to the city centre, so that to-day they lie in the heart of the city and are hindrances to the rational planning of its lay-out and traffic-net. The second feature is that the railways naturally radiate outwards from these terminals. The third is that practically every large city during the past fifty years has acquired some kind of a circular or belt line either as a freight or passenger line or as an underground or overground system for urban traffic. The railway is concerned with the transport of both passengers and freight. As regards the first, it must provide (*a*) transport for the suburban worker to and from the city, with marked concentrations of traffic at two specific "rush-hour" periods; and (*b*) main-line services to other cities, the main stations being either at "dead-ends" or located on routes which pass through the city. From this point of view, the railway plan should consist of a series of routes, radiating from the city, with branches on its outskirts serving the suburbs. Ideally, there should also be transport facilities between the great terminals in great cities—either as a surface or, preferably, as an underground railway—though this, in fact, is seldom the case. As regards the second, freight traffic, the main problem is to provide efficiently for the import of supplies of food and raw materials for the daily requirements of population and industry. There will also be a great traffic of goods through the city, either for direct transmission or for storage and redistribution. It is necessary, therefore, for the city to have—(1) sorting facilities where traffic from all directions can be assembled for their destinations, that is, marshalling yards for such entrepôt traffic, and goods yards for distribution within the city; (2) adequate connections from one radial main line to another; and (3) intra-city connections and yards which should as far as possible serve the compactly built-up areas. Thus, the belt or girdle line on the outskirts of the city becomes an almost inevitable feature of the railway pattern.

[1] See S. H. Beaver, "The Railways of Great Cities", *Geography*, Vol. XXII, 1937, pp. 116–20.

Marshalling yards will be built on the outskirts of the city near to the belt line, while goods stations lie in the vicinity of the terminals in closer touch with the urban areas of the city to and from which goods are distributed and collected.

Thus, the normal historical development of the railway net of a city, varying with its size and location, results in routes through the centre of the city—that is, routes built on what was open land bordering the town, to which the major industrial and built-up areas have largely gravitated ; terminals built at the former outskirts of the city and radiating from it ; and the final emergence of an inter-connecting belt-line. " The ideal railway plan for a large city thus somewhat resembles a wheel ; the city is the hub, the main lines are the spokes, and the circumference is the belt line" (Beaver). The symmetry of the railway plan varies, however, according to the conditions of its development and to the topography of the city, its historical development and its present size.

3. Examples of Urban Growth : Chicago, Paris, Berlin

Two modes of urban growth are well illustrated by Chicago (Fig. 24) and Paris (Fig. 25). The metropolitan district of Chicago had four and a third million inhabitants on an area of 1,119 square miles in 1940, while Chicago proper had three and a third millions on an area of 202 square miles. The Department of Seine, in which Paris is situated, has an area of about 192 square miles and a population of five millions, whereas the city of Paris has an area of about 32 square miles with a population of 2,830,000 in 1936.

Chicago has grown radially from its nucleus on the south side of the mouth of the Chicago river, the down-town skyscraper district known as the Loop (so called since it is a rectangle enclosed by an overhead railway). It has a grid plan of streets and radial boulevards and railways. The first railway companies carried their lines to the heart of the city and to-day they form a band of steel around the Loop and, together with the river to north and west, prevent its lateral expansion, and increase congestion through the concentration here of almost the whole of the passenger traffic in transit. After 1885 decentralization set in, culminating in the opening of three belt lines and marshalling yards that allow the peripheral localization of industry and relieve the centre of some traffic congestion. Industrial areas are situated on the north and south Chicago rivers. The vast complex of the Union Stock Yards, which was sited on the southern outskirts of the city when erected in 1864, is now in the heart of the city, the centre of a great area of squalid slums. Since 1905 a vast new industrial complex has developed on the open land south of the city with its chief centre in Gary, producing to-day one-fifth of the iron and steel supply of the United States. Suburban settlements extend like beads on a chain along the railways. The arrangement of the residential zones within this framework is shown on Fig. 27 and is summarized on p. 113 below. Numerous modern

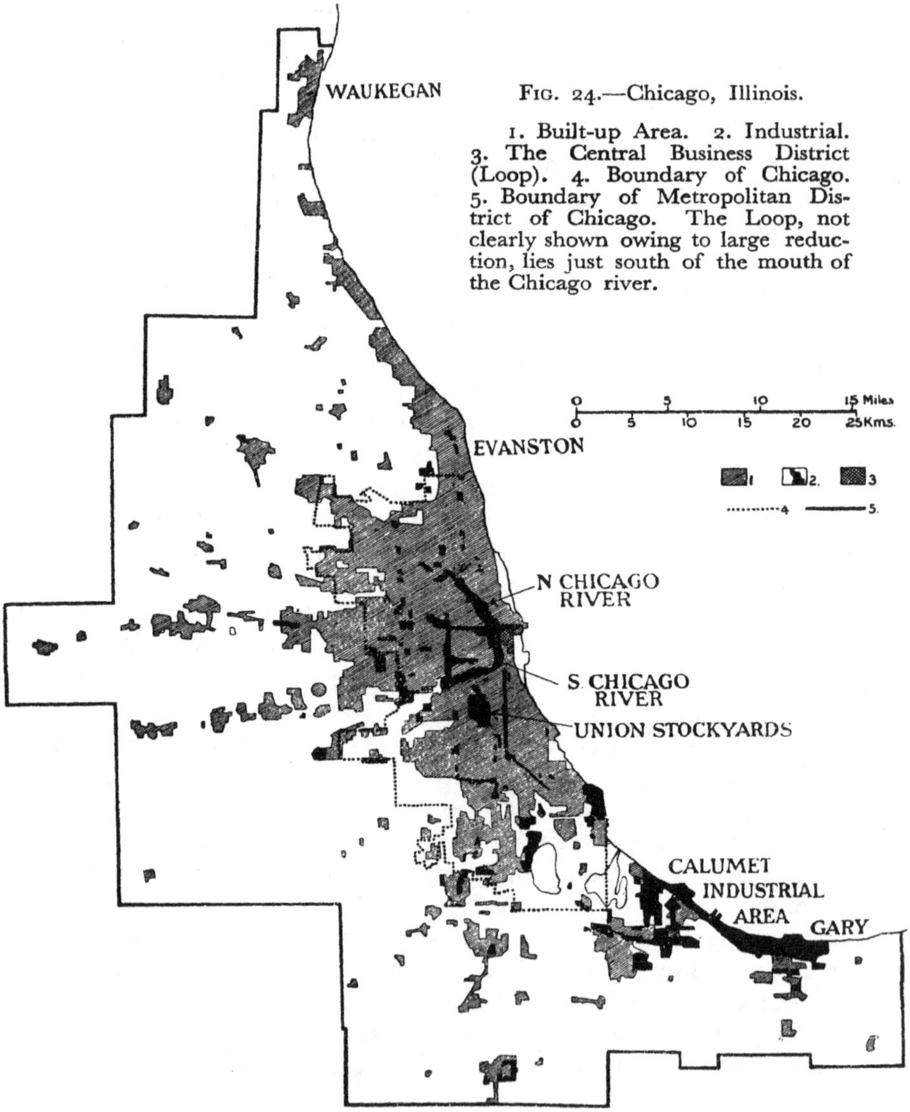

FIG. 24.—Chicago, Illinois.

1. Built-up Area. 2. Industrial. 3. The Central Business District (Loop). 4. Boundary of Chicago. 5. Boundary of Metropolitan District of Chicago. The Loop, not clearly shown owing to large reduction, lies just south of the mouth of the Chicago river.

plants, specializing on miscellaneous light industries and using primarily road transport, lie along the main roads on the margins of the city.

Paris has grown from a medieval nucleus that consisted of

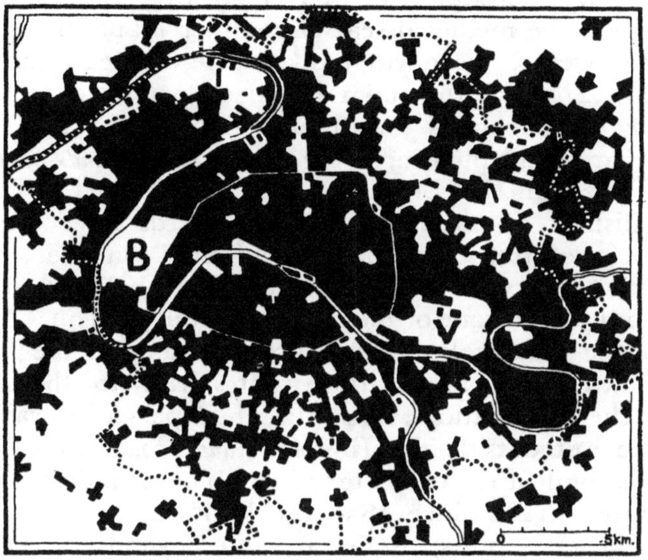

FIG. 25*a*.—The Built-Up Area of Greater Paris.

Note.—The white line is the boundary of the City of Paris ; the dotted line is the boundary of the Department of Seine. B=Bois de Boulogne, V=Bois de Vincennes. (After A. Demangeon, *Paris, La Ville et sa Banlieue*, Paris, 1933.)

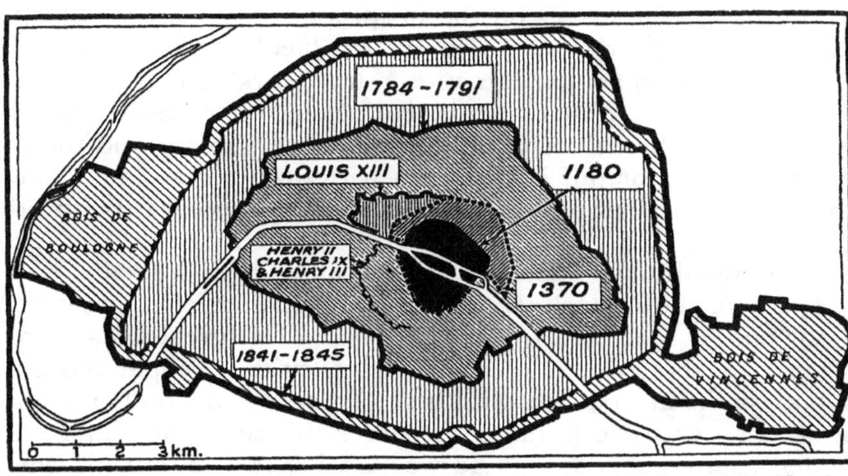

FIG. 25*b*.—The Topographic Formation of Paris, showing the dates of fortifications and the present limits of the City of Paris (by a heavy black line). (After Demangeon.)

three parts : the *cité* on the island in the Seine, the site of the Roman and of the early medieval settlement, where are situated to-day the Palais de Justice and the cathedral of Notre Dame ; the *quartier latin* on the rising ground south of the Seine, overlooked by the church of Ste. Geneviève : and the *ville*, the town proper, the medieval seat of commerce and crafts, situated along the river on its north bank. These three quarters were walled in 1210 and the town has been extended at successive stages by fortifications eventually demolished and replaced by boulevards and buildings (Fig. 25). The *grands boulevards* lie on the walls built by Charles V and Louis XIII. This is the central district of to-day. New boulevards have been built, with blocks of flats and public buildings, on the site of the nineteenth-century fortifications that were abolished after 1919, and upon the open zone of 200 metres in front of them. The inner circle of the metropolitan underground runs on the inner side of the latter belt and radial routes run to the old gates through the fortifications. All the chief passenger railway terminals lie just outside the (inner) *grands boulevards*. This is a compactly built-up area of tenement blocks built for the most part before 1914. Since 1850 the built-up area has extended beyond the city limits, as far as, and beyond, the boundaries of the Department of Seine. Here are the greatest concentrations of the heavy industries, working-class and suburban housing. Engineering, chemicals, and public utility plants (gas and electricity) lie especially on the flat land adjacent to the Seine westward, and to the canals on the north and north-east of the city, that have direct contact with the coalfield of the Nord. Industrial development, with the usual working-class quarters, appears in the south-east at the confluence of the Seine and the Marne. South of the city, building is discontinuous and there is little industry, for this is a plateau with deep picturesque valleys forming a residential district expanding westwards in the direction of Versailles. While the City of Paris lost over 75,000 inhabitants between 1921 and 1936 (2,906,000 to 2,830,000), the Department of Seine, excluding Paris, grew from 1,505,000 in 1921 to 2,133,000 in 1936. This great expansion is typical of the growth of all modern cities. People move to suburban homes on the outskirts. Factories are built on open sites, less expensive than those nearer the city centre and with better communications. Old-world villages and historic small towns lie cheek-by-jowl with, and are often transformed by, new buildings. Control is needed to direct such development and in 1932 the Paris Region was created by law, planning proposals being enacted to cover the whole region within 35 km. of Notre Dame.

Berlin (with four and a third million people before the 1939 War on about 353 square miles) has been carefully mapped according to its main types of building,[1] and Fig. 26 is interesting as a demonstration of the geographer's techniques in urban study, as well as of the outstanding features of the build of the continental city, all of which are very similar to those of Paris. It is the only European city that has been mapped in this way, and geographers would find it worth while to apply the technique to other cities. The author, Herbert Louis, framed the map on the historical growth of the city in so far as the growth is reflected in the present build. Important in this respect is the role of the fortifications. This zone, on the outskirts of the city, was normally more open than the central parts of the town and here were laid out such "peripheral" buildings as factories, hospitals, cemeteries, military establishments, new public buildings, and places of amusement. Often it marks a change in the street pattern and street names. Such a zone appears when the fortifications are still in being, but it persists when the city expands beyond as a distinct belt clearly recognizable in the build. There are two such belts in Berlin. The circular belt of the ring railway (*Ringbahn*) formed a barrier to urban expansion and growth outside it has taken place mainly since the end of the nineteenth century (Zone IV on Fig. 26) around old town centres and near to new factories.

4. The General Build of the City

The structure of the modern city, with reference to the extent and pattern of its built-up area, the distribution of urban land uses, and the social and economic structure within it, shows a broad zonal arrangement focused on the hub of its life and organization, the city centre. Peculiar conditions of topographic and historical development in individual cities may cause marked divergences from this pattern, for the tendency to zonal arrangement is only one tendency, although the most important one common to all urban development. But in studying the build of any Western city one can always discover a broad pattern

[1] H. Louis, *Die Geographische Gliederung von Gross-Berlin*, 22 pp., 2 maps, Stuttgart, 1936. Fig. 26 is a reduction of the original on a scale of 1 : 100,000.

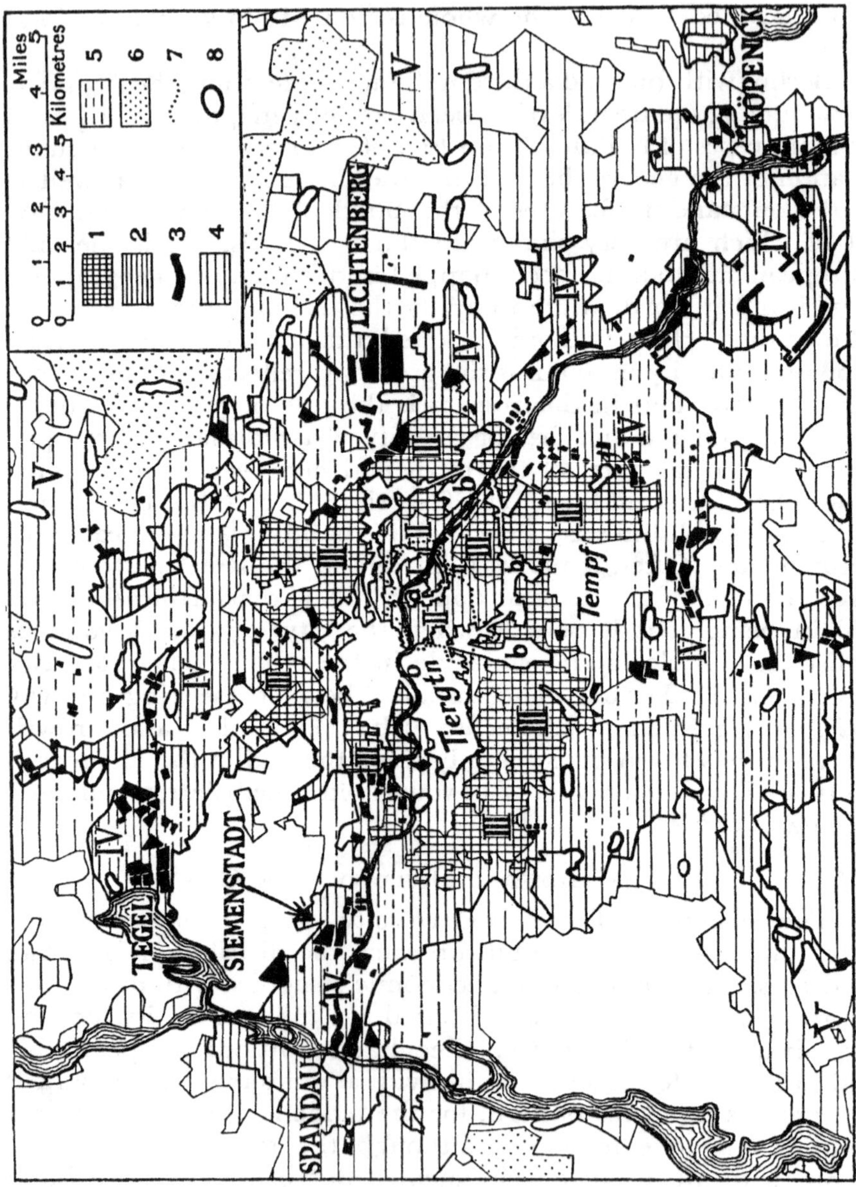

FIG. 26.—The Build of Greater Berlin (1935). (Adapted from H. Louis.)

and process conforming to this design, as is well illustrated by the three cases of Berlin, Paris and Chicago. The zones may be briefly characterized as follows :

The Central Zone is the hub of the city. It includes the medieval town and the pre-modern extensions. In it there is the maximum " friction of space ". All services requiring central locations compete for accommodation, with consequent vertical and lateral building expansion and great congestion of traffic. It includes the retail, wholesale, administrative, commerce and business districts ; markets, hotels, residential enclaves for both the *élite* and the poor ; " rooming houses " and segregated ethnic districts ; large public buildings that cater for the community as a whole ; and railway goods and passenger terminals. It is fully built up, and its vertical expansion reaches the extreme in the down-town skyscraper district of the American city, with its island of skyscrapers in a sea of relatively uniform built-up areas.

The Middle Zone in all cities is the zone that was built up during the nineteenth and early twentieth centuries. During this period most building was effected by private enterprise and there were few restrictions on the density or type of building —town-planning measures being normally devoted, at best, to questions of street lay-out and street widths. This zone is fully built up, the building blocks having little associated private open space, and has few public open spaces. In England, we find the terrace and back-to-back house ; on the Continent, as in Paris and Berlin, the tenement ; in the United States, the tenement and the timber shack. Mainly residential, the Middle Zone contains numerous scattered small factories, and old sites of early industrial development concentrated on the flat, low-lying land near the river or canal, beside the railway tracks and round the goods and railway stations. This zone, as we shall see later, is usually in a state of deterioration, with resultant social

---

FIG. 26.

1. Uniform building, four storeys and over ; mainly tenements. 2. Varied building in the Central City and Outer Areas. 3. Factories. 4. Villas, single-family housing estates, with gardens. 5. Allotment areas (*Laubenkolonien*). 6. Irrigated meadows (*Rieselfelder*). 7. Limit of the central city area. 8. Villages.

Roman numerals refer to zones. Inner unshaded belts marked *a* and *b* are former zones of fortifications.

I. Medieval town nucleus surrounded by (*a*) belt of open spaces and public buildings on the site of old town walls. II. Seventeenth- to eighteenth-century extensions (*Vorstädte*) surrounded by (*b*) belt of open spaces, public buildings, stations and railway tracks on site of eighteenth- to nineteenth-century fortifications. III. Tenement areas of the Wilhelmian period. IV. Industrial-residential zone. V. Outer (residential-suburban) zone.

degeneration and disorganization on the borders of the city zone. It coincides with Zone III on the map of Greater Berlin (Fig. 26).

The Outer Zone is often referred to as the Rural-Urban Fringe or the Suburban Zone, and all three terms are expressive of its character in respect of position, build and function. It is the zone that has emerged mainly during the past twenty years (1919–39), although its development began at the end of the nineteenth century with the advent of the electric tram, electric railway and (rather later) the automobile. It is characterized broadly by the intermixture of urban and rural uses. And is mainly in rural use, but social and economic conditions are dominated by the urban areas. The latter may be new residential or industrial clusters situated on rail, water or road ; or they may have grown around a pre-existing nucleus, village, old town market centre or old industrial centre. Thus, the lay-out and growth of this zone varies very widely according to physical build and the mode of urban expansion. Its rapid recent growth and its virtual absorption into the city (if not in its administrative area) is a general feature.

These three zones taken together form, in respect of built-up areas, density of population, work, play, interests and organization, a single social and economic unit.

There are certain broad correlations in the demographic, social and economic conditions of the population between the central, middle and outer zones of the city. These features are generally self-evident, but in their broader aspects they may be briefly summarized. The distribution of population in these zones, and the shift outward from the centre to the periphery, are accompanied by selective processes in the composition of the population. Thus, in the United States, it was found that for 96 metropolitan districts in 1930 the proportions of females, foreign-born whites and Negroes were higher in the central cities (central and middle zones), and that the proportions of children, males, and native-born Americans were higher in the outer or suburban zones. In all but four cases, the percentage of children under 15 years of age was higher outside than inside the central cities.[1]

Further, " the selective processes involved in city expansion seem to operate in a somewhat orderly and typical fashion from

[1] McKenzie, *The Metropolitan Community*, p. 180. The term central city refers to the nucleus of the wider urban aggregate (metropolitan district) used for purposes of the U.S. Census. See p. 198.

the centre to the periphery of the community ".[1] On the basis of small census tracts this has been demonstrated in the case of Chicago to be true of age-sex composition : " an excessive proportion of adult males is found in the area lying close to the Loop —the main business centre—with a tendency towards a more even age-sex distribution as one proceeds out toward the fringe of the metropolitan area ".[2] It is also asserted that " weaker and less stable elements " of the population lie close to the business centre and more " substantial elements " lie in the outer zones. The incidence of juvenile delinquency and adult crime is highest in the central zone. But many of these traits of the central zone occur particularly in the middle zone, between the city centre and the periphery, and especially in the parts of it that border on the former. In this intermediate area, closely built up in the nineteenth century with sordid residential areas intersegmented with industrial areas, and tenanted mainly by working-class folk, " old buildings, old institutions, and old ways of doing things " are characteristic. In the rapid building of centre and periphery this intermediate zone has been neglected and allowed to decay, and newer institutions have been located in the other two zones. This applies, for instance, to chain stores, banks, and cinemas, which are the " key sub-centre institutions " in the commercial structure of the city. The same trends, it is claimed, hold in greater detail inside the built-up area of the American city, as illustrated in particular by very detailed analysis of data in selected census tracts in Chicago.

Towards the centre of the city where hotels and rooming houses abound, gravitate the transient adults, mostly males. Out along the radial transportation lines, where family hotels and fashionable apartments tend to group, the proportion of women in the total population is abnormally high. In the intervening pie-shaped areas, and in the industrial suburbs, excessive proportions of children are almost invariably found.[3]

While these characteristics are true in general of the broad structure of the city, both American and European, within the compact built-up area, there are variations from district to district, based primarily on the economic level of the inhabitants, that bear little relation to the concentric zonal trend. It is certain that the " orderly and typical fashion " of operation of selective processes from centre to periphery, though real, is not nearly so important in the European as in the American city, as compared with contrasts district by district. To this we shall return later.

[1] Ibid., p. 180. [2] Ibid., p. 182. [3] Ibid., p. 245.

This structure of city populations has received relatively little attention in Europe. Sound research demands statistics for very small areas comparable with the United States census tracts. The large administrative and statistical units into which cities are divided are varied in character geographically and obscure significant differences within their limits. Nevertheless, a broad picture of the structure of the city can be obtained, even from such crude data, in line with the general treatment mentioned above of American cities. This has been done recently for Prague, and its characteristics are common to the structure of Continental and British cities. Twenty-seven districts were used for the whole city area of 43,000 acres,[1] which is divided into three areas, central area, transitional area, and peripheral area. The first is the historic core and the modern business heart of the city. The second includes the main nineteenth-century tenement areas. The third covers the outlying recent residential and industrial areas, the latter lying in particular to the north of the city. On the basis of these districts it was found that the proportion of children to the resident population decreases outwards from the centre,[2] irrespective of social and economic conditions, and that this arrangement is closely related to the density of the built-up area, the density of population per unit of area, and, most important of all, to the age of the population. If all males over 25 years old are considered, the age of the adult population increases from the periphery to the centre as the proportion of children decreases. The number of domestic servants to a hundred homes is taken as an index of social structure and the district figures show " minimal indices definitely increasing from the centre to the periphery, maximal indices increasing from the periphery towards the centre ", although the differences are not so marked as in the case of children.

5. The Concentric Pattern of Urban Growth : Chicago

A hypothetical pattern of urban growth was described by E. W. Burgess, the Chicago sociologist, in 1923[3] (Fig. 27). This pattern has been accepted by American researchers as

[1] Moscheles, Julie, " The Demographic, Social and Economic Regions of Greater Prague ", *Geographical Review*, Vol. XXVII, 1937, pp. 414–29.

[2] It was found that in the seven districts of the central area the proportion of children under 14 years of age to *resident* population was 10·3 to 12·9 per 1,000, in the transitional belt, 14·7 to 17·9 per 1,000, and in the peripheral area 20·2 to 23·6 per 1,000.

[3] E. W. Burgess, " The Growth of the City ", Chapter II, in *The City*, by R. E. Park, E. W. Burgess and R. D. McKenzie, Chicago, 1925 ; originally published as an article in *Proceedings of the American Sociological Society*, Vol. XVIII, 1923, pp. 85–9.

established by its author, and has been used by some of them as a basis of investigation of particular cities. The essence of this hypothesis (determined by special reference to Chicago) is that any town or city tends to expand radially from its centre so as to form a series of concentric circular zones as follows: (1) the central business zone; (2) a zone of transition and social deterioration, which is being invaded by business and light

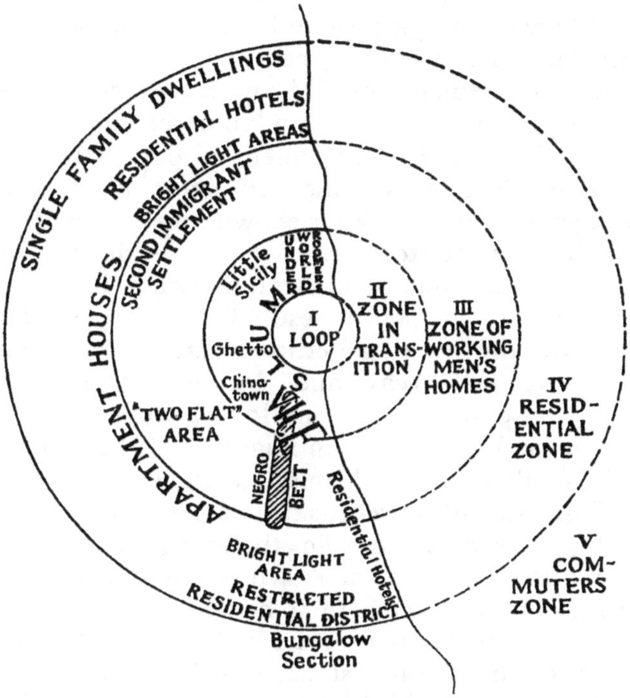

FIG. 27.—Urban Areas in Chicago (after R. E. Park and E. W. Burgess, *The City*, University of Chicago Press, p. 55). This diagram has particular reference to Chicago and should be compared with Fig. 24. The wavy line represents the shore of Lake Michigan.

manufacture; (3) the workers' housing and factory zone; (4) the residential zone of high-class apartment buildings or single-family dwellings; (5) the commuters' zone of suburban areas and satellite cities within a journey of thirty to sixty minutes of the central business district. The main feature is "the tendency of each inner zone to extend its area by the invasion of the next outer zone". This pattern of urban land use is reflected, continues Burgess, in the social structure. "In the expansion of the city a process of distribution takes place which sifts and sorts and re-locates individuals and groups by residence and occupation ", and " this differentiation into natural economic and cultural groupings gives form and character to the city " since " these areas tend to accentuate certain traits, to attract and develop their kind of individual and so to become further differentiated ".[1] Burgess went further and showed that distribution patterns outwards from the central business zone take the form of a series of gradients, such phenomena as delinquency rates, sex ratio, percentage of foreign-born individuals, and poverty, tending to decrease outwards from the centre.

Population movements in Chicago have been the subject of careful examination from the standpoint of concentric growth. The process of outward displacement of population is part and parcel of the whole process of modern urban development. The zonal arrangement in Chicago is shown schematically in Fig. 27. The oldest immigrant stocks are widely distributed throughout the urban area—Germans and Scandinavians in the north and north-west, Irish in the south-west. On the other hand, recent immigrants of the first generation are more concentrated near the industrial areas and in the zone of transition around the Loop—for example, Italians, Czechs and Russians. Second-generation immigrants of the same stocks are dispersed in smaller groups throughout the zone of working-class homes. This distribution results from a general process by which recent immigrants first settle in the congested rooming-house districts near the Loop. Then, when they have found more or less settled jobs and taken their bearings, they tend to shift to better living-quarters farther out from the centre, first segregating into communities of the same stock, and then, finally, after several moves over varying periods, depending on the fortunes of the individual, to the outer periphery, where they are completely dispersed and absorbed into the general " American " community life. The peripheral zones, in particular on the main lines of communication to the north and south along the lake front, to the west of the Loop, and exclusively in the suburban areas, are occupied mainly by people of American origin. Here are also the best-class residential areas.

There is thus a gradual decrease in the concentration of each stock outwards from the Loop to a radius of five miles. The distribution of the Negroes, who have entered Chicago since 1918 and numbered 234,000 in 1930, is a notable exception.

[1] Burgess, op. cit., p. 56.

They are clustered in a narrow north-south belt in south Chicago that cuts across the pattern of concentric zones, and have pushed southwards along Michigan Avenue, ousting the settled white residents, with whom they came into open conflict in the riots of 1918 and 1919. The fine houses of three to five storeys (some formerly the residences of well-to-do and old-established families, like the Armours, the Swifts and the Rockefellers) have been converted into congested Negro rooming and apartment houses, very overcrowded with a dilapidated external appearance through lack of proper upkeep. The Negro continues to press southwards and has reached the vicinity of the University of Chicago, where he is held at bay by tenants' associations and the like.[1] First-generation immigrants settled first near the Loop in Zone II and, differing greatly in culture and economic status, have moved out along the main avenues of communication, displacing other groups, who in turn have continued the outward trend. P. F. Cressey, by calculating the distance of specific population groups in zones from the city centre in 1898 and 1930, was able to demonstrate this centrifugal tendency. It is impossible to say how many moves were made by migrants in the city, but the general result reveals a tendency for the original segregated areas to disappear and their population gradually to be dispersed throughout the city area, except for distinct ethnic communities such as Chinatown and the Black Belt.

This concentric zone theory, formulated with special reference to the sociological structure of the urban community of Chicago, has come in for a good deal of criticism from those who have made subsequent studies of individual cities under the stimulus of the new concepts and technique of the Chicago school. Criticism of Burgess's hypothesis has been directed against the assumption that the city centre alone is the hub and driving-force in the whole process. It takes inadequate account of industrial and railway utilization, of the social disorganization associated with port areas, of the radial extension of commerce and in consequence of high land values on radial thoroughfares ; and topography may

[1] See P. F. Cressey, *The Succession of Cultural Groups in the City of Chicago*, Ph.D. MS., University of Chicago, 1930, and " Population Movements in Chicago ", in *Journal of Social Forces*, Vol. II, 1924. Also C. S. Johnson, *The Negro in Chicago*, Chicago Commission on Race Relations, 1922. P. F. Cressey, " Population Changes in Chicago, 1898–1930 ", *American Journal of Sociology*, Vol. XLIV, July, 1938, pp. 59–69. The same process has taken place in New York with the spread of the Negro from his Harlem district towards the Central Park and Morningside Heights. It is also found in the outward spread of Jews and foreigners in the districts of north-west London, as well as of the Jews in such cities as Manchester and Leeds.

indeed break up the whole "ideal" pattern. It has been recently stated in the light of these new researches that cities in their expansion patterns tend to assume an "axiate design", with the population and institutions (by this is simply meant types of land use) pushing out along main thoroughfares to constitute the "prongs" of a star-like formation, and

the population and institutions lining the sides of main arterial highways that cut squarely across the various areas of the city are frequently so different from the people and institutions located in the same general area but somewhat removed from the thoroughfares that the principle of concentricity seems to be seriously violated.[1]

An examination of twenty city land-use maps by Davie in the United States and Canada showed :

(1) a central business district, irregular in size but more square or rectangular than circular [2]; (2) commercial land use extending along the radial streets and concentrating at certain strategic points to form sub-centers ; (3) industry located near the means of transportation by water or rail, wherever in the city this may be—and it may be anywhere ; (4) low-grade housing near the industrial and transportation areas, and (5) second and first-class housing anywhere else. These seem to be the general principles governing the distribution of utilities. There is no universal pattern, not even an "ideal" type. Low economic areas are characterized by smaller incomes, fewer radios and telephones, fewer home-owners, fewer one-family dwellings, more two- and multi-family dwellings, more murders, houses of prostitution, juvenile delinquents, dependent families, unemployed, illiterates, and higher birth and infant mortality rates in proportion to population. Such areas, while in general near the center of the city, are by no means confined there, but are found in any zone. They are generally adjacent to industrial and railroad property.

This refinement of the conception of the "natural area" with which we shall deal in the next chapter, and the process of its segregation in no wise refutes the validity of Burgess's generalization regarding the broad basic background of urban growth. Study of European cities from the standpoint of their historical development, their density of built-over land, and the movement of population substantiates his general theory.[3]

[1] Maurice R. Davie, "The Pattern of Urban Growth", in *Studies in the Science of Society*, edited by G. P. Murdock, New Haven, 1937, pp. 133–61.

[2] This is certainly associated with the rectangular street pattern dominant in these cities, a circular shape being more common in European cities, reflecting the circular or oval shape of the old city walls and the modern boulevard that has displaced them.

[3] In a thorough study of the changes in the distribution of population in the city of Stockholm between 1880 and 1930, the geographers, whose work is noted

6. Areas of Social Disorganization and Blight

Social disorganization is associated with the outworn, obsolescent areas which are found in areas of high land values and old properties. These are termed by sociologists disorganized areas. Many studies have been made of the exact distribution by place of residence of the victims of such aberrations as delinquency (both adult and juvenile), vice, suicide, mental disorders, alcoholism, divorce, desertion, poverty, mortality and disease, and these traits correlated with community structure as determined by ethnic types, densities of population and income levels.[1]

Similar trends are found in English and continental cities, although they are less obvious owing to a more homogeneous ethnic composition. There is, of course, in the first place, the general centrifugal drift of people from the older, less healthy, inner housing areas, to the new outer housing areas. This is the most fundamental and widespread feature of the demography of the modern city, and it is probably most characteristic of the British city. It is both a contributory cause and an effect of urban blight in the inner areas. On the ecological side, the Merseyside survey showed that in the inner districts of Liverpool where the slums are, there is a coincidence of high birth-rates, overcrowding, poverty, physical and mental defects, alcoholism, chronic destitution, immorality and criminality, whereas the outer districts show much lower rates. The former phenomena coincide markedly in Black Patch Areas, defined by the Survey as " streets in which at least one family in every five contains individuals of the chronically unemployed and destitute kind. Such families will not infrequently be resident in one-room tenements or slums. They will be often in receipt of public assistance and not getting unemployment benefit or transitional payment." [2] The Survey concludes : " convinced environ-

above, concluded that " the growth of the town was practically concentric " and " distribution changed from high density in the centre and low density on the periphery to the reverse with the formation of a central ' city ' district ". W. William Olsson, " Stockholm : Its Structure and Development ", *Geog. Rev.*, Vol. XXX, 1940, pp. 420–38. Interesting observations on this mode of urban expansion are made by Griffith Taylor with reference to Toronto in " Environment, Village and City ", *Annals of the Association of American Geographers*, Vol. XXXII, 1942, pp. 1–67.

[1] These many studies are now briefly summarized and general conclusions drawn from them in several American textbooks on urban sociology, notably *Urban Society* by Noel P. Gist and L. A. Halbert (1941), *The Sociology of City Life* by Niles Carpenter (1931), *Urban Sociology* by Nels Anderson and E. C. Lindeman (1928), and *The Metropolitan Community* by R. D. McKenzie (1933).

[2] D. Caradog Jones (ed.), *The Social Survey of Merseyside*, Liverpool U.P., 1934, Vol. III, Ch. XIX, p. 485.

mentalists will probably claim that the congestion is the cause of the trouble. The protagonists of heredity will assert in reply that congestion, so far from being the cause, is rather the effect . . . No doubt on both sides there is some truth, but the question how much remains unsettled." [1] "Heredity, illness and 'culture shock' are all important elements in the problem. But in spite of that, poverty does appear as the environment in which these disease germs grow most readily." [2] Black Patches, or Plague Spots of this kind, have recently been studied as to their exact areal distribution by the Hull Survey group.

One of the most characteristic and serious features of the modern city is the zone of deterioration, to which attention has already been drawn (p. 96), that surrounds normally the city centre, and is also related to areas of industrial slum. These areas usually contain the oldest buildings of the city: unplanned, congested and obsolescent by all modern standards of living. They are affected economically and socially by the expansion and mere proximity of the city centre, with its high land values and its high mobility. Their inhabitants, offered since 1920, in increasing measure, chances of living in modern houses at reasonable rents on the outskirts, have been moving outwards, deserting their former dwellings. A large central zone becomes derelict and goes on the "down-grade". It is also socially disorganized. It is a slum. These various traits are referred to as constituting urban blight, which, beginning on the edge of the central business district, spreads outwards like a creeping paralysis to cover a very large part of the whole urban area—that part mainly with pre-1900 houses. Probably the commonest and best indicator of this phenomenon is the fact of both a high density of population (overcrowding) and a decreasing population (removals). A blighted area is thus one in which

> as a result of social, economic, or other conditions, there is a marked discrepancy between the value placed on the property by the owner and its value for any uses to which it can be put, appropriate to the public welfare, under existing circumstances. Old buildings are neglected and new ones are not erected, and the whole section becomes stale and unprofitable. In other words, blight is a condition where it is not profitable to make or maintain improvements.[3]

[1] Op. cit., p. 489.
[2] D. V. Glass, *The Town and a Changing Civilization*, London, 1935, p. 88.
[3] Mabel L. Walker, *Urban Blight and Slums; Economic and Legal Factors in their Origin, Reclamation and Prevention*, Harvard Planning Studies, XII, 1938.

Such an area is an economic liability to the community. A slum connotes an extreme condition of blight " in which the housing is so unfit as to constitute a menace to the health and morals of the community ". It is thus essentially of social significance, the blighted area being primarily of economic significance.

Blight has as its common characteristics :

high but falling land values ; congested but decreasing population ; obsolete and unfit housing ; a large proportion of abandoned buildings and of rental vacancies ; heavily mortgaged property ; excessive tax delinquency ; low average rentals ; generally low economic status of inhabitants ; excessive crime, mortality, and disease rates ; high per capita and per acre governmental costs.[1]

The pattern of urban blight has been summarized by Walker on the basis of Burgess's concentric zones of urban expansion and invasion outwards from the centre.[2] The causes of blight, as defined above, are various and often complicated, but it " is probably the result chiefly of loss of population, particularly as the more prosperous groups move away ", and whether a decreasing population is a cause or effect of blight, it is certainly (if examined in sufficient detail, block by block), the best general criterion of the existence of such a condition.[3]

Traits of social disorganization have been exhaustively studied in Chicago. Investigations of the distribution of delinquency rates (school truants, juvenile delinquents, adult offenders, etc.) resulted in important findings—that crime and delinquency tend to be concentrated in specific areas of the city, not evenly distributed according to the density of population ; that the incidence was highest near the central business district, and decreased outwards, with higher rates also contiguous to industrial districts such as the stockyards and steel works ; and that the incidence of delinquency was a reliable measure of social disorganization and the breakdown of community life and standards.[4] The gang and the gangster thrive also in such areas. They are found in Chicago in a semicircle around the Loop in the zone of deterioration (as named by Burgess), an area of slight social control where there is lack of law and order.[5] The distribution of vice and of

[1] Walker, op. cit., p. 36.

[2] Ibid., pp. 8–11, quoting Henry Wright in *The Survey Graphic*.

[3] H. Hoyt, *One Hundred Years of Land Values in Chicago*, University of Chicago Press, 1933.

[4] Clifford R. Shaw, *Delinquency Areas*, University of Chicago Press, 1930. Many aspects of this subject in American cities are summarized in the textbook by Gist and Halbert, *Urban Society*.

[5] F. M. Thrasher, *The Gang*, University of Chicago Press, 1927.

homeless men show similar traits since these owe their location to similar causes and are associated with the same unsettled social pattern. Suicide was found to be particularly prevalent in the central business district and the cheap hotel districts near it, in the rooming-house districts of the lower North Side, and in an area between the Loop and the " black belt ".[1] Cases of insanity psychoses were found to be clustered around the central business district, with a sharp decline of incidence outwards, and in the former there were found interesting associations of particular psychoses with different districts.[2]

7. The Rural–Urban Fringe

On the outer borders of the city, between the areas of urban and rural land use, there is an intermediate zone which shares the characteristics of each. This fringe is invaded by urban uses—by the extension of housing estates, of buildings along the main arterial roads, and by the location of new factories, as well as many other urban features which are excluded for one reason or another from the compact urban area—golf courses, water-works, cemeteries, destructors, parks, allotment gardens and the like.

Motor transport, far more than the railway, has caused the American city to expand, and to embrace outlying farms, new daughter settlements, and new land uses within its orbit.

As a result, cities have not merely expanded, they have " exploded ". " Realtors " began to lay out subdivisions. The open country features of suburban living, lower taxes, fashion, and prestige, not merely attracted the surplus population, but began to invite and allure the established residents of the city itself. This has tended to leave behind empty houses, if not " blighted areas " certainly lower and lower land values. This almost self-evident trend has only recently attracted the attention of planners, leading agencies and city officials. It was noticeable before 1930, but the 1940 census showed that while the 92 largest cities of the United States gained 1,600,000 people between 1930 and 1940, the remainder of the counties in which they are located gained almost the same, or 1,500,000. While the central cities gained by 4 per cent., the suburban areas of these cities gained by 14 per cent. Some of our largest cities, including Philadelphia, Boston and Kansas City, actually lost population.[3]

[1] Ruth S. Cavan, *Suicide*, University of Chicago Press, 1928, p. 81.

[2] R. F. L. Faris and H. W. Dunham, *Mental Disorders in Urban Areas*, University of Chicago Press, 1939. See map reproduced from this work in Gist and Halbert, *Urban Society*, showing the distribution of schizophrenia cases from 1922 to 1931, p. 221.

[3] G. S. Wehrwein, " The Rural–Urban Fringe ", *Economic Geography*, Vol. XVIII, 1942, pp. 217–28. For the meaning of " central cities " in this context, see p. 198.

This " wild expansion " or " explosion " of the city produces premature subdivisions for urban development, in the shape of vacant derelict land, reserved as residential sites and provided with paved streets, water, sewers, gas and electricity at private or public expense. The result is what Americans term tax delinquency ; the land lies idle and becomes what has been called " an institutional desert ". New residential areas must rely on rural forms of government to supply their services, though usually they are incorporated as a village or a city, raise their own taxes, and look after their own services. In this way satellite dormitories and factory centres arise upon the fringe of the city, just beyond the limits of its jurisdiction. These find no place in Christaller's theoretical pattern of the distribution of centralized service centres (see Chapter 3). The land use of the fringe is also affected by the recreational needs of the city folk—parks, playgrounds, and golf-courses—and also by institutional and legal factors, since obnoxious industries excluded by zoning ordinances from the city itself must perforce be dumped on its outskirts—slaughter-houses, oil storage depots, noxious industries, junk yards, dance halls—while some city utilities are also given a peripheral location—waterworks, sewage plants, airports, and cemeteries. Thus, the rural–urban fringe is really " an extension of the city itself, present and potential ", and " since the city or cities of a metropolitan area and the suburban or fringe area are a unit economically and sociologically, the entire area should be thought of and planned as a unit ".[1]

The British city exhibits the same tendency to extend and explode. Britain is the most urbanized country in the world. The census in 1931 classed 80 per cent. of the population of England and Wales as urban. One-third of the population lives in the six greatest urban agglomerations (Greater London, Birmingham, Manchester, Merseyside, Leeds–Bradford–Huddersfield–Halifax and Tyneside), and over half in or near the fourteen chief cities. But there is a large, essentially urban, population living outside the limits of the " urban " (administrative) districts which is not classed as " urban " for census purposes, and the total population which lives and works in urban areas is well over four-fifths. The large number of cities and the recent spread of the suburban fringe over large surrounding areas also mean that a large and dangerously high proportion of land is built over.

[1] Wehrwein, op. cit.

Expressed briefly, the uses to which the 58,340 square miles or 37,133,000 acres of land in England and Wales were being put in 1937 were as follows: some 82·1 per cent. of the total area was in agricultural production (including rough grazings); 1·1 per cent. was open land of various kinds not being used for agriculture but of potential agricultural value; 5·5 per cent. was woodland; and the remaining 11·3 per cent. was covered by buildings, roads and various other forms of constructional development, or was otherwise, unaccounted for in agricultural returns.[1]

In 1937, 41 million people in England and Wales occupied about 4·16 million acres with the various structures associated with their homes and work, so that there were 10 people per acre of land under urban use (built-over).

The impact of urban land uses on the countryside during the last forty years has thus assumed alarming proportions, and especially during the last twenty years. Without any overhead planning control, a colossal redistribution of homes and factories to the borders of the great cities has been going on—due to the shift of factories outwards from the centre of the cities, as well as to the establishment of new factories to accommodate entirely new industries. During the 1919–39 period four million houses were built, and most of them on agricultural land.[2] The great majority of these lie on the suburban fringes of the great cities, beyond their administrative limits, a fact which is demonstrated very clearly by the map showing the changes in population in the period 1931–8 reproduced in the report of the Scott Committee. This map shows clearly how the central areas of the cities have in many cases actually declined in population or have managed to maintain a very small increase well below the average for the country as a whole; whereas the outer suburban areas, frequently in non-urban districts and planted in the midst of agricultural land, have increased very rapidly. With a few exceptions the main areas of increase are in a belt from south Lancashire to the south-east coast—notably Greater London and the Home Counties and west Lancashire and Cheshire.

The impact of urban land uses on the countryside has produced a wide fringe of land which in its uses and the life of its people is neither urban nor rural, but "sub-urban". This development in Britain takes many forms—council-housing estates, private building estates on new land or clustered around an existing village or small town, occasional good-class villas scattered

[1] Report of the (Scott) Committee on Land Utilization in Rural Areas, Cmd. 6378, 1942, p. 2. [2] Ibid., p. 23.

irregularly along country roads, usually with an attractive view, or rows of semi-detached houses and bungalows strung along the arterial roads. This is the notorious ribbon-building, frequently accompanied by new factories, which all enjoy the cheapest lay-out of utility services, gas, sewage disposal, water and electricity, and accessibility for transport. Such building disfigures the countryside, and impedes the development of communal living and community consciousness. Factories, lines of houses, wayside cafés, road-houses, garages, filling-stations, bill-boards, and the like, stretch for miles along the arterial roads leading out from the cities, and sheltered, pretty nooks along the coast harbour masses of jerry-built shacks and bungalows.[1] Between the roads there are wide stretches of farm land, but much good land has been sold for houses and factories, and much farm land lies derelict, since many farmers and landowners have held their land for sale to builders, and pending such sale have allowed it to lie idle. Under such circumstances land is farmed badly, since the tenant farmer is under constant threat of eviction. This explains the frequent spectacle of derelict land on the borders of our cities with hoardings facing the roadside announcing sites for factories, or desirable residences—which strike one's eye many miles before one reaches the built-up area of the city. "The threat of the builder overshadowed and sterilized it." [2] This is a repetition of the "institutional desert" around the American city and arises from the same causes. With reference to the 1931–8 period, it is reported that :

> In every one of the great industrial cities of the central belt there is a *decrease* of population in the centre where commercial buildings replace dwelling-houses, and a huge *increase* in the surrounding fringes, where country is replaced by suburb or town. The continuance of this process we believe to have been temporarily arrested by the war, and we would call attention to the extreme urgency of legislative action to give effect to our recommendations. We are convinced that otherwise the old unregulated sprawl of towns into country with its attendant evils will recommence immediately men and materials are released for the work of physical reconstruction.[3]

8. The Journey to Work

The separation of workplace and dwelling-place, rendered possible by the development of cheap and rapid transport, is one of the most fundamental problems of the modern town.

[1] J. A. Steers, "Coastal Preservation and Planning", *Geographical Journal*, Vol. CIV, 1944, pp. 7–27.

[2] Scott Report, pp. 28–9.

[3] Ibid, p. 6.

This subject has recently been examined by Dr. K. Liepmann in a volume in this series, and to it the reader is referred. The following summary comments are drawn from it.[1]

The daily flow of passenger traffic in all its aspects follows a maze of routes and cross-currents. The most marked trend is the movement from residential suburbs or dormitories to business and industrial districts. These streams, and their reverse in the evenings, have been described as " tides of daily ebb and flow ". Workplaces are usually in the centre of the urban complex and residential districts on the fringe so that the tide is centripetal in the morning and centrifugal in the evening. But there are numerous other outstanding flows that cut across these. Thus, with the growth of factories on the fringes, residential districts have grown next to them. But in many cities, especially on the Continent, workers still live in the city and travel daily to the factory on the outskirts. These are *counter-currents* that are the exact reverse of the centripetal movements. This was true, for instance, in large measure of Berlin. There are also *cross-currents*, especially where there are several urban centres within daily travelling distance of each other ; these are especially characteristic of the big conurbations. These movements, when considered from the standpoint of the dwelling-place and workplace respectively, may be regarded as movements of *dispersion* from the former—the inhabitants of a neighbourhood leaving each morning in various journeys of very different lengths—and as movements of *conflux* to the latter.

> Since the needs of industry and commerce have so far been best met by a concentration of the day population, while for residence and domestic purposes converse conditions are preferable, industrial areas are generally associated with a larger number of dormitory areas, and such intersection as takes place is characterized by a daily pulsation between the common industrial centre and a larger number of surrounding residential areas.[2]

While this is still generally true, in the last twenty years dormitories have been built for much more numerous populations, and workplaces have been more decentralized within the urban area, and there are now many more works which employ thousands of workers. These developments produce a more marked conflux towards the single works. Finally, the daily movements

[1] K. Liepmann, *The Journey to Work*, The International Library of Sociology and Social Reconstruction, Kegan Paul, London, 1944, pp. 3–6.

[2] Census of England and Wales, 1921, *General Report*, p. 193, quoted by Liepmann, p. 5.

do not form regular patterns, for some currents are short, and these are usually the most voluminous, while others are long and of less volume. Moreover, the currents are constantly changing in direction and strength.

The effect of this mobility on the structure of the city is to permit the separation of residential from industrial and business districts. It has caused the " spread and sprawl " of our towns, which is so much deplored, a development that has been encouraged particularly in this country by the adoption of a low housing density standard, as compared with the higher density standard (based on the flat) on the Continent and in the United States. The daily journey also permits the greater mobility of labour. For good or ill, the house of the employee can be up to a journey of one hour or more, and a distance of ten to fifteen miles, from the workplace. In an area of closely clustered towns, factories can theoretically be spaced between them and draw upon their reservoir of labour. New industrial plants must either draw their workers from a wide surrounding urban area or effect the construction of entirely new settlements. The main result of this mobility of labour and the separation of workplace and dwelling-place is to create a " polarity " in the urban structure, and, among its social and economic consequences, a split of the community interests and financial obligations between the working district and the dwelling district. This is one of the most serious problems of municipal finance.

9. The Location of Industry

The growth of factories on the city outskirts is a second main cause of the expansion of the modern city. Industries are affected differently by centrifugal and centripetal forces according to their character. The *Regional Survey of New York and Its Environs* distinguishes light manufacturing and heavy industries, and, though this distinction is open to varied interpretation, the Survey based it on the following considerations. Heavy industries are large in size ; the time factor—that is, the factor of immediate accessibility to the market—is unimportant ; they demand a large ground area ; they frequently have nuisance features such as noise, odours, pollution, fire hazards ; they have a serious problem of waste disposal ; they require a large plant lay-out ; they require large quantities of fuel and water ; their products as well as their raw materials are bulky so that they require extensive and contiguous railway or water-transport facilities. All these conditions make the siting of such plants unsuitable for the central districts, and more suited to the fringes, of the urban area, whether they be shifted outwards from the central areas or established in the first place as new plants. Such industries are meat-packing, petroleum storage, smelting, automobile manufacture and assembly, lumber and flour mills, power plants and gas works. The shift outwards lags behind the need for the move, and plants established originally on the town outskirts are slow in being shifted and continue, through the inertia of fixed capital as going concerns, to function on their original sites, though unsuited in every way to the well-being of the urban community as well as to the organization of the industry. Many instances of this kind are apparent in Chicago, the outstanding example being the stockyards, which, though originally well sited on the southern outskirts of the town, are now in the heart of it in the midst of an appalling slum.

Light manufacturing industries have the following characteristics. They do not always require buildings of special construction, the time or service factor is of great importance (immediate contact with related industries, transport facilities), they often require highly skilled work, they require little ground space per worker, obsolete buildings are often suitable, they are carried on in relatively small business units, they are often seasonal in character and have a fluctuating labour demand, and they are often conditioned by the vagaries of style, especially in clothing. Finally, their products usually require materials small in bulk often going through several processes, demanding much labour, but little loss of material or bulk in the process. Such industries are garment-making, printing, cigar- and cigarette-making, instrument-making and the manufacture of cosmetics. They usually seek a few rooms or an obsolete building or warehouse on the borders of, or as near as possible to, the central business area and draw much of their labour from the poor working-class districts in and around the central business area. Their manufactured goods find their markets in the warehouses and retail stores in the neighbouring wholesale and central business districts.

This broad contrast between the two major sets of industries is sound in considering the growth of the city, but while the heavy industries have definitely this character, many of the so-called light industries may be pursued on such a large scale as to demand, like the heavy industries, an outside location. This process is well illustrated in the case of the new industrial districts of north-western London. All modern cities have had large industrial plants established on their outskirts—either as entirely new industries or as old-established ones shifted (often in stages) from the congested city sites. Such plants are situated, according to their needs, on road, rail and canal, and become in themselves the centres of crystallisation of housing developments and the foci of daily streams of traffic carrying their workers to and from their homes.

10. The Distribution and Movements of Population

The interplay of the centrifugal and centripetal forces in the modern city is most clearly reflected in the distribution and movements of population.

Overall densities of population by districts, as studied in many cities—probably the picture most accessible to the general reader will be the Ordnance Survey map of Greater London on a scale of half an inch to one mile—reveal that in all Western cities there is a fairly general pattern of distribution, though densities vary from one city to another.[1] In the very large city, densities closely fit with the concentric zones outlined above. First, the city centre has a relatively low density. It is the heart of the urban organism, throbbing with activity in the daytime, deserted at night, except for its caretakers, its slum pockets, and its hotel and apartment blocks (for both rich and poor). In smaller cities this district is not large and only detailed large-scale mapping would bring out this feature. For this reason, in cities with under, say, 750,000 inhabitants, the population of a central census district (such as a ward) includes both the business centre

[1] See also the detailed study of Merseyside, and particularly of Liverpool, by Wilfred Smith in *The Distribution of Population and the Location of Industry in Merseyside*, 1942, and the study of Birmingham in *When We Build Again*, published by the Bourneville Village Trust, 1941.

Density of Population (Persons per Acre by Concentric Zones)

Zone	*Liverpool* (1931) *Acres*	*Persons per acre*	*Birmingham* (1938) *Acres*	*Persons per acre*
Central Zone	1,600	59	3,000	62
Middle Zone (Inner)	3,300	96	9,000	32
Middle Zone (Outer)	2,800	52		
Outer Zone	17,000	17	39,000	15
Total	24,700	34·5	51,100	20
Total Population	855,700		1,050,000	

and adjoining slum districts in the transition belt, so that its density of population may be exceedingly high. The general trend, however, is for the slums to be cleared or abandoned and for their inhabitants to shift, or be shifted, elsewhere. Secondly, around the core is the mostly closely built-up residential area, and this normally has the highest overall densities of population. Farther out, land built up at the end of the nineteenth century until 1914 has more open space and lower densities. Finally, in the suburban outskirts the density—both overall and per acre of built-up land—is lowest.

The Ordnance Survey population map of Greater London shows these features clearly. The central districts—the city of London and the city of Westminster—have about 20 and 50 residents per acre respectively; then comes a surrounding zone, including such boroughs as Stepney, Lambeth, Finsbury and Paddington, with between 100 and 200 persons per acre (maxima are between 200 and 250).[1] Farther out, a wide encircling belt has 50 to 100 persons per acre, and the outlying suburban districts have under 50 persons per acre.

A further main point to be emphasized is the redistribution of population that is taking place, the most remarkable feature of urban growth during the last fifty years. It is marked by the de-peopling of the centre, owing to the ever-increasing demand on space for business, the demolition of slums, and the voluntary choice of new homes on the city outskirts in pleasant surroundings at small rents. This displacement has its opposite poles in the city centre and the suburban estate, although much growth takes place in the vicinity of new industrial plants on the city outskirts. A slowing down of population growth, and, indeed, a decrease, has also been noticeable in the very large fully built-up zone—the middle zone—around the city centre, spreading from its inner to its middle and outer margins. It is fully built-up and new families or immigrants must live there in overcrowded conditions or shift to the outskirts of the city into the new suburban areas.[2]

It is extremely important to realize that this decrease of population, while it started in the heart of the city through the expansion of business space, has spread in more recent decades to the zones around it—the zone of transition and even into the

[1] In Paris, probably the highest densities of all Western cities are found. Two working-class quarters near the centre have densities of 432 and 410 persons per acre. See J. L. Sert, *Can Our Cities Survive?*, Harvard U.P., 1943, p. 20.

[2] Reference should be made in this connection to the thorough study by K. Liepmann on *The Journey to Work*, 1944.

middle residential zone of the nineteenth century growth—through the bad living conditions in these areas as compared with those in the extensive housing schemes that have been completed in the suburban outskirts during the inter-war period.

In the city, as it emerged in the nineteenth century, the density of population was normally greatest in the centre and decreased progressively outwards to the periphery, a phenomenon that was closely related to both the percentage of the land actually built on and to the amount of cubic living space or the numbers of storeys. During the last hundred years, however, and especially since 1900, the growth of business offices and public buildings in the centre and other causes have resulted in an outward shift of people, so that, in consequence, it has a low residential density and the areas of highest densities are to be found in the congested and oldest residential areas around it. Nevertheless, in fairly small cities such as Frankfurt, for which data are available, the densities of population in concentric belts from the city centre show a

DENSITY OF POPULATION IN FRANKFURT ON THE BASIS OF CONCENTRIC RINGS (after W. Gley) [1]

Ring (km.)	*Population*		*Density per acre*	
	1890	1933	1890	1933
0–½	39,700	25,000	200	121
½–1	44,600	40,700	76	69
1–2	77,300	160,000	33	68
2–3	28,400	95,400	7·2	25
3–4	20,300	68,400	3·7	12
4–5	10,400	48,100	1·5	7
5–6	47,800	86,600	5·5	10
6–7	11,300	54,100	1·1	5·2
7–8	10,000	25,100	0·8	2·1
8–12	45,000	97,800	0·7	2·5

steady decrease outwards. The decrease, it will be noted, was most marked in a radius of one kilometre (say half a mile) of the city centre. The bulk of the population, however, is found in the 1–2 km. zone with 160,000 inhabitants out of a total of about 600,000 inside a circle of 8 km. radius.

Fig. 28*a* shows the distribution of population in the old town of Berlin (*Alt-Berlin*), prior to its extension in 1920 to form Greater Berlin. Though now obviously incorrect for a city laid in ruins,

[1] W. Gley, "Grundriss und Wachstum der Stadt Frankfurt a.M., Eine stadtgeographische und statistische Untersuchung", *Festschrift zur Hundertjahrfeier des Vereins für Geographie und Statistik zu Frankfurt a.M., 1936*, p. 87. For the most central area of Frankfurt, within a radius of half a kilometre, the density in 1930 was *121* as compared with 200 persons per acre in 1890. The increase of population in the 5–6 km. ring is due to the inclusion of the adjacent town of Offenbach.

Fig. 28a.—Density of Population in Berlin (*Altstadt*), 1925 (boundaries as before 1920).
(After Leyden.)

Persons per acre calculated on the basis of small police districts. These very high densities are common to large European cities and are, in general, lower than the densities in American cities, such as Chicago (Fig. 29a).

the features are very characteristic of the pre-war European city and it is one of the few detailed population maps that is readily available.[1] The densities are calculated for small police districts and in the same source from which this is taken there is a similar map for 1875. Comparing this map with that of built-up areas as shown on Fig. 26 (p. 108) one notices that the areas of lowest

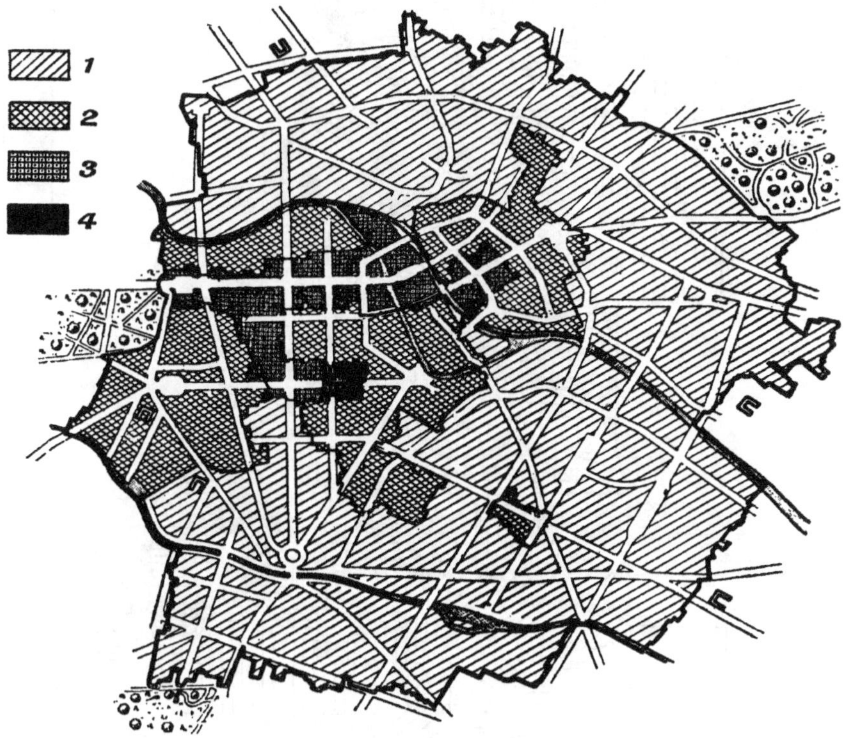

FIG. 28*b*.—The Decrease of Population in Central Berlin, 1885 to 1930, from *Berliner Wirtschaftsberichte*, 1932. (After Leyden.)

Decreases from (1) 100 to 50–100 persons, (2) 100 to 20–50, (3) 100 to 10–20, (4) 100 to less than 10. The area is bounded by the parks of the Tiergarten to the W., Friedrichshain to the E., Kreuzberg to S.W., and the Stettiner *Bahnhof* (Station) to the N.W., and Schlesischer and Görlitzer *Bahnhöfe* to the S.E.

density (10–40 persons per acre) are in the heart of the central business zone and the "West End" around the large open space of the Tiergarten, while the maximum densities (160–320 and, in patches, over 320 persons per acre) are found in the congested tenement areas built during the nineteenth century, the latter being densities that are far higher than in American cities, but are

[1] F. Leyden, *Gross-Berlin : Geographie der Weltstadt*, Breslau, 1933, p. 95.

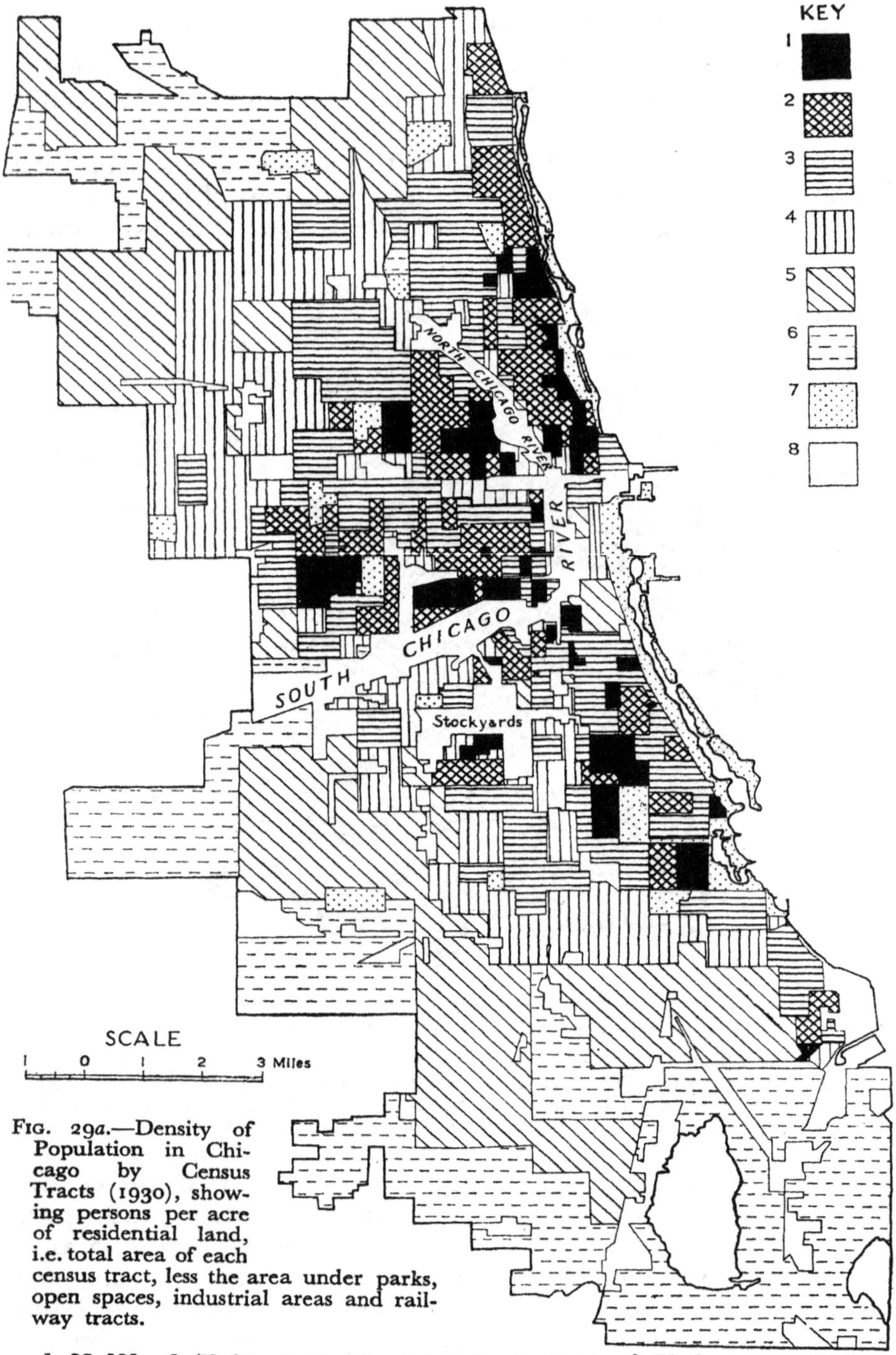

FIG. 29a.—Density of Population in Chicago by Census Tracts (1930), showing persons per acre of residential land, i.e. total area of each census tract, less the area under parks, open spaces, industrial areas and railway tracts.

1. 90–120. 2. 70–90. 3. 50–70. 4. 30–50. 5. 10–30. 6. Under 10. 7. Parks and open spaces. 8. Industrial areas and railway tracks.

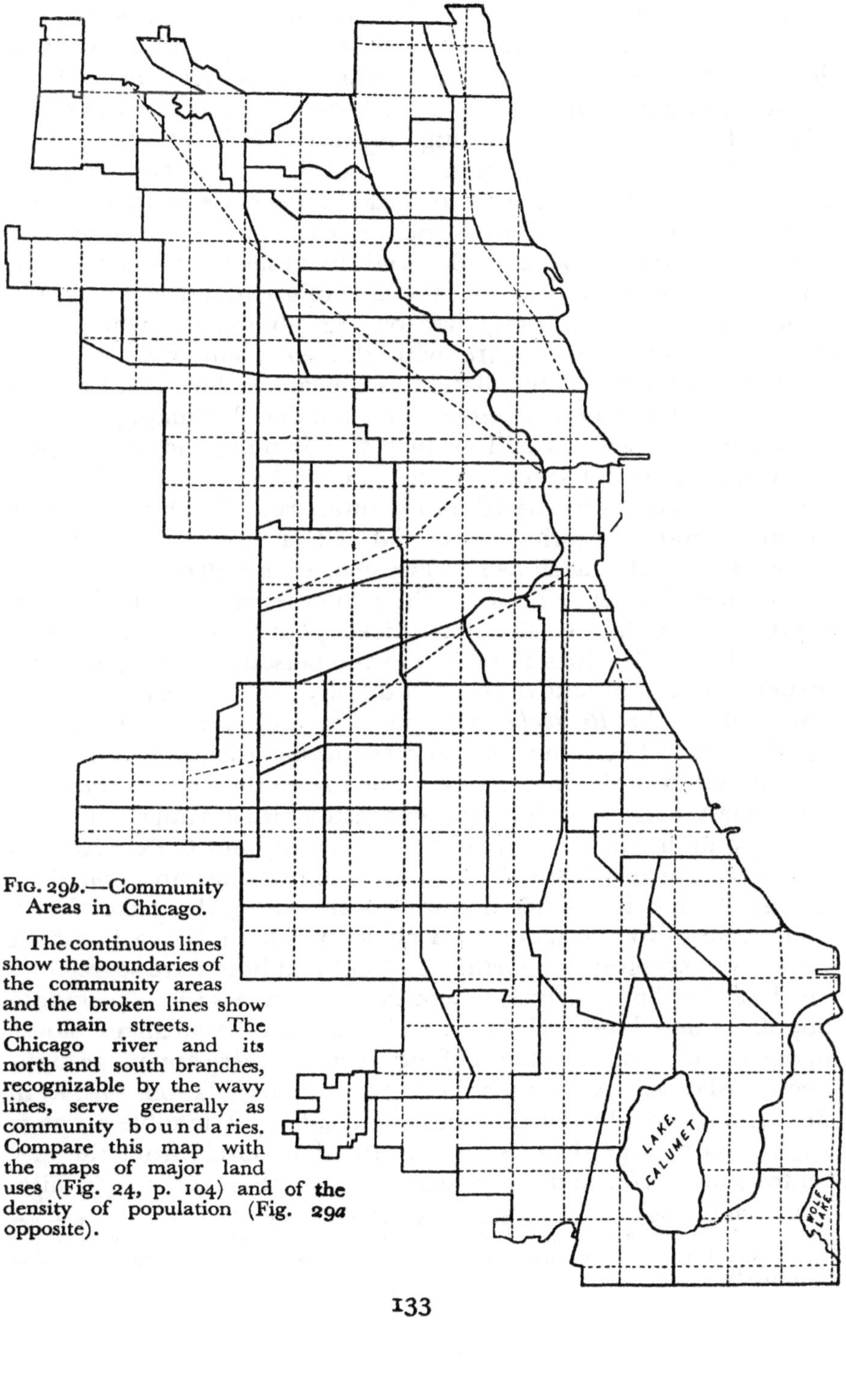

FIG. 29*b*.—Community Areas in Chicago.

The continuous lines show the boundaries of the community areas and the broken lines show the main streets. The Chicago river and its north and south branches, recognizable by the wavy lines, serve generally as community boundaries. Compare this map with the maps of major land uses (Fig. 24, p. 104) and of the density of population (Fig. 29*a* opposite).

typical of the European city. The whole of *Alt-Berlin* has an overall density of just over 100 per acre.[1] Fig. 28*b* shows the decrease of population in the centre of Berlin since 1885. Decrease began in the 'sixties in the heart of the city, accelerated after 1890, and spread throughout the whole of old Berlin, which reached its population peak of just over two millions in 1910, while the whole area recorded a decrease in 1925. The outer districts that had shown rapid increases after 1870 (tenement areas) reached their peak in 1890 and then they decreased like the centre. It will be noticed that the area of greatest decrease corresponds with the central city area on Fig. 26 and the wide areas of small decrease since 1910 with the tenement areas.

The distribution of population in Chicago is shown on Fig. 29*a* as typical of the American city. This is a detailed map, prepared originally on a scale of half an inch to one mile, and based upon 499 census tracts of several blocks each. Further, the densities are *residential* densities, based on the total area of each tract minus the areas under factories, parks and railway tracks, whereas the densities for Berlin, as shown on Fig. 28*a*, are *overall* densities. The remarkable difference between the two cities is thus all the more striking, for in Chicago the most thickly populated group of tracts has residential densities of only 90 to 120 persons per acre; whereas in Berlin this is almost the overall density for the area of the old town inside the *Ringbahn* with its two and a quarter million inhabitants. The same general distributional features, however, are evident in both cities, namely, the lowest densities in the core, the highest densities in the inner congested districts adjacent to the riverside industrial areas and the stockyards, and the decrease of densities outwards from the centre. Chicago has an anomalous feature in that its higher densities occur along the Lake Front, a fact due to its attraction as a choice residential area in which sumptuous skyscraper apartment blocks and hotels are segregated north of the Chicago river, where the zone carries the appropriate popular title of the " Gold Coast ". Isolated apartment blocks in open land also occur south of the river and these appear as high-density islands. As regards the decrease of population, this, as in Berlin, has spread from the centre outwards. The table on page 135 shows a large decrease in the two-mile circle from the city centre since 1910, and the absence of change in the 2 to 4 miles

[1] *Alt-Berlin*, consisting of the six central administrative districts, had a pre-war population of 1,790,000 on an area of 17,000 acres, with a density of 104 persons per acre. The densities of the Tiergarten and Wedding districts were 63 and 88 per acre respectively.

zone (the areas with greatest densities) in the 1910–20 decade, followed by a considerable decrease in the 1920–30 decade.[1]

The centrifugal drift is illustrated very clearly by comparative figures for American cities. The figures for 85 out of the 96 metropolitan districts show that their rate of increase from 1920 to 1930 was a little more than twice as great as that in the central cities (see p. 198), and the rate differentials tend to increase with the size of the districts. Moreover, it is worth noting that 31 per cent. of their total population lies in these outside districts. Closer study of population changes within the city itself reveals the process even more clearly. A grouping of small census districts in concentric zones from the city centre for New York, Chicago, Cleveland and Pittsburgh shows that not only is the population

POPULATION DENSITIES IN CONCENTRIC ZONES IN AMERICAN CITIES.[2]

	Population			Increase or Decrease per cent.	
	1910	1920	1930	1910–20	1920–30
New York 4-Mile Zones					
0–4 miles . . .	2,200	2,054	1,538	− 7	− 25
4–8 miles . . .	1,925	2,413	3,181	25	32
8–12 miles . . .	568	1,021	1,814	80	78
12–16 miles . . .	64·5	102	386	58	278
16–20 miles . . .	8·7	9·7	11·1	12	14
Outside City . .	1,800	2,290	3,083	27	35
Chicago 2-Mile Zones					
0–2 miles . . .	359	276	216	− 23	− 22
2–4 miles . . .	732	737	657	1	− 11
4–6 miles . . .	617	861	969	40	13
6–8 miles . . .	282	496	751	76	51
8–10 miles . . .	199	354	749	77	112
Outside City . .	266	477	829	79	74
Cleveland 2-Mile Zones					
0–2 miles . . .	162	163	119	Nil	− 27
2–4 miles . . .	300	388	373	29	− 4
4–6 miles . . .	89	192	288	117	49
6–8 miles . . .	11·6	45·4	92	290	103
8–10 miles . . .	8·7	14·6	28·69	69	95
Outside City . .	54	129	259	140	101

[1] For a map of the changes of population, 1920–30, by census tracts in Chicago, see McKenzie, *The Metropolitan Community*, 1933, p. 177.

[2] Ibid., p. 176.

decreasing in the centre but that there is a widening range of declining population from the central districts.

The main features of these trends have been summed up for American cities as follows : [1]

1. There has been an exodus of population from the central portions of the city. This exodus is caused in small measure by an increase in the size of central areas utilized for business purposes, but much more largely by the progressive deterioration of structures in large portions of the central areas. The effect of this deterioration is a creeping paralysis, commonly referred to in the United States as " blight ".

2. The exodus from the central area, together with the settlement of new population in suburban areas, has caused a drift of the masses of the population outward radially. Much of this drift is due to the promotional and sales efforts which have been made in connection with the development of new estates.

3. With the outward radial drift of population has come a recentralization of outlying district business centres, so that they sometimes reach the proportion of satellite business communities.

4. The provision of rapid transit facilities has tended to aggravate further the decline of the central area by providing non-stop express service through such areas and non-stop highways for automobile traffic.

The study of distribution of land values and of the changes that have taken place in relation to these trends centre on three phenomena :

the increase in values in the strategically located restricted central areas ; the increase in values in the outlying district subcentres ; and the decline of values in other portions of the inner area. On the decline of values in inner areas of the city not included in the intensively developed portion, data are not available. It is a common observation, however, that this decline has been great and is probably proceeding at an accelerated rate. So rapid has the decline been that in many cities the revenues from taxation of real estate have been shrinking to an alarming extent.[2]

[1] McKenzie, op. cit., p. 233. [2] Ibid., pp. 232–3.

CHAPTER 5

REGIONS WITHIN THE CITY : THE NATURAL AREA

Special attention was given in the last chapter to the city as a whole and, in particular, to the process of concentric growth within it. This process is evident not only in the general formation of broadly concentric zones, named the central, middle, and outer zones, but also in the detailed movements and segregations of population and in the succession of land uses and population groups inside the compact built-up area. It has already been indicated, however, that, important as this process of concentric growth is in the differentiation of the structure of the city, much more important is the formation of distinct districts, whose functional character depends not only on position in relation to the centre, but also on conditions of topography and of historical development. The character and arrangement of these functional regions, while differing in detail from one city to another, have certain general features, which have been noted already with reference to representative American cities (p. 116). Before proceeding to discuss particular aspects of the functional regions of the city, let us briefly examine the structure of one city, ignoring as far as possible the detail that is dependent on its special topography.

1. Functional Differentiation in a Large City : Stockholm

Stockholm has been thoroughly studied and reveals admirably the process of urban growth and the tendency to the segregation of functions in separate districts, and the succession of functions in them.[1] The inner town area covered by this detailed survey covers the site of the medieval town (Staden Mellan Broarna) on an isthmus between two east-west stretches of water (Mälaren lake and the Saltsjön) and the compact built-up area to the north and the large island of Södermalm to the south. This is a densely and

[1] W. William-Olsson, " Stockholm : Its Structure and Development ", *Geographical Review*, Vol. XXX, 1940, pp. 420–38. Also *Stockholms Inre Differentiering* by H. W. Ahlmann and associates, Meddelande från Geografiska Institutet vid Stockholms Högskola, No. 20, 1934, and Sten de Geer, " Greater Stockholm ", *Geographical Review*, Vol. XIII, 1923, pp. 487–500.

continuously built-up area, with buildings of four to six storeys, with extensive suburbs around it that are divided into eleven

FIG. 30*a*.—The Distribution of Population in Inner Stockholm in 1930. Each dot represents 25 persons.

On the isthmus is the old town of Staden Mellan Broarna, with the island of Södermalm to the south and the main part of the city to the north. Norrmalm, the central business district, lies in the centre of the latter, and is indicated by its relatively small population. The whole area represented on this map roughly corresponds to the compact built-up area of the city as a whole.

sectors by belts of water. Out of a total population of over 500,000, four-fifths live in this inner area, although the suburbs are increasing rapidly. The distribution of population in this area is shown on Fig. 30*a*.

At the beginning of the century the medieval town on the isthmus was the centre of business and administration and from it radiated the steamboat lines. But to-day the dense net of trolley lines centres on Norrmalm, a flat area on the mainland immediately north of the isthmus, which has become the business centre and the terminus of the suburban railway lines, and the focus of the chief traffic and business arteries.

A thorough study was made of the exact location of functions in 1880 and 1930 and the main conclusions, apart from local detail, are of general significance in the process of urban growth.

The inner town of Stockholm, as indeed in any large city, is used alike for residence, work and recreation. The authors classify as places of work : government offices, public institutions, business offices, retail shops and industrial establishments. Each group may be subdivided and each subdivision has its special situation requirements in accordance with which significant changes in its position may be traced between 1880 and 1930. A composite map showing the location of the functional areas in the centre of the city is reproduced in Fig. 30*b*.

Government offices have long been concentrated in the northern part of Staden Mellan Broarna, the cluster forming a national administrative centre around the Royal Palace. Many other government offices are located peripherally where they have freedom of choice of location, since they have no competition to meet and are necessary to the public. Those buildings that serve only part of a town—churches, elementary schools, police and fire stations—are normally conveniently located in the part they serve, while those that serve the whole town or State strive for a central position.

Business offices direct production and distribution. In 1880 the main cluster was south of the administrative quarter, occupying the broad street along the quays and avoiding the narrow alleys of the centre of the island. By 1930 the number of offices had increased many times and they had all shifted to broad streets and open sites in Norrmalm. In 1880 the banks and stockbrokers' offices—most sensitive of all business activities to situation—were not concentrated ; by 1930 they were clustered in the south-east of Norrmalm and formed the true " core " of the city. Offices require free accessibility between each other and the residential district, and they are situated by preference in broad streets, level

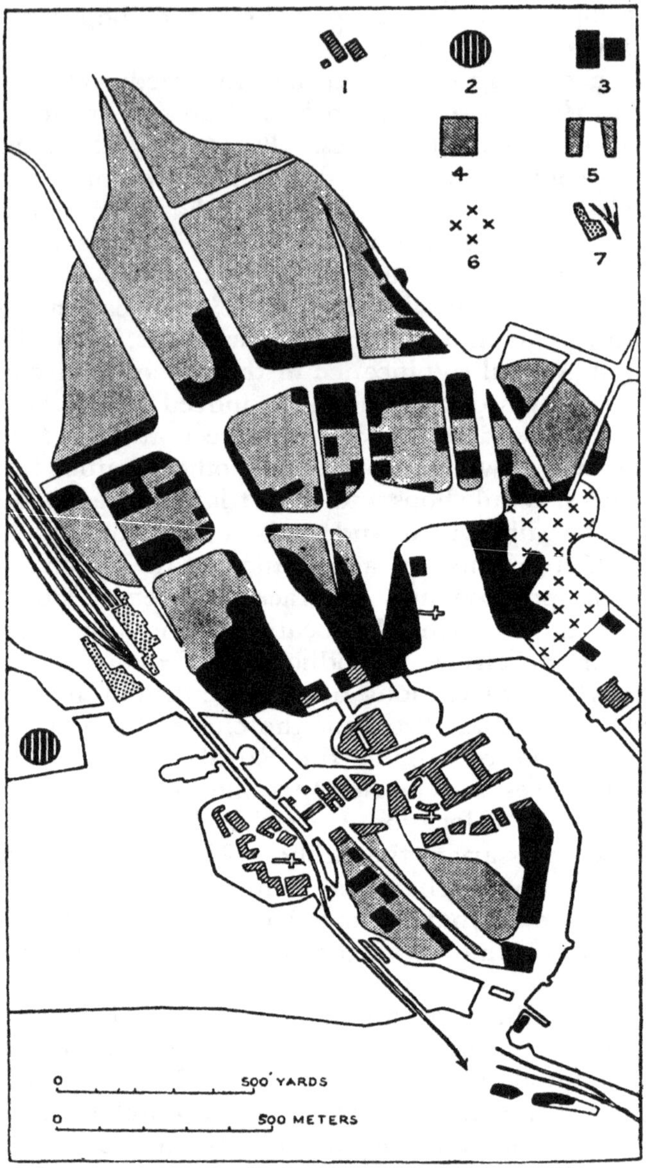

FIG. 30*b*.—The Functional Areas of Central Stockholm in 1930. For the location and extent of the district, see Fig. 30*a*.

1. Government offices and public institutions. 2. Town Hall. 3. Offices. 4. Printing and clothing industries. 5. Shopping streets, the width being proportional to the intensity of shopping. 6. Amusement centre. 7. Railways and railway stations. In 1880 these activities were much smaller in importance and areal extent and were concentrated almost exclusively in the old town of Staden Mellan Broarna on the isthmus.

ground, and high-class buildings. The great growth of business demanded space such as the congested and historically and culturally valuable medieval town of the Stadsholmen could not provide ; moreover, the administration was already there as a competitor for sites. Thus, the business area shifted to the southern part of Norrmalm, where communications were best, thence spreading northwards into the area of steep, narrow streets on higher land in southern Norrmalm through which the Kungsgatan (a main street) was cut.

Retail shops are the last link in the chain of activities by which goods are transferred to the consumer. Distribution—wholesale and retail trade—according to this interpretation includes not only dealings in goods, but also provision of meals, amusements and lodgings. Shops, to be successful, require good contact with the public and are highly sensitive to location. Neighbourhood shops sell goods for daily household consumption and are evenly spread over the town in accordance with the distribution of population. City stores have a specialized character for periodical purchases, with a margin of choice as to qualities and prices. They therefore require a central situation for a widely dispersed clientele. The importance of the shop as a distributive organ is most clearly measurable not in the shop-window frontage but in the *rent*. By adding the shop rents of a street-frontage and dividing the total by the length of the frontage there is obtained the shop rent per unit of frontage ; and this can be taken as a measure of the shopping intensity of the street. Maps prepared on this basis for 1880 and 1930 reveal the following facts. The nucleus of the shopping area—the Stadsholmen and the Drottinggatan running north-west to south-east in the centre of Norrmalm—lies in the centre of the town, and in 1930 remained, as in 1880, the main shopping area, although extension had taken place to the north-east towards the wealthier residential districts. The highest shop rents are located on centrally situated street intersections. Since shops prefer level streets of moderate width, the chief shopping streets coincide with the chief pedestrian thoroughfares, for heavy vehicular traffic is detrimental to shopping streets. Shops at the southern end of Norrmalm have been partly ousted by the banks, and indeed as a rule offices and shops have avoided each other, since their demands on location are different, the first demanding good communication, the second demanding good local contacts with the visiting public, so that competition between the two uses is rarely keen. Shop rents are normally much higher than office rents, and there is a segregation of different kinds of shops in different streets. In 1880, theatres, places of amusement, hotels and restaurants were mainly in the south-east of Norrmalm, but in 1930 they were found in all parts of the town, hotels being especially localized near the railway stations.

Industrial establishments [1]—both factories and small workshops—include in Stockholm mainly metals, paper and printing, textiles and clothing, foods and tobacco. Establishments are mapped in this study according to exact location and by a circle proportional to the number of employees. In 1880 and in 1930 industries are found both in the central fully built-up area and the outer, more open area situated especially near the quays and railway sidings. But in 1930 the latter, which included the largest units, were more widely distributed and farther on the outskirts, while the ring of industries of 1880 to the west, south and east had been replaced by new residential districts expanding from the central town. Industries highly localized in the centre of the town are printing, garment and needlework trades. The last, depending on the changing fashions and nearness to the shopping streets, made up, in 1930, 80 per cent. of the industries, and were housed in obsolescent buildings in back streets. Food products and small metal-working industries are widely distributed over the greater part of the town. The large establishments seek peripheral locations—engineering works, shipyards, etc.—though a few large concerns remain in the centre as relics of the past, and sooner or later must shift elsewhere.

Residential conditions are examined first by two maps showing the distribution of population in 1880 and 1930 respectively by absolute numbers, indicated by dot symbols. These maps reveal that " the growth of the town was practically concentric " and " distribution changed from high density in the centre and low density in the periphery to the reverse with the formation of a central ' city ' district ". The character of residential districts is determined by the type of building and the social and the economic status of the inhabitants.[2] The latter is best assessed from income

[1] Industries occupy 70,000 or 43 per cent. of persons gainfully employed. The chief group is engineering (30,000—machinery and electrical apparatus), the plants now being located almost entirely on the outskirts of the city. The other industries are those typical of all great capital cities—printing and publishing (12,000), clothing (10,000) and food (13,000). See Ph. Rebeyrol, " Les Industries de Stockholm ", *Annales de Géographie*, Vol. XLIX, 1940, pp. 35–43.

[2] This study should be compared with a similar recent investigation of Prague by Moscheles in the *Geographical Review*, Vol. XXVII, 1937, pp. 414–29 (see also p. 112 for a summary of the findings of this study), and of Budapest by Beynon, ibid., Vol. XXXIII, 1943, pp. 256–75.

returns, but as these were not available, resort was made to the distribution of ages and sexes, using the census, which gives data for each building block—a procedure quite out of the question in similar studies of British cities. In districts of well-to-do families there is usually an excess of women (due mainly to the relatively large number of domestic servants, daughters and unmarried relatives) and fewer children. Statistics by wards show that the wards having the most children per 1,000 women, the largest number of occupants per 100 rooms, and the highest ratio of manual workers to total population, are in general those in which income per able-bodied person is lowest, though it should be noted that this ratio does not apply to the central business district owing to its special demographic character. Taking the ratio of children to women as a measure of social and economic status, southern Norrmalm was the wealthiest residential district in 1880, and the poorer people lived on the periphery. But since business has invaded the former, the good residential quarter has moved to Östermalm (immediately east of Norrmalm), displacing the poorer residences there. To-day, Östermalm is the most exclusive residential district, Vasastaden and Kungholmen (adjacent to the west side of Norrmalm) are the middle-class districts, and the outskirts of Södermalm (the large island to the south of the old town) in particular are the working-class districts.

The limits of Stockholm, defined so as to include all those inhabitants who "take part in the life of the town", are determined by the population trends in the surrounding municipalities. Areas with decreasing numbers are, as far as the city is concerned, agricultural, stagnant and untouched by its growth and functions. Others with moderate increases have been slightly affected by urban influence. But nearer the city, large increases reflect "explosive" expansions due to the opening of new communications and the development of new estates. Greater Stockholm on these lines had a population of about 656,000 in 1937.

The changes in the inner town from 1880 to 1930 included no change of relative positions or internal structure, but reflect a development which "may be likened to a process of growth in which one organ, in expanding, crowds out another, and this in its turn takes up unoccupied space". The changes in the 1880–1930 period are summed up as follows :

> The administrative centre retains its situation but has been enlarged somewhat to the north ; office and shopping centre have been developed greatly ; and so have the centrally located industries ; and all these—

which together constitute the working core of the town, its "city"—have spread northward. The best residential district of 1880 has been forced to move farther to the north-east, and lately also to the west, and the better-class dwellings have displaced the poorer residences. These in their turn have pushed the 1880 circle of suburban industries farther out into the country. Finally, owing to the improved communications, large residential suburbs have grown up far from the centre of the town, encroaching on agricultural regions.

2. The Natural Area

In every urban complex there are well-defined geographical areas, called "natural areas" [1] by the ecologist or "natural" or "functional regions" by certain geographers, which are akin to the "natural formations" of the plant ecologist. Let us be clear as to the meaning of this term "natural". It does not mean an area that owes the unity of its human occupance to adaptation to similar site conditions—geographical conditions in the narrow sense—but to the operation of natural or uncontrolled forces. The natural region in the city is, according to Professor R. Blanchard (a chief exponent of this aspect of geography), the culminating aspect of the geographer's study of the city. It involves, he says, the study of its "*quartiers*, their *raison d'être*, their aspect and their population, the study of the streets that characterize the quarter or interconnect several quarters. To these topics, that represent the morphology of the city, should be added that of circulation, which is, as it were, the hydrology of the city—the traffic of its streets and places and the daily and hourly rhythm of their activities." The geographer is thus primarily concerned with the structure of the city and the functional differentiation of its parts. The study of Stockholm, summarized in the previous section, is an example of the scope and techniques of the geographical approach.

The sociologist, on the other hand, is primarily concerned with the behaviour of the individual as a member of society, and the concept of the "natural area" as developed by him appertains above all to the idea of the spatial distributions of human groups and associations and behaviour traits in the urban area. This was the concept that lay behind Burgess's concentric zone theory of urban growth and the subsequent studies of single cities. Individuals and institutions, argue the sociologists, tend to become segregated in areas populated by persons with a similar background or institutions of the same kind—"birds

[1] Refer to the footnote on p. 2 in Chapter 1.

of a feather flock together ". The individual selects a residential area in which he can pay his way. The builder or the landlord builds or lets a building where he can make the maximum profit. In consequence of this process the city tends to become a " mosaic of cultural and social islands ". Such islands are termed " natural areas " since they grow up by natural (or uncontrolled) processes, and " the natural area " represents a " type of individual and collective adjustment which the urban population has made to its social and geographic milieu ".[1] " Each area tends to select certain population types, this selection being based on economic status, racial characteristics, religious beliefs, moral codes, and the like." [2] In the American city, with its large foreign-born element, language and customs are especially potent in the formation of such natural areas, but as the immigrant stocks are absorbed, and their own language and customs disappear, " social differentiation will manifest itself in other ways—through differences in religion, occupation, education, income ".[3]

Now, it should be emphasized that the standard of relevance of the facts with which the geographer is concerned is determined by the efficaciousness of one fact or set of facts in accounting for other facts with a similar distribution. In an urban area, the most efficacious set of facts accounting for its areal differentiations, and therefore the basis of any geographical study, is the distribution and movements of the people themselves, and the buildings (or, as we may call it, the habitat) in which they live and work. Facts of individual and social behaviour are least efficacious in accounting for areal differentiations and are therefore of least relevance to geographical investigation ; but this is precisely where the sociologist begins. Clearly, the sociologist requires to know the structure of the city, as defined from the geographical point of view, as a framework for the study of social space patterns. There is no clear boundary between the two fields and disciplines. The sociologist needs base maps of land use and socio-economic conditions. The geographer, having determined the location and character of the natural functional regions, is led to consider sociological phenomena that characterize each region and to indicate contrasts between them.

There are, then, many kinds of natural social groupings in the build and social structure of the city, but the most significant are : land use—requiring a carefully considered scheme of classification ;

[1] Gist and Halbert, p. 176. [2] Ibid., p. 176. [3] Ibid., p. 179.

social and economic groups—the chief of these being based on the density and ethnic character of the population ; and community space relations.

The investigation of the natural areas must be built on the detailed analysis of the facts of land use and of economic and social life, and, above all, on the presentation of these facts on maps. In many cases the areas will in large measure be coterminous with marked breaks in the physical structure—e.g. rivers, railroad tracks and factory belts—which isolate, and foster relations within, the areas they enclose. This mode of approach has been adopted for many American cities by American sociologists. But the idea of the socio-economic unit as a geographical region seems to have been neglected in the social surveys of English towns. In Chicago, for instance, the " social base " map showing industrial and transport properties, parks, and residential areas, has served as a framework for the understanding of the pattern of social behaviour of the individual. The geographical incidence of juvenile delinquency, prostitution, criminality, the social pattern of " hobohemia " and the " rooming districts " have been examined within this framework. Social surveys of English towns are primarily concerned with the facts of economic and social structure. They accept the local government unit as the smallest unit of investigation, and little attempt has been made to map the data in greater detail or to work out the character and extent of socio-economic units within them. The reports of Regional Planning Committees give nothing but passing reference in brief introductory chapters to these basic considerations, but it is on such a basis only that the town can be understood and rebuilt.[1]

The general procedure in the detailed diagnosis of the structure of a city is well established by many American and continental studies, though it is beyond the scope of this work to deal with the detailed problems involved in such a survey. We may instance the general procedure adopted in a recent American survey of New Haven in Connecticut.[2] First, land use was plotted in the field for every building lot, the classification including single-

[1] An interpretation on these lines will be found in D. Caradog Jones (ed.), *The Social Survey of Merseyside*, 1934, Vol. III, Chapter XIX. In the Inner Districts, with their slum areas, there is a coincidence of high birth-rates, overcrowding and other social phenomena, such as alcoholism and criminality. See p. 117.

[2] Maurice R. Davie, " The Pattern of Urban Growth ", an essay in *Studies in the Science of Society*, edited by G. P. Murdock, New Haven, 1937, pp. 133–61. See also R. E. Dickinson, " Chicago : Ville-métropole et sa Région ", *La Vie Urbaine*, Paris, January, 1934, pp. 3–28.

family, two-family and multi-family dwellings, commercial buildings, light industry, heavy industry, railroad property, parks and playgrounds, public property and open spaces. Second, these detailed base maps were generalized, on the basis of the percentage of street-frontage devoted to the major categories of urban land use—residential, commercial, industrial, transport, recreational and institutional. Third, areas which had the same predominant use or combination of uses were shown on a third map. These are the land use areas of the city. Fourth, the analysis of the socio-economic structure began with the mapping, by exact place of residence, of such facts as the density of population, nationality, income, delinquency, dependency, and names included on the social registers (indicating social and professional status).[1] These facts were first plotted separately on a series of maps, and then the maps superimposed, and it was found that " to a remarkable extent the various area boundaries coincided ". In this way socio-economic areas were defined and their boundaries were marked on the map of the land use areas. Lastly, by combining the two sets of areas, twenty-five composite " natural areas ", or, as it seems better to call them, functional regions, were discovered (excluding the central business district and the industrial areas). In nearly all cases it was found that physical barriers, such as railways, water, high land, or industrial belts, divided the regions one from another, radial streets usually acting as arteries rather than as boundaries of the regions.[2]

This example of New Haven is a study of a small town. Much more significant is the division of the city of Chicago into 75 " community areas " as prepared by the Social Science Research Committee of the University of Chicago (Fig. 29*b*, p. 133). These areas are groupings of the small census tracts of which there are 499, and they are social units that are best fitted to be used for statistical purposes and for understanding the life of the city. The areas were determined from considerations of non-residential land use, such as railway tracks, streets, major industrial areas and parks, all of which often serve as barriers between neighbouring communities. The distribution of population by density and ethnic origins, on the lines we have indicated above, as well as

[1] A further line of investigation would be to examine, as for villages and towns in the country, the service areas of subsidiary business centres, churches and schools.

[2] Of many other studies of this type, particular mention may be made of Howard W. Green's work on " Cultural Areas in the City of Cleveland ", *American Journal of Sociology*, Vol. XXXVIII, 1932, pp. 356–67.

"case" studies of social groups, served as criteria for determining these areas. Detailed analysis was finally made possible by the use of the census tract (adopted by the Census in 1920). These community areas were used by the Census in 1930 and by the Chicago Land Use Survey of the Chicago Plan Commission. Such studies are badly needed for our British cities.

The cosmopolitan American city, owing to its ethnically mixed population and its very rapid growth and expansion, reveals all these features with special clarity, and has been a good laboratory for such investigation. It is not suggested, however, that the whole of each American, still less of each British, city can be divided into clearly defined socio-economic areas, or that different social and economic facts are coincident in their limits, or that these facts necessarily coincide with types of residential land use. Indeed, it will frequently occur that large areas—as, for example, in the large estates of the modern city—have little neighbourhood homogeneity. But the discovery of this very fact in certain areas, i.e. lack of communal activity and organization, is a negative discovery of the greatest significance for the planner. There seems to be little doubt that the social unity of an area is rooted in similarity of economic level, partly in the measure of its physical separateness, but above all in the neighbourliness of its people, with respect to their social and commercial relations.[1] The last is least developed in the new built-up areas, but is frequently remarkably strong in working-class areas.

The natural area as a land-use unit and as a socio-economic unit must be clearly distinguished from the social unit that emerges through community relations. It involves the concept of the neighbourhood. This is the primary locality grouping in the town, as in the country. It has been defined by the Stephensons in their study of community centres as follows:

> Among the important things which make up a neighbourhood are, first of all, the houses in which people live, and secondly, the various community buildings which they use. These latter consist of shops, cinemas, schools, churches, public houses, etc., which are to be found in a large or small degree at strategic points in any residential area. Such public and commercial buildings tend to group themselves together into some form of shopping or civic centre

[1] These remarks apply to the large city. Lack of homogeneity does not necessarily mean lack of community sense. The small country town has as a main advantage the mixture of classes and interests coupled with civic consciousness and local solidarity. It is the diversity of function and class and interests in such a town that gives to it its balanced social and economic character.

and to be patronized by the surrounding population. The radius from which people will come to them is decided by the density at which they are housed.[1]

It is in turn the social cell of the body of society, and as such the essential group from which emerge the developing folk-ways, modes and social institutions, the forms and rights of property and of government, the activities of work, of recreation and of learning, of religious ceremonies and beliefs, and of ministrations to the welfare of the community. The locality or neighbourhood group is likewise primary, along with the family, as a basic unit of social control. It is the first grouping next to the family through which the individual is inducted into social life.[2]

Modern city life has spelled the breakdown of such personal relations. Neighbours, signifying intimate association, have been replaced by nigh-dwellers, this designating adjacent residence coupled with anonymity. While this is true of great cities it has also been shown to be true of smaller towns near to them.[3]

Although an entire system of government is based on the assumption of the locality group as the unit of representation and administration and that the people living in the same locality have enough interests in common that they may be relied upon to act together for their common welfare, the assumption turns out to be invalid for large cities. . . . We expect concerted action from people who are strangers to one another. Mobility, lack of home ownership, and social distance, all operate in the disappearance of the neighbourhood as an entity possessing social and political values.[4]

The problem of city planning is to rebuild for the people so as to give a personal neighbourly framework of service as opposed to the impersonal services of the city centre.

3. Services and Service Centres

Community groupings in the city are effected through service institutions. The distribution and grouping of service institutions in urban areas are determined by the same basic factors as for service centres over the countryside. In the urban area these services may be classed broadly, from the standpoint of their clientele, as regional, city-wide, district, and local or neighbourhood services. The chapel, the club, the public-house, the small retail shop catering for immediate everyday requirements, and, above all, the elementary school, tend to be scattered about the residential

[1] F. and G. Stephenson, *Community Centres*, Housing Centre, London, 1942, p. 42.

[2] W. Russell Tylor, "The Neighbourhood Unit", in *Town Planning Review*, Vol. XVIII, 1939.

[3] Ibid., and E. R. Roper-Power, "The Social Structure of an English County Town", *Sociological Review*, Vol. XXIX, 1937, pp. 391–403.

[4] W. Russell Tylor, op. cit.

areas without segregation, although local clusters may occur at street junctions, or, as in the case of shops, along main roads. Neighbourhood shops repel, city shops attract, one another.[1] The elementary school, in particular, should be located, though it very often is not, so as to be within about ten minutes from the homes of its pupils without their having to cross any dangerous main thoroughfares. This, at least, seems now to be accepted as a first general principle of future planning, though it was advocated some years ago in the plan for New York and its Region. Above these local services there are those of a higher order, which the consumer requires occasionally and with a wider choice, and for which the maintenance of the institution, in view of its more specialized character, demands access to a larger clientele. Such services tend to segregate, normally at fairly regular intervals in the residential districts of the middle zone, though they are fewer and inadequate in the newer suburban areas. In the former, they are found at strategic road crossings and these are often old town or village centres with a parish church near by, while in the American city, the " bright light areas ", as they are called, are situated at regular intervals of about a mile at road crossings in the rectangular network of the street pattern. Here are banks and offices, multiple shops, cafés, cinemas, and so on. Above these again are the biggest and most exclusive concerns in all these categories ; these are located in the central business zone. The big department stores, newspaper offices, banks and offices, including city and regional head offices, concert halls and theatres, civic buildings and wholesale concerns all compete for space. There is a marked tendency for the central shopping area to shift in the direction of the better-class residential districts, as in the case of South Kensington in London ; and for more popular shops not merely to be segregated in separate streets, but to cluster near the centre on the side of the working-class districts. Amusement districts may also have their offshoots, as in the case of the Reeper Bahn in Hamburg and the Tauentzien Strasse in Berlin. Many instances are found in American cities where the expansion and shift of the central city or down-town district can be clearly traced within the last generation.

In a separate category must be placed those institutions that provide special services that are essential to the civic life but are

[1] W. William-Olsson, " Stockholm : Its Structure and Development ", *Geographical Review*, Vol. XXX, 1940, pp. 420–38.

not in everyday demand. Such are the civic buildings, the university and technical schools, hospitals and museums. Accessibility, plenty of space, and seclusion are the requirements of such buildings. In this country, as is well known, most of them were built at a time when there was little evidence of civic pride in the latter half of the nineteenth century, and they are in consequence situated in what is now the dilapidated middle zone. In continental cities, on the other hand, such buildings are clustered together in a district near to the city centre but secluded from it. Such an arrangement of precincts has also been achieved in the building of American cities during the last twenty years. In this country, a similar regrouping and rebuilding of public buildings is a matter for future planning.

The location and size of all such service institutions depend upon the density of population, the social character, and the standard of living of the areas they serve. In the American city, where variations in population density and the segregation of ethnic groups by districts give rise to great differences in these respects between one district and another, such considerations are of practical importance in the provision and siting of service institutions. In the English, as in the continental, city such variations, though less pronounced in ethnic structure, are much greater in the densities of population as we saw in the last chapter. The private bodies which control, for example, cinema and theatre circuits, multiple shops, restaurants and offices, and public bodies who must allocate space for commercial premises and public buildings, should have thorough information on the relation of these institutions, in size, staff and capacity, to the density of population and its social structure. In the United States large sums of money have been spent in the detailed mapping of cities, showing such data as the density of population, income levels, and social make-up, upon which basis commercial centres and individual concerns may be rationally sited. Such surveys and estimates of the purpose and capacity of the individual concern—whether it be the capacity of a municipal restaurant, an elementary school, a clinic, cinema or a bank—should form the basis of rational planning in our cities. Numerous services in the old middle zone, in many cases now blitzed, are redundant and should be more rationally distributed. On what principles must this be done and what shall be the relation between the distribution of such services and the future distribution of population? These are among the questions of planning which demand detailed civic surveys. Detailed maps of an urban area would indicate which areas are inadequately served, the nature and volume of the demand for a service, and the best site on which to establish the buildings to perform the service. In the building of new residential areas and community centres, such matters must be considered jointly as part of the problem of build and lay-out of the planner and architect.[1]

The lack of accord between the distribution of services and the distribution of population brought about by the rapid outward expansion of the urban area is well known. One of the main problems of planning is to remedy this. The great concentration of public-houses in the centre and the older districts around it and their virtual absence in the outer, newly built-up areas, has been shown convincingly in a map prepared by the Hull Regional Survey group. This applies to most other service institutions. The church may be taken as an example. Here there is a very interesting contrast between the British and the American city. As Douglass points out,[2] the church in the centre of a growing city can do one of three things—and this applies equally to any kind of service institution : it can follow its congregation to a different part of the city ; remain in the same locality and draw on its old visitants who are scattered in various parts of the city ; or endeavour to maintain and build up a congregation from its parish, though its population may indeed be quite inadequate to support it. This is the problem before many British churches. A church, however, in this country, is usually an historic edifice which cannot be moved. New churches of all denominations and of little architectural merit are

[1] The National Council of Social Service considers that at a density of 12 families to the acre, 2,000 families can be effectively served by a Community Centre. It will be noted that the basic criterion in recent studies of this kind is *accessibility* to the Centre, since the single family houses cover large areas : the block of flats obviates this distance factor, and there is no reason why the fundamental size of a small basic community should not be used in all urban areas, irrespective of density. The needs of every individual in the community are the same, irrespective of dwelling densities. On the question of the optimum size needed for a community centre, the Stephensons consider that—

" The question cannot be answered by a hard and fast rule which says that to so many thousand families there shall be one Community Centre. The rule would be unsatisfactory at once if applied to different densities of population, that is, to urban, sub-urban and rural populations. While a Centre might satisfactorily serve several hundred families on a cottage housing estate at 12 families to the acre, the same facilities would have to be duplicated many times at close intervals if they were to be used by only the same number of persons in a large town. A standard arrangement would prove impracticable in a sparsely-populated rural area where the great distance to be travelled would make the constant use of a Centre by the given number of families impossible." (F. and G. Stephenson, *Community Centres*.)

[2] Douglass, H. P., *The St. Louis Church Survey*, 1924, p. 76, quoted by Gist and Halbert, op. cit., p. 166.

built in the new residential areas, but naturally they are late in appearing and meanwhile there is absent one of the chief integrating forces of community life. But in the United States, where there is less sentimental attachment to historic buildings, the churches move outwards, quickly adjusting their location to population movements. In St. Louis, Douglass found that churches have moved consistently westward with the expansion of the city from the old town centre. One Protestant church moved three times—four blocks in 1850, twelve blocks in 1890, and two and a half miles in 1912. Many Jewish churches have moved out to the west side in Chicago, and in Minneapolis there has been a " a retreat before the expanding business buildings and a pursuit of the retreating residential districts ".[1]

4. Commercial Centres : Retail Trade in Baltimore

The pattern of business distributions in the city is of particular importance to knowledge of the community structure, but unfortunately there is no published study of a British city on these lines. Baltimore has been studied in detail and may be taken as an example of the position which is more or less common to all cities of Western civilization.[2]

The shopping area of primary importance in every city is in the central zone. Here the shops are few in proportion to the city as a whole, but, in Baltimore, they did 28·1 per cent. of the total business transactions. General merchandise, clothing, furniture and household goods, dominate the retail services in this district. With less than a quarter of the clothing stores in the central zone, these accounted for more than half of the total sales of clothing in the city. Though there were relatively few furniture stores, these did more than half of the business of the city. The jewellery stores, one-fourth of the total, did three-fourths of the business. General merchandising accounted for two-thirds of these sales in the whole of the city. Other retail services (food, automobiles, timber and building, restaurants and chemists) appear more frequently in the outer districts where most of their business is transacted.

[1] Schmid, Calvin F., *Social Saga of Two Cities*, Minneapolis Council of Social Agencies, 1937, p. 51.

[2] See " Nucleation : The Pattern of Retail Marketing ", by Inez K. Rolph in McKenzie, *The Metropolitan Community*. See also " The Population Pattern in Relation to Retail Buying ", in the *American Journal of Sociology*, Vol. XXXVIII, 1932, pp. 368–76. Rolph divides Baltimore for the purpose of her study into three areas, the Central Shopping District, the Mid-City Shopping District and the Outlying Shopping District. These correspond to what we have described as the Central Zone, the Middle Zone, and the Outer Zone.

Around the city centre is the Middle Zone, or, as Rolph calls it, the Mid-City District. This belt accounts for a quarter of the land area of the whole city and two-thirds of the population (522,000), and in general it is an area with a high proportion of foreign-born population of low economic status, with a small income level. The total retail businesses represent 80 per cent. of all shops and do 60 per cent. of the business. Establishments are scattered throughout the area, but nucleations are of a number of types. String-street-developments are those in which shops are aligned along main thoroughfares, usually radiating from the city centre. These shops draw custom from people using the thoroughfares and serve only to a minor extent as centres for the residential areas behind them. They depend like the central city district, of which they are appendages, upon transient rather than neighbourhood custom. Second, located at strategic points at the junction of main routes are sub-centres, small editions of the central city district, serving as centres for the neighbourhoods around them. These sub-centres are graded in four groups according to the number of establishments out of nine commodity groups—food, general merchandise, apparel, automotive (automobile parts), furniture, and household timber and building, restaurants and eating places, drugs and jewellery. The class is based on representation by one shop of each kind and not by quantity or quality of service. There are 18 of these sub-centres, with one-sixth of all the shops, chief of which are food shops, with clothing shops second in order of numbers, and general merchandise third.

About 60 per cent. of the shops of the whole city, doing 50 per cent. of the business, are scattered, with an average of two or three per block, over the whole area and are outlets for the local neighbourhood. More than half of these deal in food.

The outlying shopping district is essentially residential, forming separate blocks and growing into about 18 distinct and physically separate communities, far removed from the central city district. It comprises about three-quarters of the land area of the city and a third of the population (282,000). The sub-centres, of which there are 20, are the main type of retail development, with the same characteristics as those of the Middle Zone. Owing to the smaller density of population they are smaller than those in the latter area, with 4·3 per cent. of the city's stores : a quarter in the food group doing half of the business. Chain stores are particularly important. " Neighbourhood development " of small groups of shops (dominantly food) is probably more important here than in the Middle Zone.

The principal factors in determining this pattern of retailing developments are population density, income, topography, the transport net, and (especially in the American city) racial and ethnic composition of districts. Among these, population is the main factor. In the Middle Zone of Baltimore, with over 15,000 persons per square mile, there is the highest concentration of business as well as the most widely scattered services, and the greatest number of full-fledged sub-centres meeting all the essential needs of a community (7 of the 18 sub-centres are of this type). Shops selling only convenience goods, on the other hand, appear in the communities with less than 5,000 persons per square mile and between these two extremes there are gradations of retail service centres. There is a close correlation between retail service and the density of population.

> Convenience goods usually make their appearance first. Then as population increases, restaurants, in some one of their several forms, and general merchandize outlets, usually small, follow. Of the principal commodity groups, the last to enter a sub-centre are clothing apparel, furniture and jewellery. This appears to be the normal development of a sub-centre. Likewise it is possible to correlate types of operation with population density. Where density is inordinately high, one finds the greatest number of chain-operated stores. . . . In all sparsely populated communities, the independent store predominates, as it does in the least populated communities at the other extreme. It appears, therefore, that population density is significantly correlated with concentrations of retail business, with the kinds of business represented in these concentrations, and with the types of operation that prevail there. (McKenzie, op. cit., p. 264.)

5. Planned *versus* Natural Areas

The most significant development in the controlled direction of city growth is the practice known as zoning, which, through various kinds of ordinances, is now common to most cities in western and central Europe and America. The areas of the urban complex have been discussed above as they have emerged naturally, that is, without public control. Zoning is an attempt to preserve the existing character of certain reserved areas in the city and to reserve other un-built-up areas for specific types of use, such as residence or industry. Zoning is carried out according to functions : heavy industry is segregated, residential districts are preserved from the encroachments of industry and commerce, and residential districts of different types are kept separate. "Zoning ordinances represent the legal recognition that the community as a whole has a right to protect land against encroachments that would result in the depreciation of property values or be inimical to the health, morals, or safety of the residents of a given area."[1] But such zoning can only be really effective in its aims and permanent in its usefulness if based upon careful study of the structure and extent of the natural areas of the city which reflect the social and economic differences of its inhabitants. And this is quite obviously not the case. Zoning ordinances are designed to regulate the use of the land (industrial, commercial, residential, mixed) ; also the height of buildings and, in some cases, the space to be covered by buildings, in the use-zone, so as to give adequate provision for sun, light, and air.

The zoning measures adopted in this country have been effected without adequate appraisal of existing land uses and the proportions of the different categories of land use to the total built-up area, and without reference to the social and economic conditions which limit the location of such uses.

For instance, in this country, we find that enough land has been scheduled for houses in town planning schemes to accommodate 291 million people—six times the present population—so optimistic and prodigal are the present zoning schemes. Again, it is found in many cities that something like six times more land is zoned for shops and business than is actually used for these purposes, whereas in some cities as many as 30 per cent. of the existing shops are not in use. There are, in fact, too many shops—or at least a bad distribution of them—usually strung in shoe-string fashion along the main roads, which are automatically zoned for shops. Now, overzoning in any form encourages urban blight, because landowners, waiting hopelessly for a gambling chance of obtaining a buyer, in desperation allow shoddy developments to take place, or, worse still, they obtain a licence for temporary advertisement hoardings or booths. So in any case the area becomes an area of decrement and obsolescence, even while it is a site only half built-up. Thus we must start with a more scientific analysis of zoning requirements.[2]

6. The Natural Area in Local Government

Every city is a bewildering mosaic of *ad hoc* local government areas for a variety of purposes—in regard to which the average

[1] Walker, op. cit., p. 17.

[2] Max Lock, *Civic Diagnosis : A Blitzed City Analysed.* An outline summary of Planning Research undertaken by the Hull Regional Survey. Published by the Hull Regional Survey under the auspices of the Housing Centre. Introduction reprinted from *World Review*, July, 1943.

citizen is ignorant and uninterested. Parishes, wards, urban districts of different kinds and election districts, antiquated in themselves as geographical units, are inadequate in respect of the practical needs of the urban community. This we learned to our cost, for instance, in the organization of the fire-fighting services during the London blitz. New local government units are needed and a co-ordination of functions with them ; and these should be based on the principles we have outlined here. For statistical purposes this has its obvious advantages and the U.S. Census Bureau has adopted community areas as its statistical units in the metropolitan districts. The idea of the census tract was first put into practice in a few cities thirty years ago, and was then utilized by the Census Bureau to try to work out a method of establishing " permanent non-political boundaries inside of large cities ". Election districts, wards and school districts are liable to change and permanent units are needed for the statistical examination of economic and social data. These tracts, finally checked by the Census Bureau, are decided on by public and private agencies in the cities themselves and as for April, 1940, 71 cities have established census tracts for their own areas. These units in general correspond with the type of socio-economic unit we have identified above,[1] and Chicago has already been cited as an example.

7. The Natural Area in Planning : The County of London Plan

The idea of the natural community area is developed in the recently published *County of London Plan* by Forshaw and Abercrombie. The *existing* urban structure, that has emerged without external interference, is referred to in this Plan as the " natural " structure, so that the writers refer to the " natural (as opposed to the planned) use zoning and community structure ". On this functional and social basis, four main areas are recognized :

(i) The Central Area comprising the West End and the City with the adjoining areas of mixed general business and industry :

(ii) The Port and the Thames and Lea-side heavy industrial areas ;

[1] A. W. von Struve, " Geography in the Census Bureau ", *Economic Geography*, Vol. XVI, 1940, pp. 275–80.

(iii) Central Residential Areas comprising the Western housing districts coming within the influence of the West End, and the Eastern working-class district north of the river from Poplar to Kentish Town, with a high proportion of obsolescent property and with industry adjacent to or indiscriminately mixed with the residential areas ;

(iv) The suburbs outside these three areas.[1]

In the inner residential areas, community areas can often be traced back to their nuclei at important road junctions, or to the original villages, while their limits have often been determined by physical features such as railways, canals or by industry. These larger communities, such as Finsbury, Lambeth and Eltham, include smaller neighbourhood units, some well equipped, others lacking suitably located shopping, school and other communal facilities. More recent housing developments in particular lack such facilities. The residential communities form three fairly distinct groupings. The *West Central Group* is situated around the West End or comes within its sphere of influence. It is affected by the invasion of the uses of the central area and has a high proportion of large town houses and new blocks of flats, and much property of several storeys let off as flats to small families. The *East Central Group*, which includes the main areas for reconstruction, comprises the East End, Camden Town and Kentish Town and the districts on the low-lying ground in the south bank of the Thames. These contain the main areas of slum, obsolescent property and overcrowding, with a profuse " peppering " of factories. The *Suburban Group* includes the communities of recent development in which further development requires careful pre-planning so as to prevent ribbon growth along arterial roads, and to preserve open spaces between the community areas. The proposal in the Plan is

> to emphasize the identity of the existing communities, to increase their degree of segregation, and where necessary, to reorganize them as separate and definite entities. The aim would be to provide each community with its own schools, public buildings, shops, open spaces, etc. At the same time care would be taken . . . [not] to endanger the sense of interdependence on the adjoining communities or on London as a whole.[2]

After very careful consideration, it has been decided that the elementary school should be the determining factor in the size and

[1] J. H. Forshaw and P. Abercrombie, *County of London Plan*, 1943, pp. 21–2.
[2] Ibid., p. 28.

organization of the subsidiary or neighbourhood units of these communities in which large-scale reconstruction is proposed,[1]

so as to ensure a maximum walking distance from home to school.

8. Problems and Techniques of City Analysis as Evidenced by Recent Studies : Hull

It would seem appropriate to examine briefly some of the special problems and techniques associated with this particular interpretation of the city as a space structure. The scope and method of city survey in the accepted sense are sufficiently well known not to require special consideration. Classic studies are available on the social and economic structure of some of our cities, on the initial model of Charles Booth's *Survey of London Life and Labour*. There are also available briefer studies that are intended to serve as a basis for town planning. Our concern is to deal more particularly with the problems involved in the mode of approach put forward in this book, for this approach has figured prominently in recent civic surveys and demands systematic investigation by specialists for the recording of the data, their cartographic representation, and their interpretation.

Many surveys of British cities have been undertaken both as preliminary to suggestions of regional town planning and on a more elaborate scale as surveys of social and economic data by specialists. In the former category we may instance the surveys of Sheffield [2] and Southampton.[3] In the latter category the pioneer work is Charles Booth's *Survey of London Life and Labour* and in more recent years the survey of Merseyside.[4] But the weakness of all these studies is that they do not map the individual sets of facts as a comparable series on the basis of which physical and social planning can be built. The technique of investigation has received much attention in America but had no real repercussions on British investigators until the war and the blitz made the reconstruction of our cities, and hence a fuller understanding of their structure from the point of view of land planning, a matter of high priority. One of the first results of

[1] Ibid., p. 28.

[2] *Sheffield : A Civic Survey and Development Plan*, by P. Abercrombie and R. H. Mattocks, 1924.

[3] *Southampton : A Civic Survey*, edited by P. Ford, 1931.

[4] *Social Survey of Merseyside*, 3 Vols., edited by D. Caradog Jones, 1934.

this stimulus is the work of the Hull Survey group under the direction of Max Lock, whose whole mode of approach is conditioned by the American pattern.

The field of a city survey is summarized by Lock under ten main heads.[1]

1. The geographical and physical structure of the region.
2. The utilization of land and the basic physical structure of the city (or cities) in the region.[2]
3. The location of industry and of the industrial population, the industries being examined by type of manufacture, mobility and numbers of employed, location of factory, atmospheric pollution.
4. Economic survey of trade and labour.
5. Transport facilities including the volume and flow of traffic.
6. Piped and wired public services—gas, water, electricity, drainage and telephones.
7. Investigation of housing, its location and conditions, and the analysis of urban blight.
8. The distribution of shops and analysis of retail trade, according to type of shop (food, clothing, household goods, miscellaneous merchandise), areas within half a mile of shopping centres, numbers employed per shop, annual turnover, economic type (independent, multiple, co-operative, department stores), spheres of service, shopping habit of economic groups, incidence and trends of increment and decrement.
9. Social and physical survey of cultural, religious, recreational and social services, both indoor and outdoor.
10. Administration—local government areas, local taxation, and land values.

Of particular importance under these headings are housing and the problem of the slum. The method suggested by the U.S. Federal Housing Administration in Washington is adopted of analysing the geographical distribution of urban blight, that is, areas of obsolescent housing. This is done by mapping separately on tracings, block by block, the worst areas of several distributions and then by superimposing the tracings the worst areas of urban blight are determined. These distributions show the following factors—rateable value or rental value of housing (which is the best guide to social and economic level of their inhabitants) ; age of structures ; state of repair of structures ; density of dwellings per acre ; sanitary amenities (houses having no baths and no hot water) ; mixed development (factories, etc., in housing blocks) ;

[1] Max Lock, op. cit.

[2] There is clearly confusion of wording, and probably of conception in these two heads. What is meant by physical ?

amount of sunshine in dwellings ; amount of overcrowding ; atmospheric pollution by smoke and smell, noise, juvenile delinquency and infant mortality. A series of maps prepared on this basis not only permits the identification of the worst areas of urban blight [1] where the distributions overlap, but also reveals, through the coincident distribution of the phenomena, the degree to which natural community areas exist throughout the city, as well as those social and physical conditions the planner requires to know when rebuilding cities or reallotting land uses according to social and economic requirements of the community.

9. Current Surveys in the United States : The City Inventory

Chicago.—The Chicago Land Use Survey, directed by the Chicago Plan Commission, has recently produced the first of three volumes on residential Chicago, which shows the prodigious amount of methodical research that is going on in that city, the laboratory in which the modern study of urban sociology has grown up.[2] The goal of this comprehensive survey is

> to provide an inventory of the physical, economic and social characteristics of every use of land in the city of Chicago, to picture the pattern of the local variations of these characteristics ; to provide an indication of city-wide and local trends—their nature, direction and rate ; to furnish a reservoir of facts which may be used in determining more precisely the fundamental principles of city growth . . .

The project is backed by large sums of money and a large staff of office and field workers. Over 300 clerks transcribed tax assessor's records for over 600,000 buildings in the city. Over a thousand field workers worked for a year in investigating and recording the character of buildings in 20,000 assessors' blocks and had a filled-up questionnaire from every family in these blocks. These investigations are tabulated, mapped and discussed in the volume. Sixteen large-scale maps in red and black show a wide range of residential conditions, and these are summarized in tables for each of the city's 75 community areas as adopted by the Census. Special mention may be made of maps showing the " decayed inner core " by means of low rent areas.

[1] A tenth of the houses of Hull (total 86,000) suffer from six of these defects. To these should be added the blitzed housing area.

[2] *Chicago Land Use Survey, Vol. I, Residential Chicago*, 1942. Directed by the Chicago Plan Commission and conducted by the Work Projects Administration, sponsored by the City of Chicago. Reviewed in the *Geographical Review*, Vol. XXXIII, 1943, p. 682.

New York City. The Real Property Inventory of New York City taken in 1934 reveals in the greatest detail the socio-economic structure of that city, and its broad features are summarized in a series of maps.[1] This Inventory has its statistics tabulated by census tracts averaging 40 to 50 acres each and in the Residential Reports 139 items are given for each tract. There are seven volumes, prepared under the direction of the New York City Housing Authority, the United States Department of Commerce, and the Mayor's Advisory Committee on Real Property Inventory. The survey covers " the physical, occupancy and income conditions of all real property " in the five boroughs of the city. The reports are supplemented by seven large coloured maps, prepared on the basis of city blocks (not tracts—eight blocks normally making one tract) on a scale of 1 : 49,000. These maps bear the following titles : (i) Predominant Use (i.e. whether residential or non-residential) ; (ii) Predominant Residential Age (i.e. age of residential buildings) ; (iii) Predominant Residential Type (i.e. single family, two family, multiple family, etc.) ; (iv) Predominant rental per family quarter ; (v) Predominant Non-Residential Type (public buildings, stores and offices, warehouses, garages, industries) ; (vi) Population (residential population) ; (vii) Day Population.

A map of Manhattan, based on this survey to illustrate its findings, shows a broad division between residential and non-residential areas. The latter are divided according to the predominant type of building into office buildings, stores and others, these being mainly docks and warehouses. The two main areas of office buildings and stores are the two down-town districts between the Bowery and Central Park that are centred on Wall Street and Times Square. Docks and warehouses lie along the Hudson River front. The non-residential buildings are of four main kinds—" loft structures ", office buildings, buildings used for " private special uses " (largely theatres) and stores. The " loft structures " are defined as buildings " suitable for light manufacturing and rented by floors or parts of floors " in which not more than 25 per cent. of the building area is occupied by offices. Of these uses in Manhattan, clothing

[1] For a summary of its findings see an article by John K. Wright in the *Geographical Review*, Vol. XXVI, 1936, pp. 620–39. A detailed survey of Manhattan in this article includes maps showing residential and business properties, the former being classified according to rentals ; a map showing daytime population (over 200 persons per acre) with stippled areas to show residential areas ; and the daytime densities per acre of non-residential structures and per acre of residential population.

manufactures are the chief. Residential buildings are classed according to monthly rentals per family quarter (a structurally separate dwelling) of 60 dollars and over, 20 to 60 dollars, and under 20 dollars. Predominantly single-family house districts and districts with over half of all families Negroes are also shown. Residential buildings range from 5 to 20 storeys in height. But most of the tallest skyscrapers are office buildings.

The inventory adopted a unique method of recording information of the daily journey to work. There is recorded the time taken by the principal wage-earner in the family to reach work. Four time-groups are differentiated—0 to 20 minutes, 20 to 40, 40 to 60 and over 60 minutes. For each census tract the percentage of wage-earners in each group is tabulated. The areas are also mapped in detail in each census tract where there is over one person to the acre. In general, the majority living in such an area in Manhattan travel in under 20 minutes to their work. This is surrounded by a 20- to 60-minute zone, and beyond this is the over-60-minute zone.

10. A Basic Programme for the Analysis of British Cities

In order to determine and characterize the areal differentiation in the life and organization of the large urban community in Britain, we require a set of maps for each city with over, say, 100,000 inhabitants (the size beyond which such differentiation becomes marked), produced on a standard scale with a standardized set of symbols. Base maps would be prepared on the 6-inch scale, but it is probable that the 1 : 25,000 would be the best standard scale.[1] Essential maps in each series would be the following : (1) Historical Map, showing the stages in the expansion of the city.[2] (2) Map of the proportion of land which is built-up, on the basis of a small unit area, such as the block. (3) Land Use Map, to show public buildings, factories, retail, wholesale and office properties, and types of residential property. (4) Population Map, to show the distribution and density of population, on a shading or symbol basis, block by block.

[1] A model example of this type of analysis is contained in *Southampton : A Civic Survey*, edited by P. Ford (1931), notably Chapter III on " Land Utilization " by O. H. T. Rishbeth.

[2] The historical period is of great importance, for while villages and the county towns and other focal points have been swallowed up by the city, their old buildings often remain ; they have often persisted as centres of integration and form local traffic nodes and commercial sub-centres, and in this and in their historical significance often form in fact, and would be suitable in re-planning to be, the centres of new planned districts.

(5) Socio-economic Map, based on key criteria, such as wage levels, age and sex composition, numbers of children and domestic servants per cent. of population. Such data are not available in this country for small districts, or even for wards, so that they would have to be obtained from sample family returns. (6) Social Service Map, to show the location of institutional centres such as churches, clubs and schools, and the distribution, by exact place of residence, of their regular members. (7) Urban Regions Map, produced on a scale of 1 : 25,000 (2½ inches to a mile). Such a map could be produced for each city, accompanied by a short monograph describing the special aspects enumerated above. The main drawback to such work is the absence of the detailed census data ; the publication of statistics for the large cities by small districts, as is done in the census of cities in the United States, would greatly facilitate such research. Moreover, a central body is required to collect and collate what material exists and then to proceed systematically to a survey of our great cities on the same lines as those on which the Land Utilization Survey has studied rural areas. The mapping of these data raises many problems of observation and recording in the field, as well as the cartographic problem of mapping the data in the office—and specially skilled workers would be needed.

PART III

THE CITY-REGION

CHAPTER 6

THE REGIONAL RELATIONS OF THE CITY

1. The City-Region

The city cannot be fully understood by reference only to its arbitrarily defined administrative area. It has to be interpreted as " an organic part of a social group ",[1] and in approaching the analysis of the four main urban functions—dwelling, work, recreation and transport—" it must be remembered that every city forms part of a geographic, economic, social, cultural and political unit, upon which its development depends ".[2] The problem of the regional interpretation of the city, of defining and analysing the functions and limits of the city and the *unifying* relationships in the surrounding area, is one of disentangling the regional component and examining the multitude of tributary areas served by and serving the city. Each group of functions has its particular extent and characteristics. Many functional areas have no close relation with each other in their geographical extent—which is often difficult to define—or in their causes or characters. But they all have a common denominator in their dependence on the city and, in consequence, in the scientific sense, we may refer to this area that is functionally dependent on the city as the city-region.

The regional interpretation of the functions of the city involves a twofold approach : first, an assessment of the effects of the character of the region—its resources, and economic production—on the character of the activities of the city ; and, secondly, an examination of the effects of the city, as a seat of human activity and organization, on the character of the region. There is also involved the question of the limits of the city, and its spheres of influence or tributary areas in its multitude of regional

[1] M. Aurousseau, " Recent Contributions to Urban Geography ", *Geographical Review*, Vol. XIX, 1934, pp. 444–55.

[2] J. L. Sert, *Can Our Cities Survive ?*, Harvard U.P., 1943, p. 10.

functions. Some attention has been paid in Part I to the question of defining the limits of the city as a regional centre, but this should be subordinated to the main aim of this approach which seeks to evaluate both the city and its region, however vaguely defined, in terms of their mutual relations and in the light of their historical development.

Settlement, route and area are the three facets of the geographical interpretation of urban economy. The commercial output of the area—farming of different types, forestry, industry, or combinations of these—calls into being centres differing widely both in their interests, their commerce and in the industries arising from the processing of the primary products marketed in them. The quantity of output that passes through commercial channels is the sum total of economic, political, and cultural intercourse.[1] It is, in effect, a measure of the nodality of the urban centre. If all such intercourse is concentrated in one city, all the commerce for the area would pass through the city ; and the sum total of this commerce would be the total of its exports and imports. This theoretical state of affairs is never reached because the degree of concentration of circulation in one city in any area depends on the suitability of the area for commerce relative to the location of the city and of its neighbouring cities, to the conditions of historical development, and to the physical build of the land, which may rigidly affect the orientation of routes. Nevertheless, the potency and extent of the sphere of influence of a city are to be measured in theory from the degree of concentration of the circulations of the area around it in the form of freight, passenger and general intercourse.

The city produces goods, and processes and stores imported goods not only for a nation-wide market, but also for the market in its surroundings—whatever it can sell in competition with its neighbours. The city, in addition to its own natural increase (by excess of births over deaths), draws the folk from its surrounding area to enjoy its special amenities—its shops, institutions, markets, art galleries, and theatres. With the great growth of cities in the early nineteenth century, the rural population has been drawn into the towns, with the resultant phenomenon of rural depopulation. The city is a melting-pot and fount of opinion. It disseminates its views on matters relevant to the life and affairs of its citizens and the people of the surrounding

[1] H. Bobek, " Grundfragen der Stadtgeographie ", *Geographische Anzieger*, Vol. XXVIII, 1927, pp. 213–24. This is called *Verkehrspannung* by Bobek.

towns through the medium of the press. It is a home of learning, culture and political life. The city must be fed, with food for its people and materials for its industry. Before the development of cheap and rapid transport, every city was almost entirely dependent upon its surrounding area for both. Distant supplies of food or materials or immigrants were brought by the only cheap means of transport—water, and it is no accident that in the past, before the railway era, the chief cities in Europe and America were either ports or riverside cities at the heads of river navigation. In the modern era, however, although the movement of foodstuffs and raw materials is world-wide, there is, in fact, a still closer relation between town and country. For all perishable goods must be delivered quickly and daily to the city consumers. Moreover, the economic factor of accessibility to the best market dragoons farm areas to supply large urban markets, so that an even closer tie-up between the great city and its environs results. Again, with the ever-increasing complexity in the social and economic structure of society, in service and organization, the city has acquired a great increase of functions as a regional centre for the distribution of both consumer goods and producer goods, and as a centre of services—social, economic and administrative. The city makes its impact on the surrounding towns and countryside, especially since the advent of the automobile, by the expansion of urban built-up land—for residence, industry and recreation. It also affects the character and structure of their social and economic life.

The question of the limit of the city when considered as a centre of regional services of collection and distribution may be approached by referring to Figs. 3*a* and 3*b* (p. 30). Suffice it to note here that this theoretical distribution of towns is based on the assumption that functions are centred in towns that may be graded according to the importance of these functions. Consequently, a city of the fifth grade in Christaller's scheme, for instance, will combine all the functions of its own and of the four lower grades, and each set of functions in each grade will have its corresponding limits as a series of concentric circles passing through the towns of the next lower grade. This scheme is most nearly approached in extensive and dominantly rural areas with an even distribution of towns and occasional, evenly spaced, large, dominant cities as in eastern England, France, or, indeed, in south Germany. But the following *conditions must* be added to this distributional pattern—quite apart from irregularities of distribution brought about by topographical and historical conditions, though these, in an evenly settled area, cause relatively small deviations.

First, the modern growth of population has been mainly in urban centres, proportional to the size of the centre. This has meant the snowball growth of existing towns, and in no way interferes with the basic pattern of distribution of service centres.

Secondly, new seats of industrial production, clustered at seats of production of raw materials (or at places of assembly), have given rise to new population clusters, which give rise in turn to central service centres.

Thirdly, the spread of population from the big city results in the spread of the urban area radially and frontally: merging with, often absorbing, pre-existing centres in its closer environs. These outlying centres, though absorbed in the urban mass, usually retain their functions as commercial sub-centres.

Fourthly, the extension of the big city results in the appearance of new settlements, budding off from it, sometimes being independent centres, both legally (if beyond the city boundary) and economically, without any relation to the laws governing the origin and growth of centralized services.

We have already discussed the general structure of the city and its fringes. These, and the outer and more widespread areas influenced by the city, are arranged into three main zones that can be described as the *urban tract,* the *city settlement area* and the *city trade area.*

The *Urban Tract* is used to define the compact and continuous urban built-up area in preference to the term "conurbation". The latter term defines the urban agglomeration that extends beyond administrative boundaries, and was in fact first defined by Geddes as a group of two or more contiguous administrative units that were urbanized. The term has been further elaborated by Fawcett, who describes it as "an area occupied by a continuous series of dwellings, factories, and other buildings, harbour and docks, urban parks and playing fields, etc., which are not separated from each other by rural land; though in many cases in this country such an urban area includes enclaves of rural land which is still in agricultural occupation".[1] This assumes that the conurbation ends with the limit of the compact built-up area,

[1] C. B. Fawcett, "Distribution of the Urban Population in Britain in 1931", *Geographical Journal,* Vol. LXXIX, 1932, pp. 100–16.

but there is invariably a fringe of rural-urban uses, a fringe that is wide and irregular in this country and still more complicated in other countries. This definition has given rise to much confusion of thought, because with it is associated the idea of several administrative units. An urban agglomeration, no matter what its extent or population, would not be counted, on this definition, as a conurbation if it were one administrative unit.[1] In such cases the administrative boundaries have extended with the growth of the urban area, or the urban area itself has grown by the coalescence of separate units—whether independent towns, villages, or satellites thrown off by a central city and later absorbed in its extension. Moreover, the peripheral rural-urban fringe is so diffuse that there arises the problem of deciding whether to include places that are cut off from the main area but sufficiently near to it to be a part of its economic and social organization. In other words, on the margins the emphasis must change from compactness to function and accessibility.

The limits of an urban tract are to be defined in the first place by mapping the land uses and enclosing those areas that are closely built-up, as suggested by Fawcett. It is of interest to note the minimum density of population for such marginal areas —though it is not suggested that the tract is to be limited on this basis. The Ordnance Survey takes 6,400 per square mile as the limit of " urban " ; Jefferson [2] suggested a theoretical limit, many years ago, of 10,000 per square mile for American cities. A recent study of Paris takes 250 persons per square mile as the extreme limit of urban influence against rural areas and 1,250 persons per square mile as the limit of the compact urban tract. A comparative study of German cities shows that in the suburbs (*Vororte*) of the cities fully urban areas have a minimum overall density of 2,500 persons per square mile and that 250 persons per square mile is the outer limit against the rural areas.[3]

The *City Settlement Area* embraces the urban tract and the outer zone or rural-urban fringe, as we have described it in a previous chapter. This fringe of settlement and city influence extends as far as communications will allow. A journey-time of

[1] This actually is the interpretation given by J. Soulas in his recent study of the French conurbations in the *Annales de Géographie*, Vol. XLVIII, 1939, pp. 466–71.

[2] M. Jefferson, " The Anthropogeography of Some Great Cities ", *Bulletin of the American Geographical Society*, Vol. XLI, 1909, p. 543.

[3] R. Clozier, *La Gare du Nord*, Baillière, Paris, 1940, and M. Reichert, *Die Vorortsbildung der süd- und mitteldeutschen Grossstädte*, Stuttgarter Geog. Studien, Stuttgart, 1936.

one hour is usually considered to be the main limit of daily travel for the city worker, and dormitory settlements lie on the main railway routes outside the greatest cities within a radius of about twenty miles. The map showing isochrones, or lines joining places accessible in the same time to and from a selected centre, is a fundamental basis for planning. This outer area, however, is not merely one of residential and industrial settlement. It supplies the city with milk and vegetables and receives many goods from its wholesale warehouses and its retail shops. It is sufficiently accessible to permit regular visits to the city. It forms a part of the labour market of the city complex and has intimate social and economic associations with the activities of the city. The area has been appropriately called by Chabot the *zone du voisinage.*[1] Its characteristics are summed up by its relatively high densities of population, intermediate between the urban tract and the country, and, more significantly, by its high rate of increase of population.[2]

The City Circulation or Trade Area is the area of wider and more extensive, more occasional circulations to and from the city, these relations being normally more intense and varied as the centre is approached through the fringe and the tract to the core. The great bulk of local circulations, such as are found in the urban tract, are directed to the local towns. The city is the head of affairs, the seat of opportunity, offering in all fields what the local town has not got. Clearly the relations with the big city are occasional and diffuse, and normally (except for through routes) do not appear on the road traffic map until the threads collect on the main roads near to the urban fringe. The city settlement area is, of course, served by a net of routes—rail, bus and tram. The outer limit of this circulation area is vague, and indefinable as a line except very diagrammatically. The ultimate limit of any particular circulation is fixed principally by the accessibility of the city relative to surrounding cities of similar status offering the same service. In fact, the limits will coincide in the peripheral towns, where such goods and services are received for distribution to their local service areas. It is normal for towns of medium size with considerable functional independence to be placed on the border of

[1] G. Chabot, "La Détermination des Courbes Isochrones en Géographie Urbaine", illustrated by reference to Dijon in *Comptes Rendus Congrès Internationale de Géographie,* Tome II, *Géographie Humaine,* Amsterdam, 1938, pp. 110–13.

[2] A particularly interesting study of the *banlieue* of northern Paris will be found in R. Clozier, *La Gare du Nord,* Paris, 1940.

the sphere of influence of two cities, with close relations with both.

2. A British Example: Leeds and Bradford

We shall now pass to a brief treatment of the regional relations of two cities—Leeds–Bradford—that jointly serve as an outstanding regional focus.[1]

Owing to the complex physical configuration of the site of the West Yorkshire conurbation, its lack of direct contact with the sea, and the rapid growth of Bradford in the nineteenth century, displacing Leeds as the merchanting centre of the woollen and worsted industries, there is in the industrial area of the Aire and Calder valleys not one principal nucleus as in other British conurbations, but two—Leeds and Bradford, and there exists a subdivision of regional functions between these two cities.

Bradford is the hub of the commercial organization of the woollen textile industries. The area tributary to it in this capacity is indicated by the distribution of members of the Bradford Exchange and the Chamber of Commerce. It is co-extensive in the main with the distribution of woollen textile workers, as given in the Census, in the Aire and Calder valleys west of Leeds. In this connection the close association of Keighley, Halifax and the Spen valley with Bradford should be specially noted. The character of the commercial activities of Bradford is dominated by the woollen industry and its broader regional activities are of little importance. Leeds, however, with numerous industries, a large population, and an excellent central location, has acquired functions of a more general and varied character. It is the headquarters of many industrial associations and societies established in the interests of the industries pursued in the conurbation, the variety of pursuits undertaken in the city itself fostering this development. It is pre-eminent in Yorkshire as a centre of branch offices and depots established by firms with their works outside the conurbation. Such offices are generally located at Leeds, Sheffield, and Newcastle, and the area in which business is transacted from the Leeds centre, though varying widely according to the nature of the business and the number of competitive branches, exhibits some consistency in that it regularly excludes Cleveland and south Yorkshire,

[1] R. E. Dickinson, "The Regional Functions and Zones of Influence of Leeds and Bradford", *Geography*, Vol. XV, 1930, pp. 548–57. If the reader is not especially interested in the three detailed city studies that follow, he may proceed direct to Section 5 on p. 192.

the former being controlled from Newcastle and the latter from Sheffield.

Leeds is also prominent as a banking and insurance centre (Fig. 31*a*), though, in contrast to what has happened at Manchester, no banking house has resisted amalgamation and there is not one insurance company with its head office in Leeds or Bradford. In the development of banking in Yorkshire, each of the West Yorkshire towns played a prominent part and no one centre was dominant. The largest private banks had their headquarters in Leeds (2), Bradford (1), Halifax (1), and

FIG. 31*a*.—Leeds and Bradford. Yorkshire Insurance Areas.

1. Original Leeds Branch area. 2. Branch areas. 3. Full Branch office. 4. Sub-branch office. 5. Branch and sub-branch areas.

York (1). To-day, though all these banks have been absorbed by the Big Five, there is still some measure of administrative control centred in Bradford and Leeds. The Stock Exchanges at both cities are negligible and are of no higher status than similar exchanges at Halifax and Huddersfield. The lack of a dominant focus is one of the most notable features of the financial organization of the West Riding, in striking contrast to the extreme concentration of the financial activities of S.E. Lancashire at Manchester. Insurance development, however, lends itself admirably to regional treatment. In Leeds there are the district offices of almost a hundred insurance companies and the area controlled approximates to that shown in Fig. 31*a*.

A regional capital functions as a general distributing centre through the agency of the wholesale markets,[1] wholesale provision merchants, distributing depots of confectionery and kindred firms and the retail shopping trade. The areas served in these capacities, based upon maps of the areas served by individual concerns, are shown on Figs. 31*b* and 31*c*—the distribution of fruits, vegetables and wholesale provisions, and retail trade areas.

The sphere of influence of Leeds and Bradford as cultural

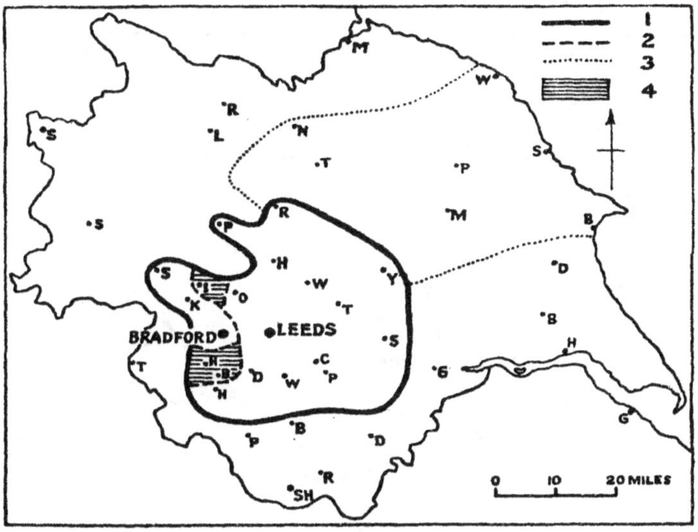

FIG. 31*b*.—Leeds and Bradford. Wholesale Provision Merchants and Wholesale Markets Areas.

1. Areas served by Leeds and Bradford with provisions, meat, fruit and vegetables. 2. Boundary between Leeds and Bradford areas. 3. Boundary of area receiving occasional deliveries of fruit and vegetables from Leeds. 4. Areas receiving supplies from both Leeds and Bradford.

centres is more difficult to define. It may be estimated from the circulation of their principal newspapers, and those of competitive regional centres, the extent of the " Yorkshire Regions " adopted by educational and political societies with branches or headquarters in either city, and the areas served by the higher educational institutions. From the evidence culled from these various sources we conclude that Cleveland and north Yorkshire have intimate relations with north-eastern England and its capital

[1] Ministry of Agriculture and Fisheries : Reports on Markets and Fairs in England and Wales, Part III, *Northern Markets*, 1928.

at Newcastle; south Yorkshire and north Lincolnshire with Sheffield. Doncaster divides its allegiance, having close relations with both west Yorkshire and Sheffield.

As an administrative centre, Leeds is of outstanding importance and ranks with the British provincial capital cities. This contrasts with the relations of Bradford, whose commercial and administrative functions are mainly concerned with the woollen textile industry. There are located in Leeds the divisional offices of government departments, railway and post office areas, and trade and professional societies.

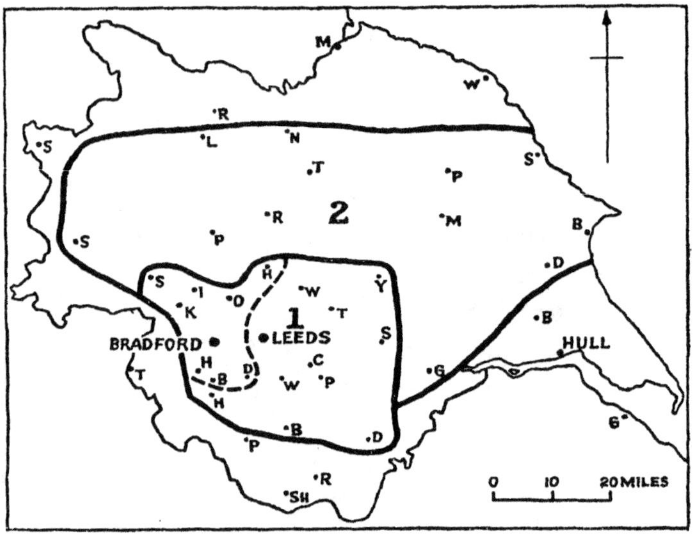

FIG. 31*c*.—Leeds and Bradford Retail Shopping Areas.

1. Market Day Customers' Areas. 2. Seasonal Customers' Area, mainly shopping at Leeds.

From a consideration of the areas served by Leeds and Bradford in their varied regional relations and of the factors which condition their extent, we have indicated the limits of their sphere of influence (Fig. 31*e*). The area thus defined we have named the "Yorkshire Region". It is served from Leeds by the district branches of many business firms and insurance companies; it is the area adopted by the "Yorkshire" trade, professional and voluntary associations; it is served by the Leeds and Bradford newspapers and higher educational institutions, by the large retail shopping firms (mainly furnishers and high-class drapers) and the regional secondary industries. Leeds is its principal administrative and business centre, and Bradford the hub of its chief manufacturing industry.

The Middlesbrough industrial area has intimate relationships with north-eastern England and is too remote to be effectively controlled from west Yorkshire ; therefore it lies definitely within the Tyneside sphere of economic, administrative and social influence. The West Riding falls into two clearly defined industrial areas focused upon the west Yorkshire and south Yorkshire conurbations and separated by the Calder–Dearne

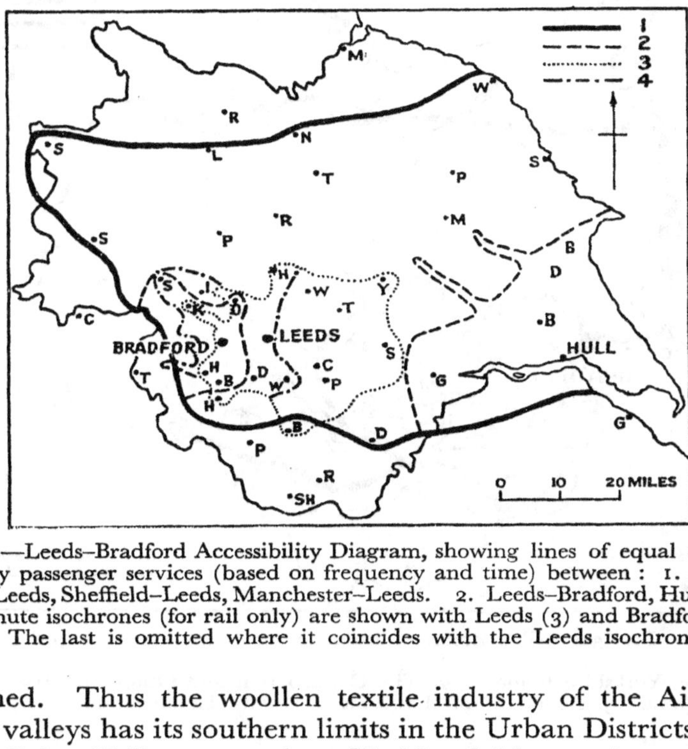

Fig. 31*d*.—Leeds–Bradford Accessibility Diagram, showing lines of equal efficiency of railway passenger services (based on frequency and time) between : 1. Middlesbrough–Leeds, Sheffield–Leeds, Manchester–Leeds. 2. Leeds–Bradford, Hull–Leeds. Sixty-minute isochrones (for rail only) are shown with Leeds (3) and Bradford (4) as centres. The last is omitted where it coincides with the Leeds isochrone.

watershed. Thus the woollen textile industry of the Aire and Calder valleys has its southern limits in the Urban Districts of the Upper Colne Valley centred on Huddersfield ; to the south the iron and steel industry is dominant. The Yorkshire Coalfield is similarly divided into two sections, the west Yorkshire field centred on Wakefield, and embracing Huddersfield, Halifax and Bradford to the west, and Pontefract to the east ; and the south Yorkshire field with its two chief centres at Barnsley and Doncaster. The orientation of these two fields towards Leeds, Bradford and Sheffield respectively is indicated, for example, by the distribution of members of the Leeds and Sheffield Coal

Exchanges and the West and South Yorkshire Coal Owners Associations. Thus, south Yorkshire, with a distinctive group of related industries, reflected in a community of economic organization and interests, is to be regarded as a portion of a separate region with its capital at Sheffield. The limit of the west Yorkshire sphere of economic influence is similarly set on the south-west borders of the West Riding by the appearance of the cotton industry around Todmorden, and in the urban

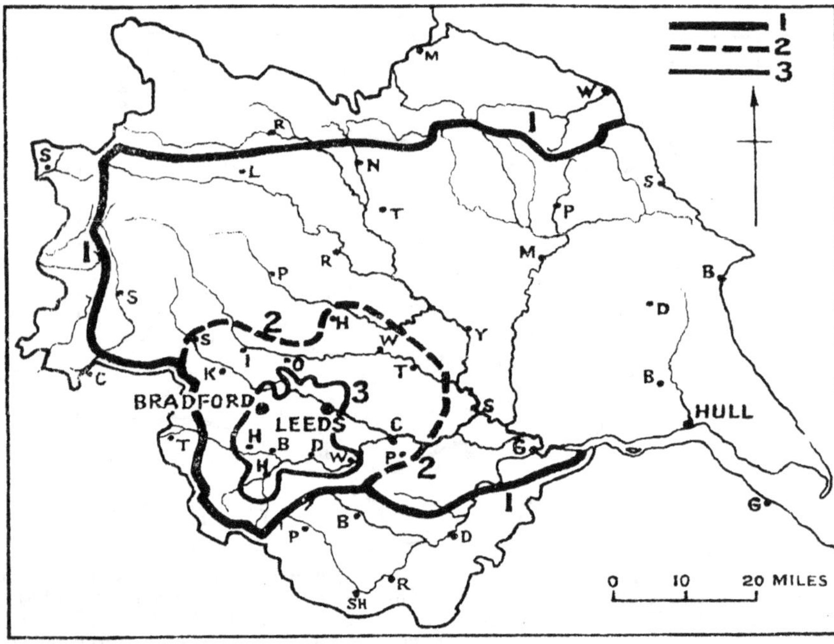

Fig. 31e.—The Zones of Influence of Leeds and Bradford.

1. The Yorkshire Region. 2. The Conurbation and Outer Suburban or Commuting Area. 3. Boundary of the West Yorkshire Conurbation.

districts of Springhead and Saddleworth contiguous with Oldham, which lies just beyond the county boundary. These areas are definitely oriented in every respect to Manchester. Hull and the East Riding are to be included in the west Yorkshire sphere of influence, for this area is regularly included in the Yorkshire districts of Leeds branch offices, insurance companies and "Yorkshire" administrative areas. In all these respects Hull possesses no fundamental regional functions. It is primarily a great port, and not a centre of regional integration. All its associations are local and related to commerce and the industries pursued in the city, and it is negligible as an administrative centre. Moreover, its very existence as a great port is dependent upon its trade relations with the West Riding industrial areas. Yet it is such a large city that a separate organization for various purposes is essential, and in the interests of economy and in view of the relative inaccessibility of most of the East Riding from west Yorkshire, the East Riding often forms a separate district with Hull as its centre. As a large city, it is also a good shopping centre for the thinly populated area which surrounds it.

Within this region there is a zone which has more intimate relations with Leeds and Bradford. This embraces two main areas. First, there is the conurbation and its associated industrial towns. This may be divided into two sections, each with its predominant industries ; the first, to the west of Leeds, concentrates upon the textile and engineering industries, and has its specialized commercial focus at Bradford ; the second, to the south and south-east of Leeds, is the principal coal-mining area (Hemsworth, Featherstone, Castleford, Pontefract, and Garforth) and has intimate commercial relations with Leeds. Secondly, there are the residential areas lying to the north and north-east of the conurbation, including middle Wharfedale and Harrogate. A conspicuous modern tendency is for the population to move outwards from the conurbation—mainly Bradford and Leeds—to these outlying residential towns.

Most of this zone is accessible to either Leeds or Bradford by rail in under one hour, and there are daily reduced fares on most lines ; all parts are also served by through motor-bus routes to one or both of the cities. That portion of the zone west of a line from Otley to Dewsbury, extending south to include Halifax and Brighouse (but not Huddersfield and its tributary area in the Colne Valley), is more accessible to Bradford by bus and rail than to Leeds. It is these travelling facilities that mainly account for the coincidence of the zone with the shopping and insurance areas of the two cities and their daily inward and outward movement of workers. The whole zone is served by the four chief towns of the conurbation, but Bradford and Leeds are the principal distributing centres, and Leeds in particular distributes occasionally throughout its whole extent.

3. An American Example : Salt Lake City [1]

Salt Lake City is " the capital of the State (Utah), the seat of a religious denomination, the Mormons, that has played a dominant role in the regional development of this country and serves as an important basis of regional integration, a nucleus of commercial and financial enterprises, a focus of transportation, a leading centre in educational activities, and the largest city in a vast section of the United States ". In many ways this city of only 150,000 inhabitants both serves and dominates an extensive tributary area that may be called its region.[2] This region is physically diverse, but owes its functional unity to the binding influence of Salt Lake City. It has an area of 185,000 square miles and a population of 790,000 inhabitants (cf. England, 51,000 square miles and 38,000,000 inhabitants). It includes Utah, southern Idaho, eastern Nevada and south-western Wyoming.

Salt Lake City has a well-balanced occupational structure. Its 54,000 gainfully employed persons are occupied as follows (figures for 1930) : manufacturing and mechanical industries 25 per cent., trade (wholesale, retail and other) 20 per cent., clerical occupations 15 per cent., transport and communications 11 per cent., domestic and personal service 12 per cent., professional services 11 per cent., public service 3·4 per cent. Harris estimated that a fifth of the total gainfully employed were employed in occupations over and above the needs of the Salt Lake City and its immediate environs—that is, that a fifth of the employed were concerned directly with meeting *regional* as opposed to *local* needs. These " represent merely the elemental occupational base upon which are pyramided many other occupations serving chiefly the inhabitants of the city itself ". On this basis Salt Lake City has 62 per cent. of the clerical occupations ; 71 per cent. of the wholesale trade ; 46 per cent. of the retail trade (which is more evenly distributed with the population in accordance with local needs) ; 56 per cent. of other trades (bankers, brokers and moneylenders, 61 per cent. ; insurance agents, managers and officials, 60 per cent. ; commercial travel-

[1] *Salt Lake City, A Regional Capital*, by Chauncy Dennison Harris, published by the University of Chicago Press, 1940, Private Edition. We summarize the findings of this work since it illustrates excellently the viewpoint and technique of our approach and the study in itself has a very limited circulation and will therefore not be generally accessible. Maps are reproduced with the kind permission of the author.

[2] The Census metropolitan district had 184,000 and the city 140,000 in 1930.

lers, 65 per cent.) ; 59 per cent. of public service (mainly military service, since the city is a garrison centre) ; 42 per cent. of professional service [1] ; 37 per cent. of transport and communication. Wholesaling is a main regional function, the 407 establishments selling $100,000,000 worth of goods and employing over 4,000 persons in 1935. Two-thirds of the wholesaling of both producers' and consumers' goods of Utah are concentrated in the capital. Manufacturing is not a dominant activity in the occupational structure of Salt Lake City. It produces about a quarter of the value of manufactured products in the State of Utah. But manufactures are none the less important, and figure large in the life and structure of the city.

The Region serving Salt Lake City has a populous core, and scattered oasis and mining settlements in the midst of grazing and desert lands (Fig. 32). The boundary of the Region is therefore a wide no-man's land in which there is really no effective competition with other regional capitals. Boundaries are, therefore, more clearly defined in the populous sectors along the main lines of communication. The Region was determined by studying the areas served by a number of important regional functions (Fig. 33). Each of these service areas covers a distinct area that is limited by different factors. Twelve areas were defined and superposed. The service areas for retail trade, wholesale grocery trade, wholesale drug trade, radio broadcasting, and generalized trade were all taken from published sources, and all of them (except the fourth) are the result of careful statistical studies by competent authorities. Newspaper circulation and the extent of the Mormon religion are also based on statistical study. The former is recognized (if carefully mapped) as a good indicator of potency and extent of metropolitan influence. The latter is a feature peculiar to the State of Utah, which gives to Salt Lake City special claims to leadership over a wide area.[2] Mormons are still to-day in a majority throughout Utah and in sectors of southern Idaho (except the Twin Falls district), and throughout a wide surrounding area make up to 10 to 20 per cent. of the total population. Interviews provided the author with the extent of telephone, bakery and

[1] Publishing 66 per cent., baking 60 per cent., clothing 58 per cent., petroleum refining 100 per cent., managers and officials 55 per cent., building (the second largest group) 38 per cent., all other (the largest group) 31 per cent.

[2] Established as the capital of the Mormons in 1847, the settlement of much of the surrounding lands was effected by the religious leadership of the Mormon community, which directed the exploration of irrigable lands, the lay-out of new villages and the settlement of European immigrants.

petroleum distribution areas. From these twelve areas the generalized boundary was drawn (Fig. 32). The retail trade

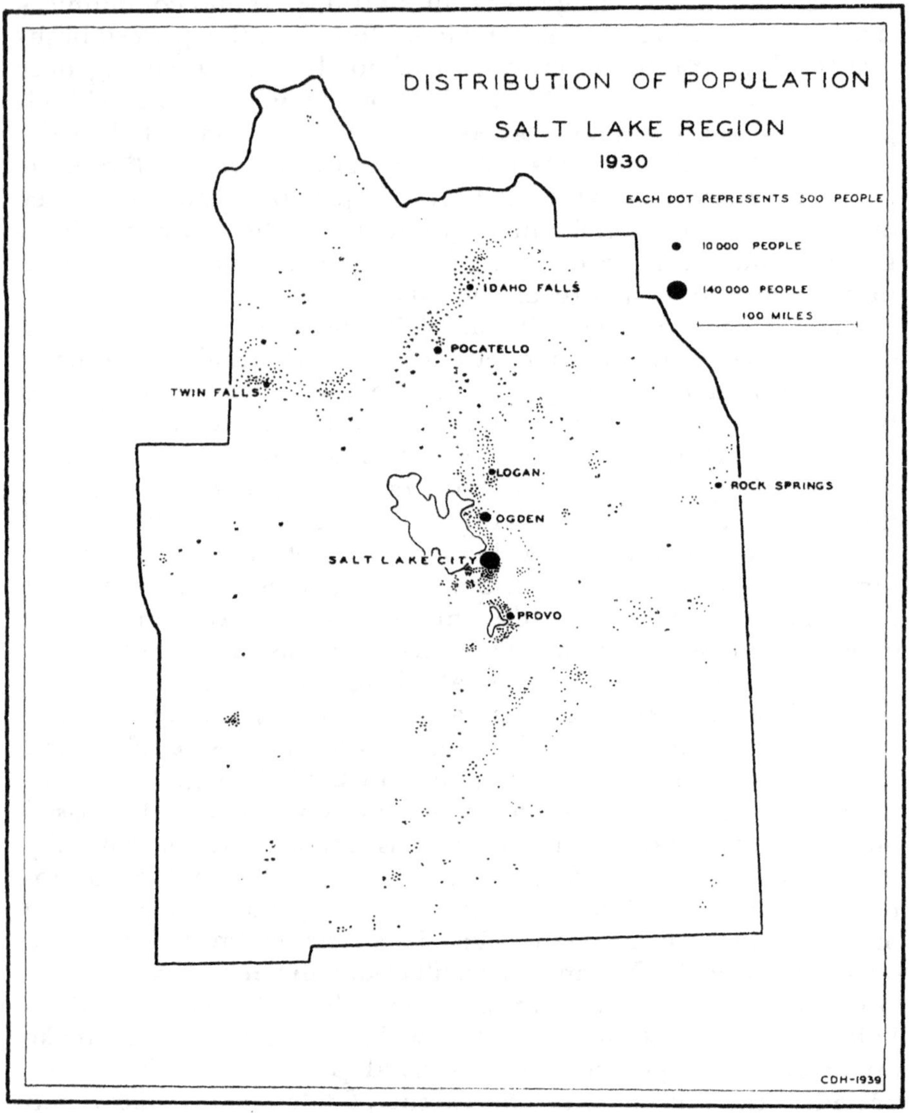

FIG. 32.—Salt Lake City. Distribution of Population in the Salt Lake Region.

area (which is prepared for department stores, clothing, furniture, and jewellery stores rather than for grocery and drug stores) differs from the rest in having a very small service area. The retail trade area is small since it is concerned with a frequent service so that customers will not travel far for it. Local

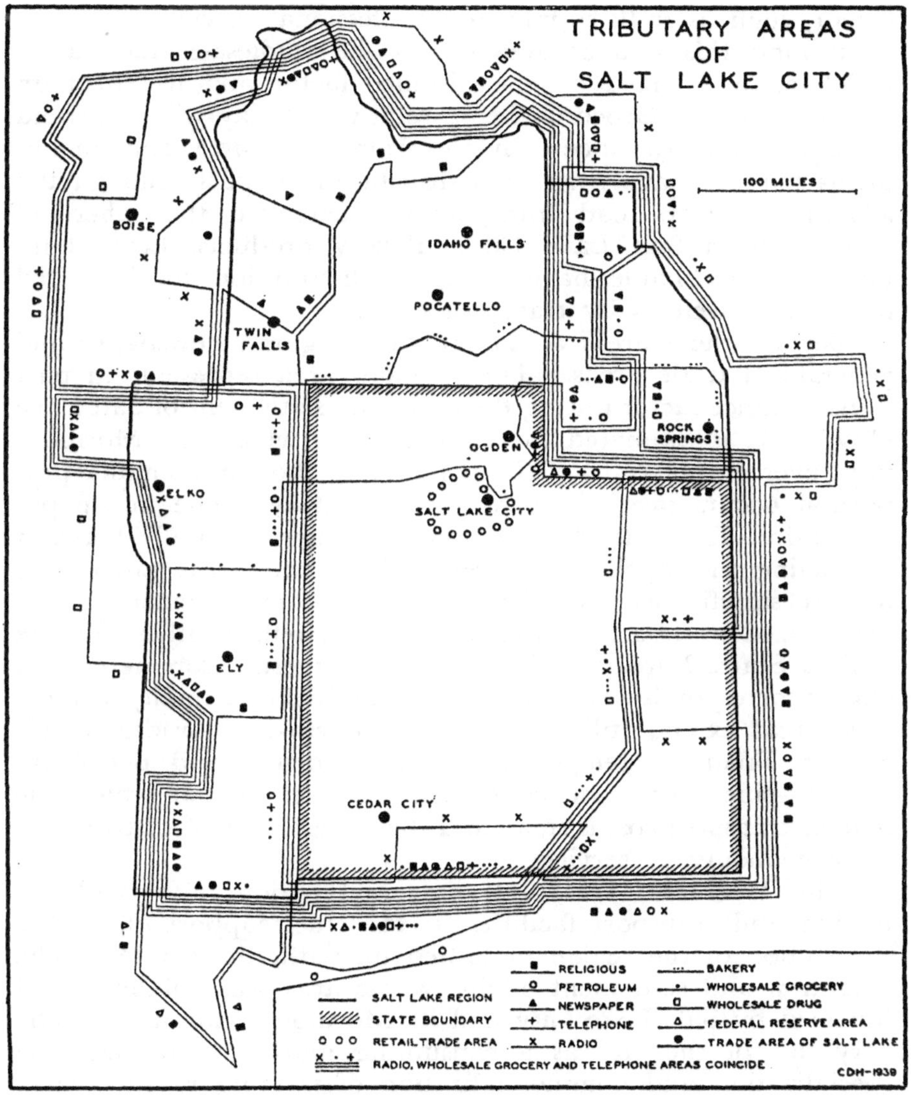

FIG. 33.—Salt Lake City. Tributary Areas.

service centres serve, for the same reason, small local areas. The higher quality and higher priced goods have larger areas —both from the point of view of the consumer travelling to the centre or the goods delivered to him. Occasional deliveries (or visits) may cover a long distance, but the area considered is that of regular service. There are other areas of similar restricted extent, such as milk supply and professional service.

Beyond this smaller area comes the wholesale trade area, which again is fixed by the distance factor and its effect on transport costs. Groceries being relatively bulky will not stand the cost of long-distance deliveries from one centre, and smaller towns tend to have their own distributing depots—though the advent of cheap road transport has extended the sphere of delivery from Salt Lake City. Bakery products, being both bulky and perishable, have a range of distribution that is limited by time as well as by transport costs.

Beyond these areas again are the religious, newspaper and generalized trade areas. The religious area represents in part " the distance factor during the early day settlement of Salt Lake City ". It is estimated from the area in which the Mormons form over 20 per cent. of the population, though a smaller proportion occurs over a much wider territory. The newspaper circulation limits tend to follow a " time divide " between competing metropolitan centres. These two areas are of the highest significance as cultural and economic agencies. The generalized trade area is the result of a careful synthesis by the United States Bureau of Foreign and Domestic Commerce and closely corresponds with the extent of the Region as finally defined. Five areas extend still farther to the north-west—radio broadcasting, financial, telephone, wholesale drugs, and petroleum service—since these are concerned with services in which the item of transport cost is small and the widest area of distribution is an economy in service.

The Wasatch Oasis, with its combination of agriculture, mining and transport facilities, is the chief support of, and is the section of the region most tied up with, the capital. The irrigation and settlement of this whole area were planned and financed by Salt Lake City and its Mormon community. The very size of the city as a consuming market encourages the specialization and commercialization of farming in the Oasis. The influence of the city in all its regional relations covers the whole of the Oasis, and the business and financial leadership of the city has been influential in the development of its transport facilities. The impact of the city on the types of commercialized farming on the irrigated land is reflected in the importance of dairying, for the Oasis produces half of the milk of the State, and 80 per cent. of the fluid milk consumed in the Oasis is produced in it. There is also a marked specialization on the production of fruit, vegetables and other perishable products.

Salt Lake City, together with Ogden, forms an important nucleus of railways, while the Oasis, in addition to being such a focus, has a closely knit net of electric inter-urban lines. In contradiction to this complicated pattern with its two foci in Ogden and Salt Lake City, the highways are centred on the latter, eight of them radiating in all directions. The business activities of the city and its region are reflected in the trade flows as canalized on rail and road—although it should be noted that such linear routes cannot be used as a means for defining the market area that they serve. Railways dominate this goods traffic. This falls into three categories, through traffic (from California to East or Middle West) making up 54·5 per cent. of the total carload traffic of the region ; inter-regional traffic which either originates in or terminates in the Salt Lake City Region, 21 per cent. ; and intra-regional traffic, which both originates and terminates in the Region, 24·5 per cent.

The role of Salt Lake City in these traffic movements may be viewed from the standpoint of either the service areas or the mapped streams of traffic flow in which the services are canalized (Figs. 34 and 35). Ogden is a transfer point between major railway systems, a dispatch centre for certain raw materials such as live stock, and an industrial satellite of Salt Lake City (with railway workshops, meat-packing plants, and flour mills). Salt Lake City handles the trade in diversified wholesale goods, and while Ogden is a through route centre for rail traffic, Salt Lake City is by far the more important terminal for goods and passenger traffic. Inbound shipment of goods far exceeds outbound shipments at Salt Lake City in contrast to Ogden, for it imports valuable and diversified products for distribution (often including processing in the city) by rail and lorry throughout the region. The carloads of goods (in and out) handled by Salt Lake City (1932) amount to a sixth of the total for Utah, and some of these products are handled dominantly by the city for distribution in the region : petroleum at the refinery ; miscellaneous products through the hands of distributors ; ores and concentrates terminating at refineries in neighbouring satellite towns ; imported products such as automobiles, furniture, and

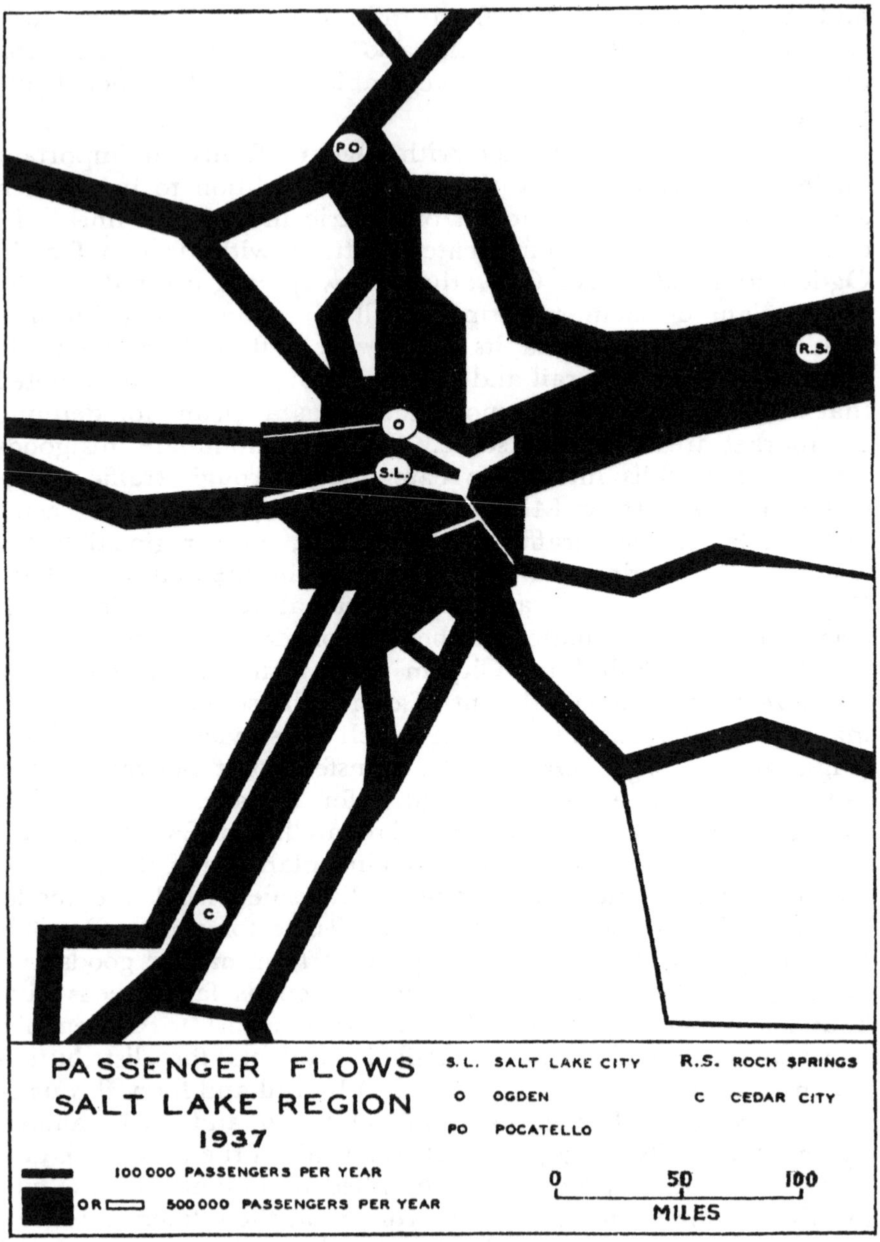

FIG. 34.--Salt Lake City. Passenger Traffic Flows, 1937 (Railroad, Bus, Air and Private Car).

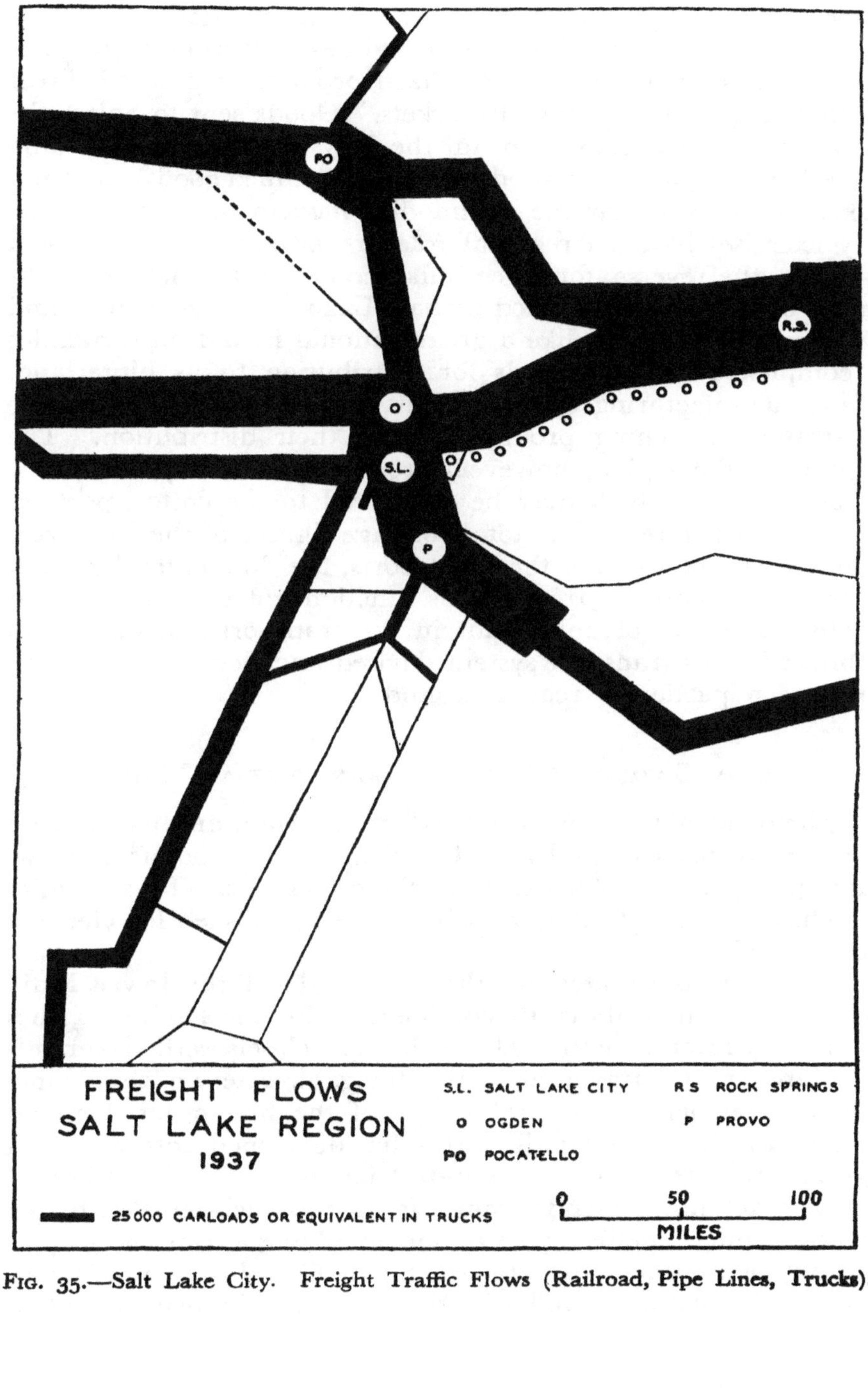

FIG. 35.—Salt Lake City. Freight Traffic Flows (Railroad, Pipe Lines, Trucks).

machinery (77 per cent. of the trade of Utah) ; and paper and paper products (83 per cent. of the trade of Utah).

The city plays a much smaller part in exports from the Region, for products from seats of specialized production proceed direct to their regional or national markets. Goods sent to Salt Lake City are mainly consumed in the city itself—milk, building materials, manufactured goods (sugar and canned goods), coal and coke. It follows that the commodity movements in which the city exercises its chief regional role are not the largest in bulk—rather the reverse, for three bulky commodities make up two-thirds of the freight handled in Salt Lake City : coal, ores and petroleum. The city, like a great national inland port, handles a complex variety of goods for distribution to its hinterland. Many manufacturing establishments have arisen at railsides to treat these incoming products before their distribution. The exports of the region, however, proceed from many small raw material ports, which may be compared to the cotton ports of the southern states. The total tonnage figures of the Salt Lake City yards, like those of the great ports, are dominated by a few bulky commodities. It owes this function not to its location at a physical break of bulk point in the transport system, but to a break in this transport system devised and developed by man to serve a particular region around it.

4. A German Example : Frankfurt-am-Main

The regional relations of Frankfurt-am-Main and its adjacent towns have been carefully studied from various points of view. Figs. 36, 37 and 38 show some of these relations. These studies are characteristic of many that have been published for German and French cities.

Frankfurt is situated on the north side of the lower Main river, a few miles above its confluence with the Rhine. It lies in the centre of an extremely fertile and closely settled agricultural area, that is traversed by the navigable artery of the Rhine, and it is a focus of natural routes of the first order from all directions. This surrounding area has developed both as a rich agricultural area and as an industrial area. A discontinuous urban tract now extends from Mainz and Biebrich-Wiesbaden (both forming, in effect, part of one area lying astride the Rhine) through Rüsselsheim to Frankfurt and Offenbach and Hanau over a distance of 25 miles, and accounting for over a million people. This nuclear area is completely encircled to the north, east and west by forested uplands, with tongues of fertile, closely settled, lowland stretching north along the Rhine gorge, south in the plains of the Middle Rhine, and east through the lowland strip through Hesse to Giessen. Here, from prehistoric times, routes have converged and to-day it is a focus of outstanding importance, with two chief centres in Mainz and Frankfurt. There converge on Frankfurt and Mainz ten important railway routes and over 700 trains normally (pre-war) enter or leave Frankfurt daily. It was planned to focus five autobahns on Frankfurt. Before the recent war 23 airways converged on the city, which was the second airport in Germany after Berlin. The nearest competitive traffic nuclei are Mannheim, Ludwigshafen and Saarbrücken to the south, Cologne and Kassel to the north and Würzburg to the east.

The interrelations of these cities with each other and with the surrounding countryside have been the subject of particularly thorough analysis, especially in respect of the food supplies to Frankfurt and the "journey to work" to the chief towns.

An analysis of the movement of food supplies to Frankfurt (Fig. 36), causes one writer to recognize two main zones around the city. The inner zone near Frankfurt is dominantly agricultural and specializes on vegetable and fruit production, owing primarily to the fertility of the soil. Large quantities of milk, wheat and potatoes also find their way to the Frankfurt market. An outer zone supplies milk, vegetables, meat and potatoes in varying quantities, the products differing in character and quantity from one part of the zone to another, for its environmental conditions are markedly contrasted. Considerable quantities of food are imported also from other parts of Germany, especially the west, and from abroad.[1]

The "journey to work" has been exhaustively studied by Hartke on the basis of the new tax, established in 1936, to be paid by workers to their residential places if they work outside them (Fig. 37). The study covers Frankfurt (28,000), Mainz (6,200), Rüsselsheim (15,000), Darmstadt (4,000), Hanau (6000) and Offenbach (6,500), the figures in brackets indicating the total number of persons recorded on this basis, as working in these places but living outside them. A separate analysis is made of some of the biggest industrial plants. The most remark-

[1] W. Brosius, Die Lebensmittelversorgungsgebiete der Stadt Frankfurt-am-Main, *Rhein-Mainische Forschungen*, Vol. X, Frankfurt.a.M., 1934.

able facts about this study are the closeness of these centres and, in consequence, the overlapping of their passenger traffic flows; the fact that most of the movement is limited to a journey-time of about one hour in one direction; and finally the considerable importance of the bus and cycle as against the train. Frankfurt,

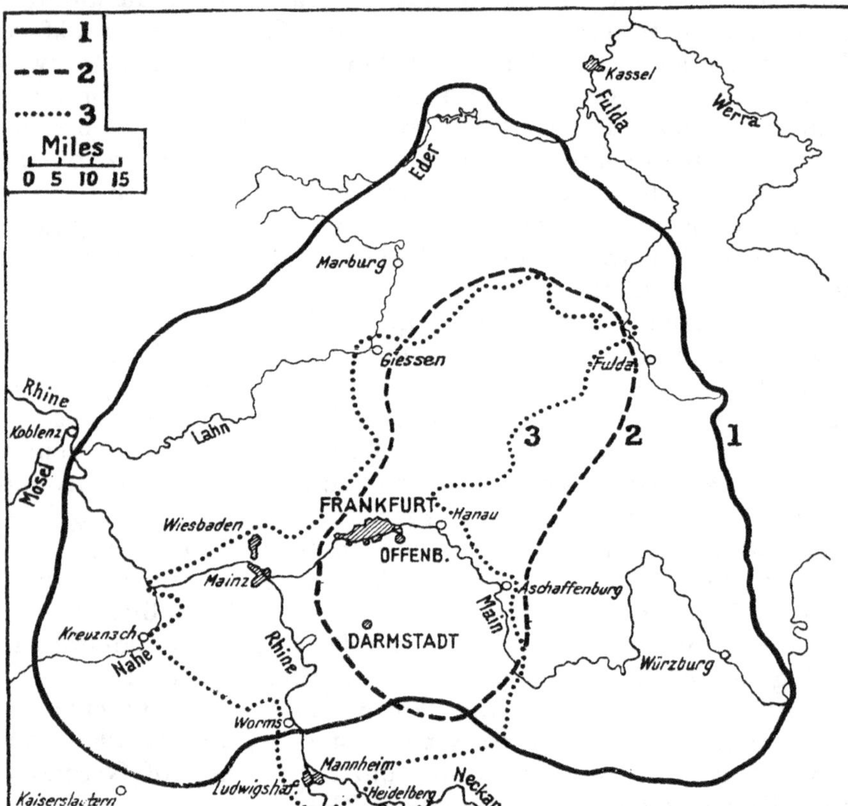

FIG. 36.—The Rhine–Main Region. Food Supply Areas of Frankfurt, Offenbach and Darmstadt.

1. Frankfurt (after Brosius). 2. Offenbach (after Ulrich). 3. Darmstadt (after Schilling). After H. Schrepfer, "Über Wirtschaftsgebiete und ihre Bedeutung für die Wirtschaftsgeographie", *Geographische Wochenschrift*, Heft 21–2, 1935, pp. 497–520.

as one would expect, is the chief centre of this movement. But most remarkable is the vast Opel works at Rüsselsheim between Mainz and Frankfurt. This place, extremely small as compared with Frankfurt, employs 15,000 people who live elsewhere, the great majority being employed in this one concern. The main residential area lies south of the Main in open country that is

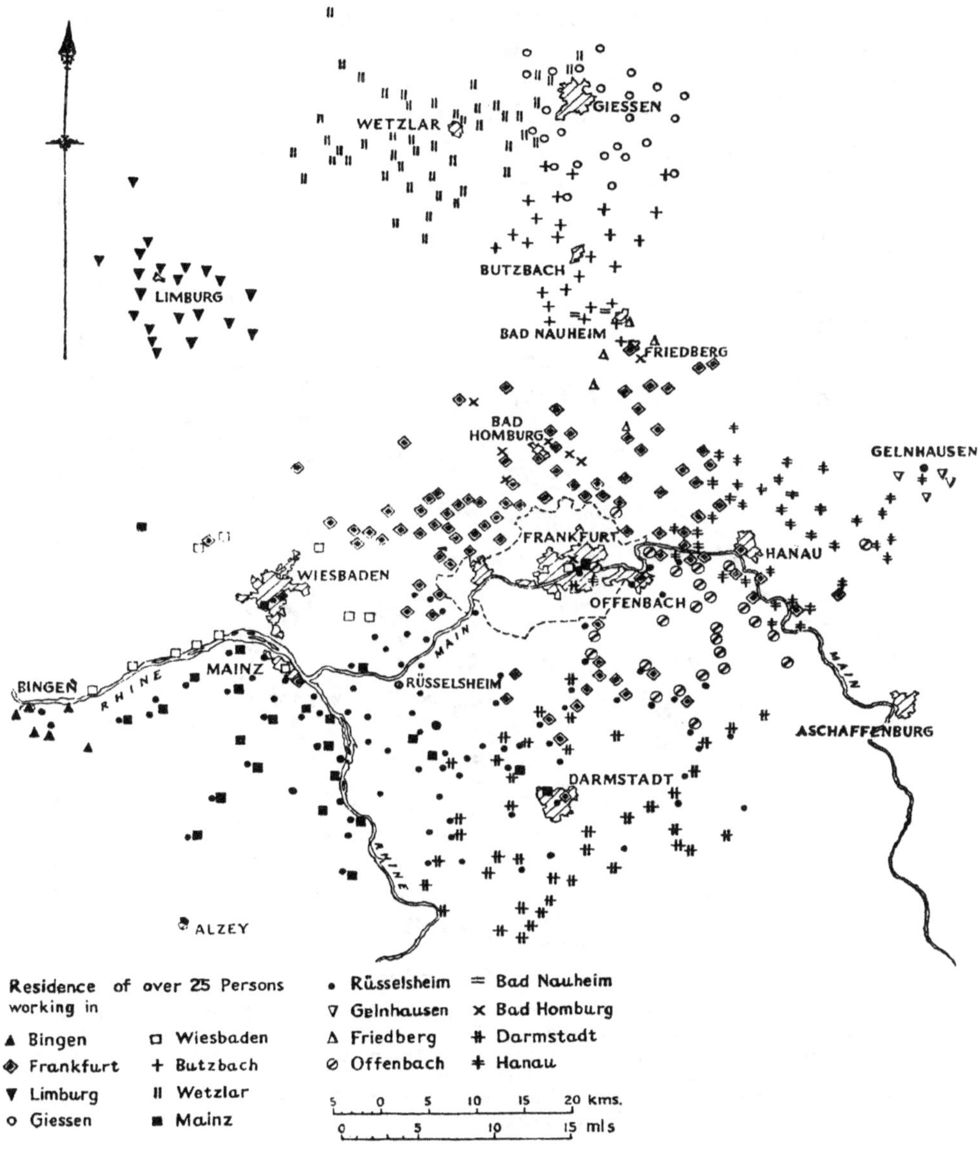

FIG. 37.—The Rhine–Main Region. Places of Residence of persons working in, but living outside, Frankfurt and the chief towns of the Rhine–Main Region. Each symbol shows the residence of over 25 persons working in the towns indicated. Based on tax returns. After W. Hartke, "Das Arbeits- und Wohnungsgebiet im Rhein–Mainischen Lebensraum", *Rhein-Mainische Forschungen,* Heft 18, Frankfurt.a.M., 1938.

served not by train but by bus and cycle. Of the towns, Mainz sends the largest daily contingent to Rüsselsheim. The aggregate of persons moving to these towns amounts to about 75,000 out of an aggregate population exceeding a million.[1]

The zones of influence of Frankfurt as worked out by Schrepfer are shown on Fig. 38. They may be described as follows:

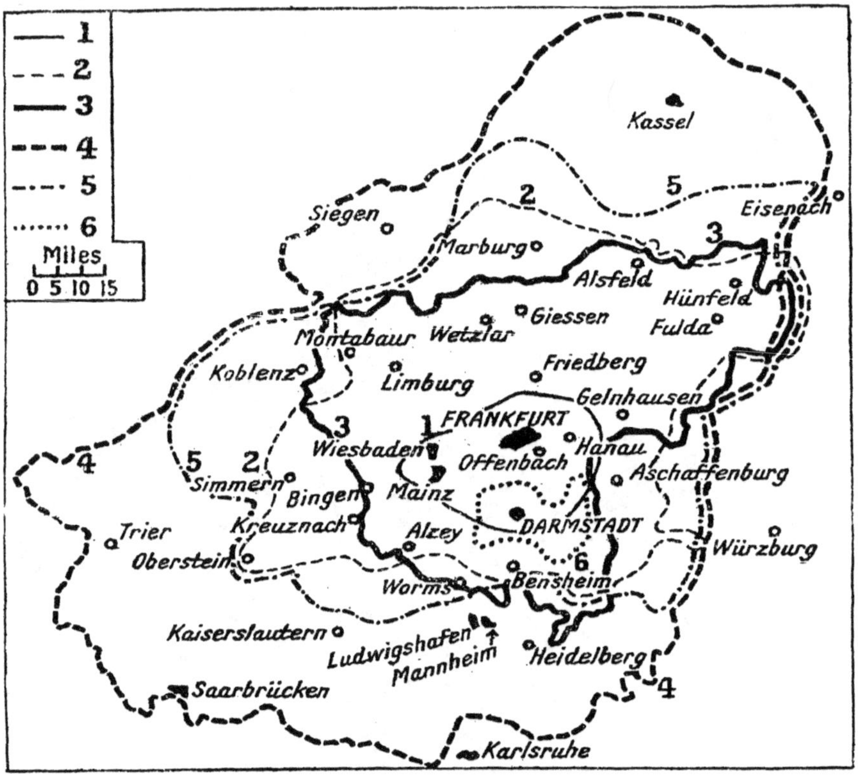

FIG. 38.—The Rhine–Main Region. Zones of Influence of Frankfurt-am-Main.

1. The Nuclear Region. 2. The Economic Region (after Schrepfer). 3. Boundary of the *Rhein–Mainischen Industrie- und Handelstages* (seat at Frankfurt). 4. Limit of the Outer Rhine–Main Region and (5) of the Inner Rhine–Main Region, after Otto Maull. 6. Sphere of Influence of Darmstadt, after M. Schilling. The towns shown have independent local spheres of influence. After Schrepfer, *Über Wirtschaftsgebiete Geographische Wochenschrift*, Heft 21–22, 1935.

1. The Nuclear Zone is formed by the triangle of Mainz, Wiesbaden, Frankfurt, and Darmstadt. It contains the areas most intimately associated with these places in respect of resid-

[1] The 1939 population figures were as follows: Frankfurt 570,000, Offenbach 86,000, Wiesbaden 160,000, Mainz 154,000.

ence, food supplies and industry. This zone ends at the foot of the surrounding wooded hills.

2. The outer zones are mainly uninhabited. They are mainly forested, and have a few small towns. Only against the region of Saar–Palatinate is there a competitive border zone. Weinheim is an economic satellite of Mannheim and Worms lies in the economic sphere of the Palatinate. The border zone against central Franconia with its capital in Würzburg (which city is an old-established economic and cultural focus with a tributary area which has been clearly defined by the surrounding wooded uplands and a forest-free fertile centre) is marked by the thinly peopled Odenwald and Spessart uplands. The Main valley as far as Wertheim is dominantly oriented towards Frankfurt. Kassel to the north is an independent centre serving northern Hesse. To the north-west, Siegerland with its iron ore mining and iron and steel-making industries, is oriented towards the Ruhr and Cologne, but the Lahn and Dill valleys lie in the outer zone of Frankfurt. The Koblenz region has close relations with both Cologne and Frankfurt and is in the nature of a buffer area.

The smaller towns within the range of influence of Frankfurt have independent local tributary areas. One of these towns, Darmstadt, has been the subject of a very detailed study, and its composite tributary area is shown in Fig. 38.[1]

Darmstadt is the administrative centre of Hesse with a population of 110,000 (1939), lying about halfway between Frankfurt–Mainz and Mannheim–Ludwigshafen. The regional relations of the city are classed as follows : milk supplies ; meat supplies ; foodstuff supplies (*Lebensmittel*), that is, vegetables, fruits, butter and cheese ; the movements of people and ideas—migration citywards from the countryside, commuting, and " cultural " movements, that is, visitors to theatres, pupils to schools, and the circulation of the Press. Each of these relations is examined in detail by mapping the places concerned quantitatively, the whole area in each case then being framed by a line. Superposition of these maps revealed two zones, an inner zone that embraces all the above relations, and an outer zone that has varied relations, with the city and the larger surrounding cities, that gradually fade off with increased distance from the city centre.

[1] M. Schilling, *Analytische Untersuchung der Wirtschaftlichen und Kulturellen Einflusssphäre der Stadt Darmstadt*, Ph.D. Dissertation, University of Frankfurt, 2 vols, Darmstadt, 1935.

5. Comparative American Studies : The Regional Influence of the City on Rural Economy

The city exerts powerful influences on the social and economic structure of the territory around it. These influences are expressed in the types of rural land use and farm economy, in the nature of urban land uses, and in the social and economic structure of the villages and towns affected.

As soon as a city arises it exerts an influence on the uses of the land around it according to a principle that was clearly expounded by Johann Heinrich von Thünen in his *Isolierte Staat*—the Isolated State—first published in 1826.[1] The principle was worked out in respect of rural land use by imagining one large city in the centre of the State in a plain with uniform soil and climate. Types of land use would then be conditioned entirely by the economic factor of distance from the city (by road) as the market centre. The State, served by the one city, is assumed to be surrounded by a wilderness completely cutting it off from the rest of the world. The type of crops grown will be determined by the price of the crop in the city market, a crop, for instance, being grown no farther from the city when the cost of transport plus cost of production reach the selling price in the city. Six concentric zones surround von Thünen's city, namely, a small zone just outside the city producing perishable commodities that cannot stand long-distance transport, e.g. milk and vegetables ; a narrow zone of forest, wood being used in this pre-industrial era for fuel and building, placed near to the city owing to its bulk and high cost of transport ; third, a zone of rotation grain cultivation ; fourth, a zone of less intensive rotation cultivation with pasture and fallow ; fifth, a wide area of widespread three-field farming, the dominant system in western Europe until the end of the eighteenth century ; and finally, a zone of cattle-raising and hunting. According to von Thünen's calculations, cereal cultivation would cease at 31·5 miles from the city. It is the competition of land uses through rent that explains these concentric belts of land use in the " Isolated State ". The use which can pay the highest rent at a particular place occupies the land.[2] In countries of the present day where road transport is the means of carrying crops to a railhead

[1] J. H. von Thünen, *Der Isolierte Staat in Beziehung auf Landwirtschaft und Nationalökonomie*, 1826, reprinted, with introduction by Heinrich Waentig, Gustav Fischer, Jena, 1910.

[2] R. T. Ely and G. S. Wehrwein, *Land Economics*, New York, 1940, p. 135.

(rather than to a city), the limit of movement before the advent of the motor lorry was 15 to 25 miles, though the bulk came from a five-mile radius. In Rhodesia the limit is 15 miles. "Roughly the zones are 'farm land' within 25 miles of the railway, 'ranch land' 25 to 50 miles, and land beyond 50 miles is of little practical value to the settler." [1]

This effect of the modern city on the rural land uses and the crops grown in its environs is related to two basic trends: first, the orientation of commercialized farm output towards the city market; secondly, the effect of the spread of the urban area on the values of open land around it. For such land is likely to be used sooner or later for urban uses and will therefore rise in value in anticipation, with the result that farming will be intensive in order to get the maximum returns. Alternatively, if there is immediate likelihood of its being sold for building, it will lie derelict in the hands of the speculative builder, awaiting a purchaser. The location of commercialized horticulture near to cities is due to the high price of land as well as to the proximity of an immediate market. For the same reason the demand of the city market for fresh milk has a marked influence on the spread of dairy farming around cities irrespective of climate and soil. In the United States it has been shown that there is a tendency to the formation of concentric zones around the city market, the sequence being milk production, feeding grains, bread grains and ranching, each fading into the next by a zone of transition from one type of land use to another.[2] The influence of the city market upon the kind of land utilization is thus based on the factor of distance from the market.

The theoretical pattern of land use trends around the great European cities has been summarized by Olof Jonasson.[3]

The same principle may be applied to the delivery of goods

[1] Ibid., p. 136, and I. Bowman, *The Pioneer Fringe*, American Geographical Society, 1931, pp. 216–19.

[2] Ely and Wehrwein, op. cit., pp. 133–8.

[3] *Horticulture.*

Zone 1. City, plus greenhouses and floriculture.

Zone 2. Truck products, fruits, potatoes, and tobacco.

Intensive Agriculture with Intensive Dairying.

Zone 3. Dairy products, beef cattle, sheep for mutton, veal, forage crops, oats, flax.

Zone 4. General farming—grain, hay, livestock.

Extensive Agriculture.

Zone 5. Bread cereals and flax for oil.

Extensive Pasture.

Zone 6. Cattle (beef and range), horses (range), sheep (range), slat, smoked, refrigerated and canned meats, bones, tallow, and hides. (*Continued on* p. 194.)

outwards from a centre. If instead of von Thünen's single city market there are several competitive market centres, then providing these centres serve as equivalent sources of supply and distribution, the price to the consumer becomes the cost price plus the transport charge. Around each centre lines (*isotims*) may be drawn to show the limits to which any commodity can be delivered for the same price and where the lines of neighbouring centres meet is a transition zone in which it is immaterial whence the product comes, or, *vice versa*, in which centres it is marketed. Given the same cost of production, the centre enjoying the more convenient, more efficient, or cheaper, form of transportation will penetrate the market area of the other. This principle is of special importance in respect of retail and wholesale distribution.[1]

6. The Regional Influence of the City on Social and Economic Conditions

The impact of urban influences on the country and smaller towns has many and deep-seated effects on their social and economic structure, such influences normally being most intense near the city and decreasing outwards from it. An obvious method, at any rate in theory, for examining the nature of the relations between town and country is to examine various conditions statistically on the basis of small administrative units in the environs of the town. No such studies appear to have been undertaken in Europe, but an attempt has been made in the American researches of Brunner and Kolb in their *Rural Social Trends* Monograph.[2] Eighteen cities were selected with populations ranging from 20,000 to over half a million. A statistical analysis of concentric zones round each city was made, taking as limits for the area the limits of the wholesale trade area for the city. The county in which the city is situated was taken as a unit, and the counties contiguous to it were called Tier One, counties contiguous to Tier One were called Tier Two, and so on up to Tier Four. The outer boundaries of each tier from the boundary of the city county are approximately 6, 12, 18 and

---

Forest Culture.

Zone 7. Outermost peripheral areas. Forests.

Olof Jonasson, "The Agricultural Regions of Europe", *Economic Geography*, Vol. I, 1925, pp. 284–7.

[1] Ely and Wehrwein, op. cit., pp. 140–2.

[2] E. de S. Brunner and J. H. Kolb, *Rural Social Trends*, Chapter V on Rural-Urban Relationships, 1933, pp. 141–2 and 151–2.

24 miles respectively.[1] There averaged nineteen counties to each city, covering 10 per cent. of the population of the country. Literally hundreds of indices were taken in these studies, and the characteristics of the tiers or zones in 1930, and the changes in these characteristics since 1910, were determined. The indices include fertility rates, sex ratios, ratio of children, proportion of land in farms, average acreage of improved land per farm, value per acre of farm property, and of farm crops, and value of dairy products. Several significant "gradient relationships" were discovered in the successive concentric zones. The ratio of children under ten years of age to women of reproductive age increased with distance in the first three tiers but declined in the fourth tier. The birth-rate tended to increase with distance of the county from the city. Distance from the city was found to have an important effect on farming. The percentage of farms devoted to dairying tended to increase with distance in the first three tiers and declined rather sharply in the fourth tier. The percentage of truck (market gardening) farms declined in the first three tiers, but showed a slight increase in the fourth. The percentage of poultry farms declined through all four tiers, while the proportion of stock farms increased with distance in all four tiers. The farms tended to increase in size with increased distance from the city. The value per acre of farm land declined consistently with distance and the value per acre of all farm products also declined with distance. The value of dairy products per acre tended to decline in the first three tiers and increase in the fourth tier. These data were then examined for the 1910 to 1930 period and the following conclusions were drawn.

The City County and, to an increasing extent since 1920, Tier One, are likely to be given over to smaller farms for market gardening, fruit growing and intensive dairying. This means more compact communities and relatively high densities of population. It means more frequent contacts of all sorts with the city centre. Demographic trends are akin to those of the city. Tier One and especially Tier Two show an increasing specialization on dairying. Tier Three has many markedly transitional features. Tier Four and beyond, until the influence of another urban centre is reached, form what is described as an

[1] Wholesale trade areas were taken from the *Market Data Handbook of the United States*, by P. W. Stewart, published by the U.S. Department of Commerce, Domestic Commerce Series, No. 30, 1929.

outer zone. It has larger farms but fewer cultivated acres per farm. Population densities are lower, communities larger, and there are fewer contacts with the city. City influence in retail sales declines beyond Tier Two, although there is some dependence on the city stores and mail-order houses for clothing, furniture and household effects. But most remarkable are the demographic trends in Tier Four that stand in fundamental contrast to the inner Tiers. There are more children under 10 to women 20 to 45 years of age and there are higher birth-rates. There is a higher proportion of people under 21 years of age, and a higher proportion of males to each 100 females, and this is in turn reflected in the marriage ratios. Children are a distinguishing feature of the farm populations removed from urban influence, and they have been becoming more so, even though the general differences between rural and urban populations in this regard were tending to decrease.[1]

Changes in farming conditions with increased distance from the city were revealed some years ago in a study of Louisville. Within 8 miles of the city market place market garden products and potatoes provided 68 per cent. of the farm income, whereas at a 15-miles radius only 20 per cent. was drawn from this source, even though the soil that is especially suitable for market gardening extends more than 20 miles along the river. At 15 miles or more dairying and general farming took the place of the intensive crops. The greater intensity of land use near the city was shown by the smaller farms, greater operating expenses, and higher expenditure on fertilizers. Gross earnings were five times as high per acre 9 miles from the city as 16 miles, and this was reflected in rents paid and land values.[2]

A similar investigation has been made by the Urbanism Committee of the National Resources Committee.[3] It allowed one tier of counties around each county that contained one city with over 100,000 inhabitants, a second tier was added to eight central nuclei with a city with over 500,000 to one million inhabitants (allowing for its greater range and potency of influence), and a third tier for the five central nuclei with a city with over one million inhabitants. On this basis, 89 regions were established around the counties containing the 93 cities

[1] Brunner and Kolb, op. cit., Chapter 5, especially pp. 141–2.

[2] Ely and Wehrwein, op. cit., p. 137, quoting J. H. Arnold and F. Montgomery, *Influence of a City in Farming*, U.S. Dept. of Agriculture, Bulletin 678, 1918.

[3] *Population Statistics, 3. Urban Data*, National Resources Committee, October, 1937, pp. 43–5.

of 100,000 inhabitants or over. Data considered were the growth and density of population, industry, income, occupations (grouped as agriculture, industrial non-service occupations, and industrial services),[1] and the percentage of total population ten years old and over gainfully occupied. This investigation showed that an increasing proportion of the total population of the country is concentrating in these regions, and that settlement is, as it were, segregating in wider areas around the central urban districts. The character of these city regions varies from one part of the country to another. In the Atlantic Seaboard, the Great Lakes and California, population is highly urbanized and concentrated and is becoming increasingly so and increasingly industrialized. In the rest of the States, the cities have a more limited range of influence and lead " a more isolated and independent existence ". They are primarily commercial distributing points to their agricultural hinterland. It was, moreover, clearly revealed that the city was predominantly engaged in the service trades and professions, as compared with less than a half of the total employed so engaged in the outlying counties, and 25 to 60 per cent. (an average of 40 per cent.) in all the outlying areas. The city is thus clearly revealed as " the centre of managerial, commercial, clerical and professional functions ".[2]

APPENDIX TO CHAPTER 6

The Metropolitan District of the United States Census Bureau

(*Fifteenth Census of the United States; 1930, Metropolitan Districts. Population and Area.* 1932, pp. 5–6).

An attempt to recognize the unity of the urban tract, at least for statistical purposes, has been made by the United States Census Bureau, by creating the metropolitan district.[3] A committee

[1] These included all those engaged in transport, trade, public service, professional service, domestic and private service, and industry not specified.

[2] *Population Statistics, 3. Urban Data*, National Resources Committee, 1937, p. 45.

[3] The metropolitan district as defined for the 1910 and 1920 censuses was a city of over 200,000 population together with every civil division in which the density of population exceeded 150 to the square mile, the whole or the major part of whose area or population lay within ten miles of the boundary of the central city. In the case of cities with 100,000 to 200,000 inhabitants, the whole of the area within this radius was included.

appointed by the Industrial Bureaus of the Chambers of Commerce reported in 1927 as follows:

> The real city to-day, because of the automobile, the telephone and other distance diminishing agencies, extends not only beyond existing city boundaries, but beyond the boundaries of any area which might be annexed. Because of this there are students of the subject who incline to oppose further annexation except in the cases of obviously misfit or irregular boundaries and to favour the creation of a new unit, the metropolitan region, which shall have charge of certain common public services—such as main highways, water supply and sewage disposal, parks systems, police and fire protection, etc., leaving the municipalities within the region autonomy as to their individual concerns. A clear definition of such metropolitan regions, capable of application to all situations, is still to be worked out. There are, however, certain considerations that should be borne in mind when drawing the boundaries of a metropolitan region, i.e. it is an area within which the conditions of manufacturing, trade, transportation, labour and living, in brief the daily economic and social life, are predominantly influenced by the central city.

Pursuing this idea, the United States Chamber of Commerce, with the approval of the Bureau of the Census, sent a circular letter to the Chambers of Commerce of towns of over 50,000 population, inviting them to supply maps and data defining a metropolitan district (or region) as determined by certain proposed tests or control factors. " These factors included telephone services, electric power service, retail store delivery, commuting service, water service, gas service, mail delivery, sewer service, residential membership in social and athletic clubs, operation of local real estate companies, and soliciting and collecting routes.[1] A subsequent study of the extent of these factors, their relations one to the other, and the general interpretation made of these by local organizations, made it evident that there was more or less inconsistency, and that if the metropolitan districts were made to be uniform and comparable with one another, some factor should be used which could be applied uniformly and consistently to every city."

It was found that the local organizations in preparing their maps were thinking of an industrial or trade area, rather than of the area which represented the concentration of population around the large central cities. The problem for census purposes was to establish metropolitan districts so as to show the principal population centres taken as a whole, " by including in a single total both the population of the central city itself and that of the suburbs and urbanized areas surrounding it—or, in some cases, the population of two or more cities which are located in close proximity and that of the suburbs ". These districts were officially defined in the 1930 census as including " in addition to the central city or cities, all adjacent

[1] It is important to emphasize that all these services refer to services received *direct* from the central city.

and contiguous civil divisions having a density of not less than 150 inhabitants per square mile, and also, as a rule, those civil divisions of less density that are *directly* contiguous to the central cities, or are entirely or nearly surrounded by minor civil divisions that have the required density ". The minimum aggregate population of the metropolitan district was taken as 100,000. The method of definition used by the Census changed the population of the metropolitan district by less than 5 per cent. for about three-fourths of the cities for which districts had been defined on the service basis.

The metropolitan district, as defined by its geographico-economic boundary rather than by its political boundary, is of great value in the United States to business and industry as affording a scientific market measurement for a given economic urban unit, and to the cities themselves for planning, regulating and welfare departments, and to state and federal governments in their assistance to cities. Fawcett bases his definition of a conurbation on the extent of predominantly urban areas, thus including much of the fringes. The U.S. Census Bureau takes an extremely low density limit so as to have a consistent statistical basis that will ensure the inclusion of most of the areas that form a part of the city as a social and economic unit.

CHAPTER 7

THE CITY-REGION IN THE UNITED STATES

1. The City or Metropolitan Region

The influence of the metropolitan city extends well beyond the limits of its settlement area to a vaguely defined tributary or trade area, or hinterland. This trade area, however vague it may be as to extent, has become much more effective and influential with the development of the complexities of modern civilization. This is mainly due to the addition of numerous nation-wide services, which demand large city centres as their seats of operations, near to the great masses of population and accessible to the whole of the surrounding territory so as to effect a nation-wide coverage for the particular needs with which the services are concerned. It is also due to the increasing concentration of the best and most exclusive commercial, cultural and social services in the greatest cities. So important are these metropolitan regions that they are used by a great variety of concerns as units and they have also been seriously put forward as potential new political units and planning units. For a nation-wide division, however, they have been rejected in recent studies in the United States as planning units. But though too specialized as social units to serve the wide range of demands as planning units, they are of the greatest importance in the United States as in other countries for two main reasons. They are the most important single regionalizing force in the social organization of modern society. All aspects of so-called regional and physical planning, all measures for dealing with social problems and all questions of the delimitation of new political and administrative areas require the recognition of these regions. This chapter will attempt to indicate their characteristics as socio-economic units.

The importance of the metropolitan region from the standpoint of business organization is strikingly evident in the business practices of the United States. Numerous concerns have sought to parcel out the whole country into areas dependent on cities for their own particular interests. In general, their investigations contain maps showing such data as the following : " newspaper circulation and trading practices, accessibility, transportation facilities and rates, customary trade practices, market characteristics of cities and counties, tax reports, wealth and consumption indexes, sales data, telephone, radio, automobile and home ownership, postal receipts, and population characteristics ".[1]

The delimitation of metropolitan regions is also a primary concern of planning agencies and of governments.

> Their activities proceed generally on the basis of regions which conform to existing administrative areas, although they, too, are beginning to recognize the significance of economic and social characteristics which have a bearing upon the functional unity and interdependence of local areas belonging to different governmental administrative units. Such items as location of industries, transportation and communication facilities, growth, distribution and movements of population, commutation, housing, land utilization, subdivision planning, public utilities, water supply, sewage disposal, public health, public safety and public welfare, public finance, constitutional and legal provision, and administrative practices are among the most important factors with which planning and governmental studies of metropolitan regions are primarily concerned. The regional plan of New York and its environs, and the regional studies of Chicago, Cincinnati, Cleveland, Detroit and St. Louis represent such efforts.[2]

2. The Growth of Urban Population

The growth of population in the United States falls into three periods. In the pre-railway era, from colonial times to about the middle of the nineteenth century, settlement was confined to the Atlantic seaboard east of the Alleghanys. The railway era, from about 1850 to 1900, was an era of great westward expansion, when cities grew rapidly, first as commercial centres and later as industrial centres, gateway cities serving as the " child and servant of expanding rural settlement " (McKenzie) ; 42 of the 93 cities with over 100,000 inhabitants were incorporated as towns in this period. About 1900, with the advent of motor transport, the third period begins. The rise of manufacturing, population and wealth, and the filling up of the easily accessible free lands in the West caused an increasing proportional growth of the big city agglomerations, which thus

[1] *Population Statistics, 3. Urban Data*, National Resources Committee, October, 1937.

[2] Ibid., p. 36.

increased their economic and cultural dominance as centres for their own inhabitants and for the rural settlements as well.

In the peopling of the United States, Benton MacKaye [1] recognizes four periods of migration, which he compares to the flow of running water. The first was a "broad-gauged" continental *outflow* covering the period of the westward movement beyond the Alleghanys in the days of the covered wagon. The second was a *reflow*, a deepening and widening of the field of migration through the advent of the railway. It established the manufacturing industry and the factory and produced the rapid growth of towns as seats of commerce and industry. The third migration is marked by an *inflow* into the great growing cities, by "a series of local in-drawing movements attracted by the city skyscraper". Here the immigrants engage not only in factory industry, but also in "industry in paper" in offices—commerce, business and administration—concentrated in the big metropolitan cities, the heads of "financial direction". Migration in the fourth period consists of "a series of local back-pushing movements impelled by metropolitan congestion and breakdowns. It is not a quest for a new source of livelihood; it is a relocation of its populations and industries resulting from the second and third migrations—the relocation, that is, of the factory worker and the office worker". The fourth migration is thus a *backflow*. A substantial part of it is made up of commuters or suburban dwellers, but factories and offices also are making their exits.

In the 1930–40 decade, for the first time in history, the great cities with over 100,000 inhabitants increased more slowly than the country as a whole, the figures being 5·2 and 7·0 per cent. as compared with 31 and 16 per cent. in the 1920–30 decade when the cities increased twice as quickly as the country. Out of 82 metropolitan districts [2] in 1940, 25 recorded decreases (almost all of these being in the north-eastern States), 43 increased by less than 10 per cent., 9 increased from 10 to 25 per cent., 4 from 25 to 50 per cent., and only one—Miami—increased by more than a half. Cities that were still increasing rapidly in the 1920–30 decade had slowed up or decreased in the 1930–40 decade. Detroit increased by 57·9 per cent. from 1920 to 1930,

[1] Benton MacKaye, *The New Exploration, A Philosophy of Regional Planning*, New York, 1927.

[2] See definition on p. 198. Figures are from Gist and Halbert, op. cit., p. 44. See also M. Jefferson, "The Great Cities of the United States, 1940", *Geographical Review*, Vol. XXXI, 1941, pp. 479–87.

but only by 3·2 per cent. in the last decade. New York City dropped from an increase of 23·3 per cent. to 6·5 per cent., and San Francisco from an increase of 25·2 per cent. to a decrease of 0·8 per cent. To put these figures into clearer perspective, the United States as a whole increased by 16·1 per cent. in the 1920–30 decade and by 7·0 per cent. in the 1930–40 decade. The main cause for this trend is that the cities within their census boundaries are fully built up, and new growth is accruing to the places contiguous with them and sharing in their life both by immigration from elsewhere and by migration outwards from the central cities.

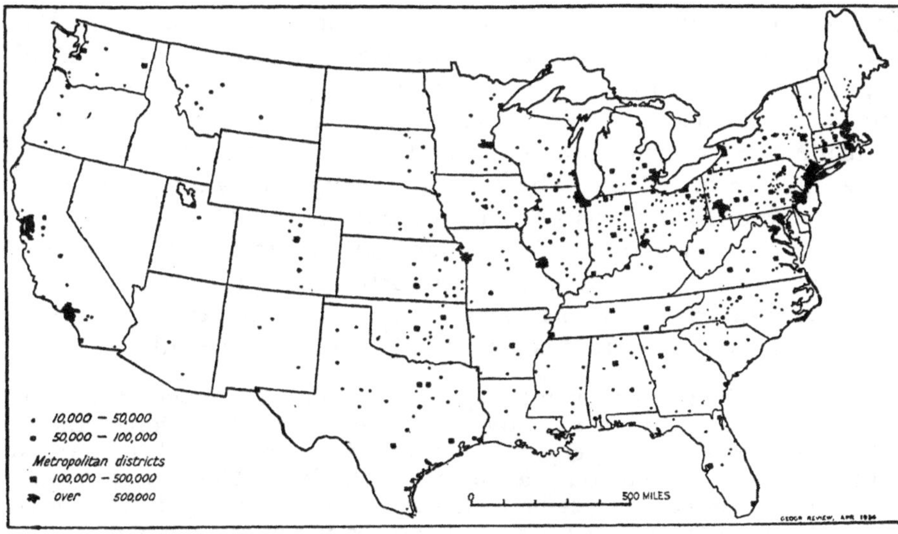

FIG. 39.—United States. Urban Settlements in 1930.

It is thus true that in the United States " the historical trend of urban expansion is apparently being reversed, and the indications are that the decade between 1940 and 1950 will witness an even greater number of cities losing population than declined in the decade just ended ".[1] But the fact to be emphasized from the point of view of the structure of the city is that while its aggregate population showed in many cases a decrease in the percentage rate of growth of its numbers, there has been going on within it and on its margins a fundamental and rapid change in the distribution of population. People are shifting outwards from the central cities to newly built-up areas, either on the

[1] Gist and Halbert, op. cit., p. 46.

outskirts of the city administrative area or beyond them into contiguous administrative districts, many of which are essentially rural, and their inhabitants, for census purposes, are counted as such.

These trends are brought out even more clearly in figures of growth for the metropolitan districts of the census when distinction is made between the chief central cities and the remaining outer areas of the districts. Such figures are available for the 1920–30 decade, but not yet for the 1930–40 decade on the same basis. In 1930 there were 96 metropolitan districts containing 44·6 per cent. of the total population in the United States. Between 1920 and 1930 these districts increased by 28·3 per cent. as compared with 16·1 per cent. for the whole country. But growth inside the metropolitan districts varied. Their central cities decreased from 72·4 per cent. in 1920 to 69·1 per cent. in 1930 of the total population, whereas the peripheral areas outside the central cities increased in proportion from 27·6 in 1920 to 32·9 in 1930. The central cities grew by 22·3 per cent., while the outside areas grew by 44 per cent. Thus, though the central cities still increased 40 per cent. more rapidly than the country as a whole, a larger proportion of the dwellers in the metropolitan districts was living in the outside areas. This trend has been much more marked in the 1930–40 decade.

> The centripetal movement of population in cities, which was overpowering until rather recently, is losing some of its force and is dropping some of the people who would formerly have been carried into the central cities outside their peripheries. This new direction in distribution cannot yet be called an actual centrifugal movement but it does suggest that such a movement is possible, that the automobile, good roads, the wide distribution of electricity, the ubiquity of the telephone, and the use of the radio are beginning to exert some influence on the distribution of population. Whether this movement will attain the proportions sufficiently large to amount to an actual decentralization, or whether it will merely mitigate some of the more obvious evils of continued concentration cannot now be told.[1]

Eighty-five great cities with over 100,000 inhabitants in 1940 are listed by Jefferson.[2] Of these, New York was the giant with 8,289,000 inhabitants, followed by Chicago with

[1] *Population Statistics, 3. Urban Data*, National Resources Committee, October, 1937, pp. 7–9.

[2] M. Jefferson, " The Great Cities of the United States, 1940 ", *Geographical Review*, Vol. XXXI, 1941, pp. 479–87. These include 18 " composite cities " each consisting of two or more political cities.

3,863,000. Philadelphia, Detroit, Los Angeles and Boston each had between one and two million inhabitants. Five cities had between 750,000 and a million inhabitants,[1] twenty-five had between 750,000 and 250,000, and 49 had from 250,000 to 100,000 inhabitants.

3. The Development of the Metropolitan City

The main phases in the economic development of the United States are clearly reflected in the growth of its great cities. In the pre-railway era, until the middle of the nineteenth century, settlement was confined for the most part to the Atlantic coastal belt east of the Alleghany Mountains, with sporadic pioneer settlement beyond these mountains along the river courses. Settlement was mainly rural. Towns were small service centres and the chief were situated on the Atlantic coast. The second period of settlement corresponds with the expansion of the railway and reaches from about 1850 to 1900. Settlement spread westwards across the Continent. The United States was primarily an exporter of foodstuffs—amounting to two-thirds of its exports in 1900. Cities grew as centres for the collection of agricultural products and as distributors of supplies to the farmer brought in from manufacturing centres on the Atlantic seaboard. The principal cities were located at the crossways of railway and water transport. They all developed as railway centres.

Towards the close of the nineteenth century three changes mark the emergence of the modern United States—the development of the manufacturing industries, the great influx of European immigrants and the advent of the internal combustion engine and the automobile. Industries spread both at the seats of natural resources and also in the great cities, which drew the resources to them so as to supply their market areas with the manufactured goods which in the earlier phase they had imported from the east. The advent of the automobile has brought about a much greater mobility and diffusion of both goods and persons than was possible on the railway, since the car has the whole of the road-net at its disposal, whereas the railway is a definite track to which movement is confined. Whereas mainly under the influence of railway transportation there was

[1] St. Louis, Cleveland, Baltimore, Minneapolis, and Pittsburgh.

an ever-increasing concentration of population in cities, the automobile has permitted the outward expansion of the city and has enabled it to establish much more intimate relations with the surrounding towns, villages and farms, so much so that the city or metropolitan community, extending far beyond the limits of the city as an administrative unit, is now a new and important fact in the framework of American society.

By reducing the scale of local distance, the motor vehicle extended the horizon of the community and introduced a territorial division of labour among local institutions and neighbouring centres which is unique in the history of settlement. The large centre has been able to extend the radius of its influence ; its population and many of its institutions, freed from the dominance of rail transportation, have become widely dispersed throughout surrounding territory. Moreover, formerly independent towns and villages and also rural territory have become part of this enlarged city complex. This new type of supercommunity organized around a dominant focal point and comprising a multiple of differentiated centres of activity differs from the metropolitanism established by rail transportation in the complexity of its institutional division of labour and the inability to get population. Its territorial scope is defined in terms of motor transportation and competition with other regions. Nor is this new type of metropolitan community confined to the great centres. It has become the communal unit of local relations throughout the entire nation. Its development has introduced a vast amount of rearrangement of populations and institutions, a process which is still far from having attained an equilibrium.[1]

It is in this final phase of metropolitan integration that the metropolitan city becomes not only a commercial and industrial centre, but also a cultural and financial centre, with an ever-increasing measure of independence of other cities as the source of services and leadership in culture and commerce for the surrounding territory which regards it as its centre.

Throughout the nineteenth century the United States was primarily a country of agricultural workers and the world's greatest producer and exporter of foodstuffs. The turn of the century witnessed the increasing growth of trade centres as seats of commerce for the distribution of farm requirements and the collection and distribution of farm products, but in ever-increasing measure also as seats of industry—both of producers' goods and consumers' goods. The last stage in this growth has been the emergence of financial centres, independent of New York, the national economic metropolis. The great commercial centres

[1] R. D. McKenzie, *The Metropolitan Community*, 1933, pp. 6–7.

of the nineteenth century have become " million " cities, seats of commerce, industry, finance and culture. Within the span of fifty years we find each of these great cities attaining metropolitan proportions, first as collectors and distributors of products drawn from their hinterlands, then as producers of those very goods which hitherto they had imported from distant producing areas through larger cities ; and finally, as independent centres of finance and culture. There is no mistaking about a dozen of these cities. There are, however, marked variations in their functions, depending upon the character of their hinterlands—whether concerned with agricultural products, with industry, the production of raw materials, such as oil, or a mixture of both. But all have the same essential characteristics. They are economic capitals for their wide tributary areas, for which they are also the leaders in economy and culture, interests and aspirations.

The idea of the economic metropolis and the stages in its development have come from the United States, for in that country, owing to its great size and the absence of tariff barriers and its rapid settlement in the last hundred years, the stages in the development of the metropolis can be traced over a few generations, the commercial or wholesaling phase, the industrial phase, the transport phase and the financial phase. In this development there has been acute rivalry between neighbouring cities—New York, New Orleans and Montreal in the early nineteenth century ; Chicago and St. Louis as rivals for the leadership of the Middle West ; the Twin Cities (Minneapolis and St. Paul) and Kansas City, with the opening of the grain and cattle-raising lands of the western states in the 'nineties and after ; and San Francisco, Los Angeles and Seattle in the Pacific West. But the fiercest competition has been between the cities on the Atlantic seaboard, old cities clustered in its north-eastern corner, near to each other, backed by closely settled hinterlands, and serving as the potential outlets of the whole of the Middle West—New York, Boston, Baltimore and Philadelphia.

> By every peaceful means possible each of these cities has endeavoured to outdo its rivals, by constructing highways, canals and railroads, . . . by establishing transatlantic lines for freight and passengers, and by getting as low land and water rates from and to its ports as could be secured. They have been rivals for grain, cattle, coal and general merchandise, and the passenger trade.[1]

[1] N. S. B. Gras, *Introduction to Economic History*, pp. 286–7.

In examining the extent of their tributary regions one sees " a see-saw of unending struggle, not marked by political elections, or military engagements, but by advertising, the circulation of newspapers, the activities of commercial travellers, the struggles of boards of trade, rate wars, and the migration of workers and business men ".[1] Gras, on the basis of his definition of a fully-fledged metropolis, considers that there are eleven metropolitan cities in the United States—New York, Chicago, St. Louis, the Twin Cities (Minneapolis and St. Paul), Kansas City, San Francisco, Baltimore, Boston, Cleveland and Cincinnati, the last being a doubtful choice, since it has been eclipsed by its competitors. In one city after another in the heart of the Middle West and Far West in particular, one can trace over a hundred years the stages in the development of each metropolitan city, which have been described above. Each began as a commercial centre, then added manufacturing to its activities, and finally emerged, in varying degree, as a dominant seat of finance.

4. Metropolitan Cities of the United States

In order to select from the United States census list those cities that are metropolitan, several criteria have been chosen. The first is the value of sales *per capita* of manufactured, wholesale, and retail goods for cities with over 50,000 inhabitants [2] (Fig. 40). Retail sales (*per capita*) do not show a very wide range, differences between cities depending on the areal extent and purchasing power of the market and the competition of other centres. The other two, however, range widely. The principal wholesale cities (over $1,000 sales *per capita*), with the exception of several specialized markets (mainly cotton and live stock), are all cities of 100,000 people. There are 37 of them and they contain a little under one-quarter of the total population of the nation and two-thirds of the wholesale trade by value.

A large number of cities are low in wholesaling (under $500 *per capita*), and many of them are also low in retailing. Cities of this category are grouped around the metropolitan

[1] N. S. B. Gras, op. cit., p. 298.

[2] Fifteenth Census of the United States, 1930 : Census of Manufactures and Census of Distribution (Wholesale and Retail Trades). Wholesale trade statistics are taken from the State reports of the Census of Wholesale Trades.

cities, especially in the north-central states. They are typically manufacturing communities. All cities of this kind are themselves tributary to metropolitan cities and may be excluded from further consideration ; examples are Akron, Canton, South Bend, Lowell, Lawrence.

In Fig. 41 the classification combines manufacturing and wholesale sales. Grouped in the north-east are the largest manufacturing and wholesale centres (over $1,000 sales *per capita*). Three of these—New York, Boston and Minneapolis—

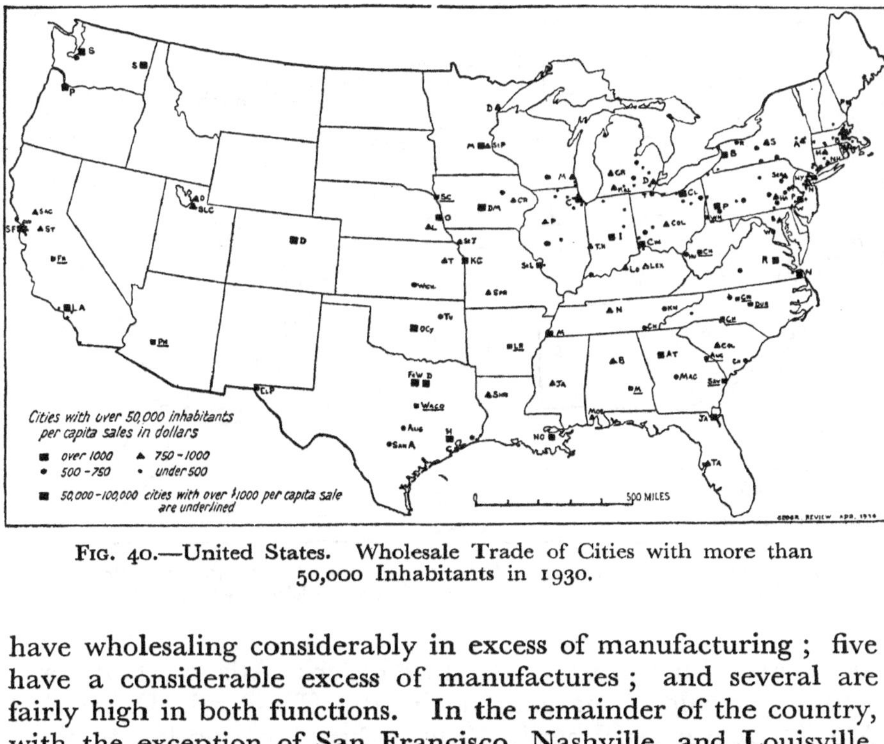

FIG. 40.—United States. Wholesale Trade of Cities with more than 50,000 Inhabitants in 1930.

have wholesaling considerably in excess of manufacturing ; five have a considerable excess of manufactures ; and several are fairly high in both functions. In the remainder of the country, with the exception of San Francisco, Nashville, and Louisville, all the large cities are high in wholesaling and low in manufacturing sales.

Other criteria will broaden the basis of and strengthen this functional classification. Thus there is the distribution of merchandizing warehouse space. It has been calculated that 90 per cent. of the total is concentrated in 26 cities.[1] Again,

[1] E. A. Duddy, " Regional Distribution of Public Merchandise Warehouse Space in the United States ", *Distribution and Warehousing*, 1928.

we may consider the sales and distribution effected through branch offices, warehouses, and manufacturing plants ; 20 large national concerns [1] have been selected and the location of their district branches noted and totalled. The location of Federal Reserve banks and branches and bank clearings has also been studied.[2] Several of the 37 cities referred to above as the principal wholesale cities do not function in any of these three capacities as metropolitan centres.

From these data it has been concluded that the following

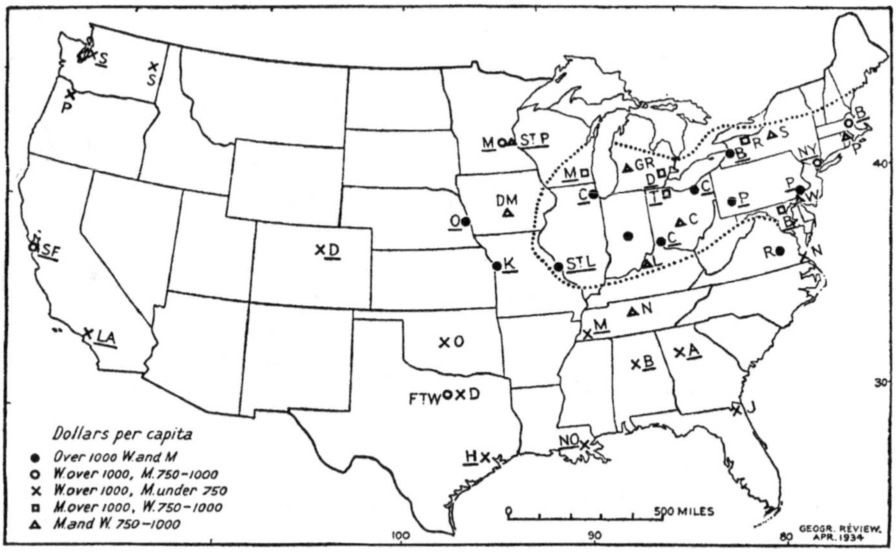

FIG. 41.—United States Wholesale Sales and Value of Manufactured Products of Cities in 1930. The dotted line encloses the Manufacturing Belt.

are metropolitan cities of the first order : Atlanta, Baltimore, Boston, Buffalo, Chicago, Cincinnati, Cleveland, Dallas-Fort Worth, Denver (?), Detroit, Kansas City, Los Angeles, Minneapolis—St. Paul, New York, Philadelphia, Pittsburgh, St. Louis, and Seattle. Metropolitan cities of the second order are : Birmingham, Jacksonville, Omaha, Richmond, Houston, Memphis, Portland, Salt Lake City, Louisville, Milwaukee, New Orleans and Spokane.

[1] *Thomas's Register of American Manufacturers*, New York. [Published annually.]

[2] Location of Federal Reserve Districts in United States, *63rd Congr., 2nd Sess., Senate Doc. 485*, 1914 ; see Gras, op. cit., p. 296, for map.

5. The Metropolitan Regions: Special Functions

The commercial functions of a metropolitan city are collection and distribution. It assembles and redistributes the primary and manufactured products of the region, and the character of its activities varies according to the character of its region. The products of distribution, however, are more standardized, except for the supply of materials to industries in tributary towns ; for example, cotton at Boston, metals and minerals at Cleveland, Pittsburgh and Detroit. With commodity distribution must be included all types of services, of which the most important is the circulation of newspapers and printed matter : printing and publishing is, indeed, one of the most important and distinctive industries of the metropolitan city. All these distributed commodities are either imported to the city or manufactured in it. Consequently a full-fledged metropolitan city is characterized by the great diversity of its regional industries and a growing metropolitan city by the rapid increase of such industries, as it strives to assert its independence of other centres.

The degree of concentration of the total commodity movement in a metropolis and the extent of the region within which the movement to and from it takes place depend on the location and site of the metropolis, on the degree of development of both the metropolis and the region, and on transport facilities, particularly freight-rate structure.

Of various service-area maps used in determining the boundaries of the metropolitan regions, Figs. 42 to 45 have been selected for reproduction. Fig. 42 shows wholesale trade areas. Many wholesale and retail trade areas have been defined for the country with a view to forming a framework and a sound economic basis for the distribution and sale of wholesale products. The map is based upon a grouping of several hundred jobbing and retail trade areas in such a way that each area is focused on one central warehousing centre.[1]

Newspaper circulation has been demonstrated to be a delicate indicator of the extent and potency of urban influence.[2] Cir-

[1] P. W. Stewart, *Market Data Handbook of United States*, U.S. Bur. of Foreign and Domestic Commerce, Domestic Commerce Ser. No. 30, 1929 ; J. W. Millard, *Atlas of Wholesale Grocery Territories*, ibid., No. 7, 1927.

[2] R. E. Park, " Urbanization as measured by Newspaper Circulation ", *Amer. Journ. of Sociology*, Vol. XXXV, 1929, pp. 60–79.

culation can be accurately determined from the records of the Audit Bureau of Circulation. Fig. 43 has been prepared on

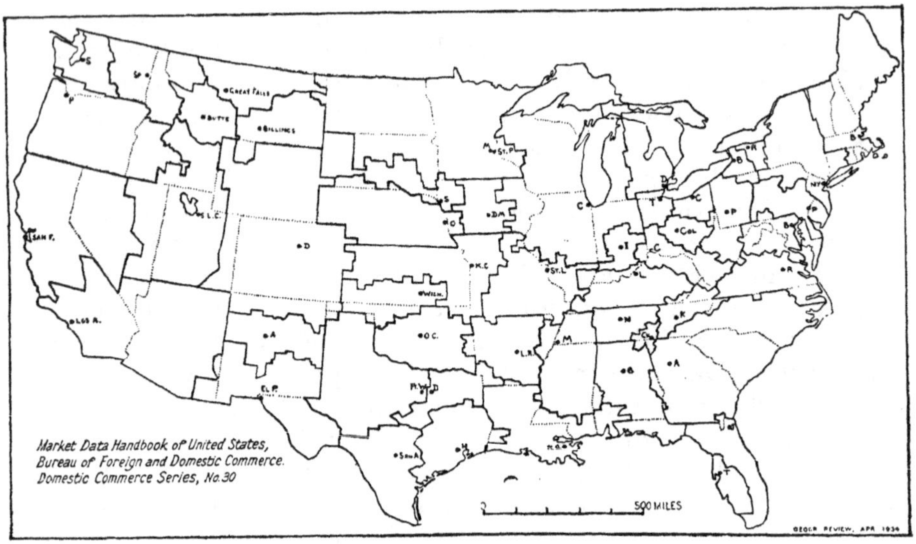

FIG. 42.—United States. Wholesale Trade Areas.

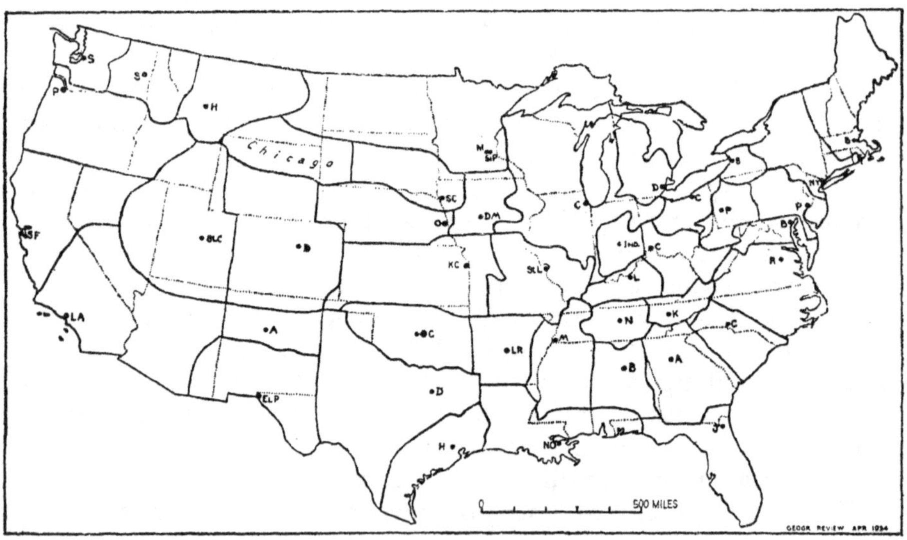

FIG. 43.—United States. Newspaper Circulation Areas.

this basis by the Social Science Research Committee at the University of Chicago.

With regard to the marketing of agricultural products, the chief livestock markets of the country fall into four groups. Salt Lake City, Ogden and Denver collect range sheep and cattle and ship them east for fattening ; the High Plains and Middle-Western group, the chief markets, are the slaughtering and packing centres ; Buffalo, Cleveland, and Pittsburgh have considerable receipts of all kinds of stock,[1] serving an area in which stock are bred and fattened and to which only small supplies come in from the West ; and the Atlantic-seaboard

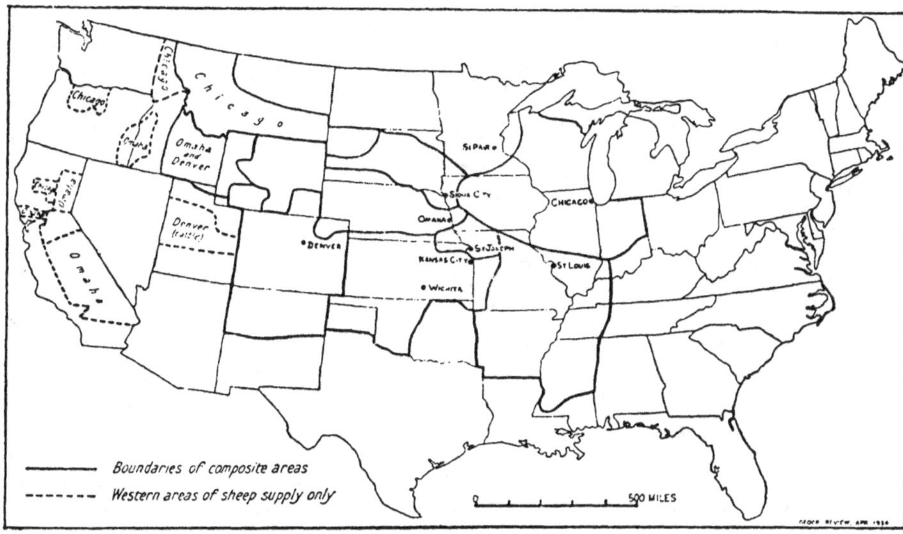

FIG. 44.—United States. Composite Live-stock areas of nine Mid-Western Markets.

These areas are based upon car load shipments on a County basis for 1927, each County shipping 50 per cent. or more of its total rail shipments to one of the nine markets.

markets, where, in proportion to population, stock receipts are small, and the large cities are principally centres for warehouse distribution by the large meat-packing companies. The South and the Pacific seaboard depend on local supplies and local distribution by the meat-packing companies.

The composite areas of livestock supply of nine Mid-Western markets are shown on Fig. 44, generalized from a series of three

[1] " Sources of Cattle Receipts, Pt. I ", *Monthly Letter to Husbandmen*, Armour's Livestock Bureau, Chicago, September 1931, pp. 8–9.

maps—for cattle, hogs and sheep.[1] The three areas closely correspond, although large supplies of store sheep are drawn from the Far West. In the northern states of the intermontane belt supplies are sent to several markets. Western Montana, indeed, has its chief market in Chicago.

The great bulk of grain shipments to primary markets is focused on Minneapolis and Chicago, and data indicate the close correspondence of the grain and the live-stock supply

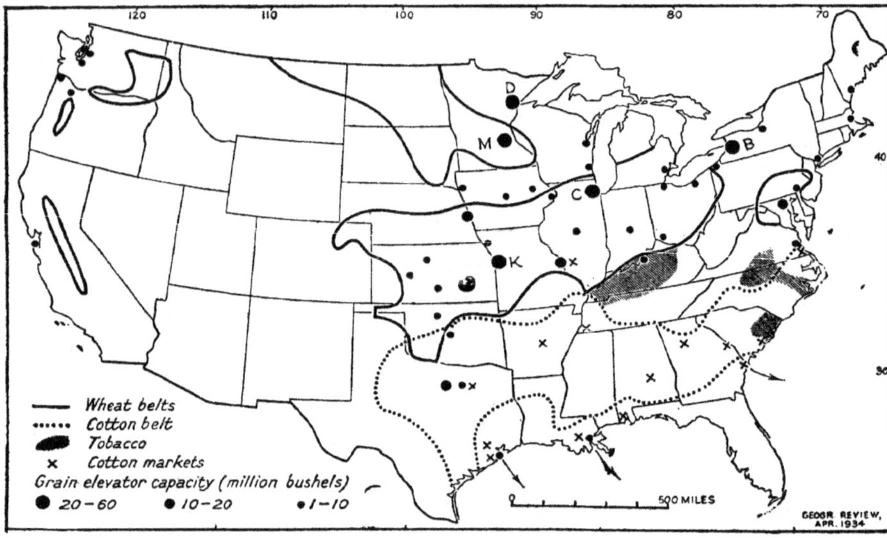

FIG. 45.—United States. Some Agricultural Marketing Relations.

areas.[2] With Kansas City, St. Louis, Duluth, Milwaukee, and Omaha, these markets handle over 90 per cent. of all grain passing through primary markets. Grain passing through secondary markets is handled mainly by Buffalo (40 per cent.) and the Atlantic-seaboard group—New York, Baltimore, Philadelphia, and Boston (55 per cent.) (Fig. 45).

[1] E. A. Duddy and D. A. Revzan, *The Supply Area of the Chicago Livestock Market*, Univ. of Chicago, School of Commerce and Administration, Studies in Business Administration, Vol. II, No. 1, 1931, maps on pp. 16, 26 and 31; E. A. Duddy, "The Potential Supply Area of the Chicago Livestock Market", *Journ. of Farm Economics*, Vol. XIII, 1931, pp. 410–25.

[2] "Analysis of Sources of Grain Receipts (1912–13 to 1916–17), on State Basis", in *Report of the Federal Trade Commission on the Grain Trade*, Vol. II, Washington, 1920, pp. 40–1; also M. L. Hartsough, *The Twin Cities as a Metropolitan Market*, Research Publs. Univ. of Minnesota: Studies in the Soc. Sciences No. 18, 1925, and information from Duddy on grain movement on county basis to midwestern and Pacific terminals.

See also "The Grain Elevator Capacity of the United States and Its Regional Distribution", by Duddy, *Amer. Elevator and Grain Trade*, 1928.

Attention should be drawn to the competitive territory between the markets of the Middle West and those of the Pacific seaboard as determined by the potential movement of hogs and grain.[1] This zone, the location of which is the expression of market prices and freight rates, is an important economic divide. It follows closely the belt between the 100th and the 105th meridians.

The marketing of cotton in the southern states is directed to a number of inland markets. Supplies for export are sent from these markets, or direct from country points, to the ports of Houston–Galveston, New Orleans, Savannah, and Norfolk, or they are sent by rail direct to the domestic manufacturing centres.[2]

6. Characteristics of the Regions and Cities: Centres of the Manufacturing Belt

On the basis of this selection of metropolitan centres and the maps of areas served by the chief cities in various functional capacities, a map (Fig. 46) has been prepared to show the composite metropolitan regions of the United States. The remainder of this chapter will be concerned with a correlation of these regions with other types of economic regions, agricultural and industrial (see Figs. 68 to 71), and, in relation thereto, the distinctive industrial and commercial activities of the metropolitan cities will be described and explained.

The main manufacturing belt, in the north-eastern States,[3] is the area of maturest metropolitan development, in which the metropolitan centres are the commercial and cultural foci for tributary industrial towns. The remainder of the country receives the bulk of its manufactured products for general distribution from the manufacturing belt. Focal distributing points are so located that they are conveniently accessible to the manufacturing belt with hauls reduced to a minimum, and to their extensive tributary service areas; hence the central location of the southern metropolitan centres and the seaboard

[1] E. A. Duddy and D. A. Revzan, "The Potential Supply Areas of Pacific Coast Markets for Hogs", *Journ. of Farm Economics*, Vol. XIV, 1932, pp. 586–98.

[2] The hinterlands of the ports as given in the *Port Series* prepared by the Corps of Engineers of the U.S. War Department have also been taken into consideration.

[3] Sten de Geer, "The American Manufacturing Belt", *Geografiska Annaler*, Vol. IX, 1927, pp. 233–359. Compare the grouping here suggested with Jefferson's regional grouping of population, "Some Considerations of the Geographical Provinces of the United States", *Annals Assn. of Geogrs.*, Vol. VII, 1917, pp. 3–15.

location of the western distributing centres receiving supplies mainly via the Panama Canal. On the Atlantic seaboard Boston, New York and Philadelphia have central wholesale markets, which distribute food products and raw materials. Each region and city has its distinctive industries, and each has the characteristics of a full-fledged metropolitan centre. Yet the similarity of urban pattern and function, freight-rate structure, and the dominance of New York give to this Atlantic group a certain measure of unity.[1]

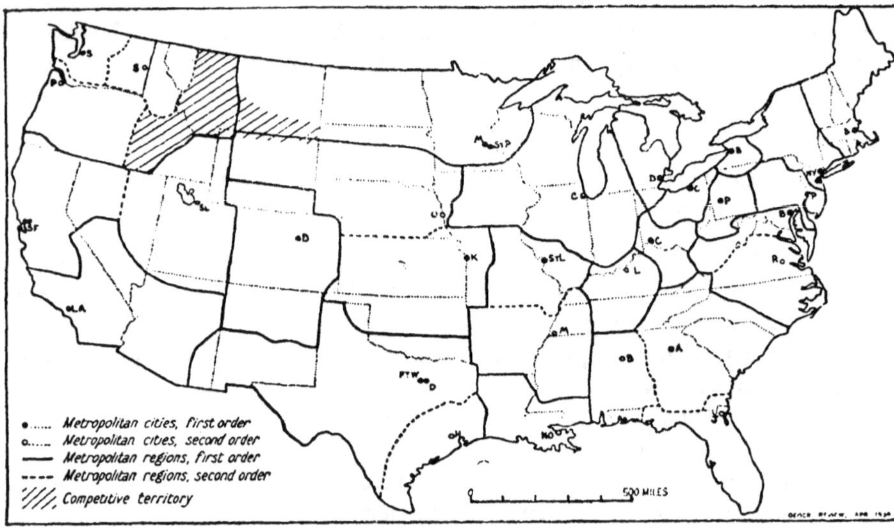

FIG. 46.—United States. Metropolitan Regions.

The central section of the manufacturing belt consists of a large number of industrial towns, which fall into four groups, each with its distinctive industries and focused upon a metropolitan city—Detroit, Cleveland, Pittsburgh or Buffalo. In each of these cities metals and minerals form the chief item (20 per cent.) of the wholesale trade. The group of four regions lies, like the Atlantic-seaboard group, in the hay-and-pasture belt, and each metropolis has a considerable live-stock market serving a local region.

[1] J. F. Dewhurst, *Commercial Survey of the Philadelphia Marketing Area*, U.S. Bur. of Foreign and Domestic Commerce, Domestic Commerce Ser. No. 1, 1925; E. F. Gerish, *Commercial Structure of New England*, ibid., No. 26, 1929: *The Port of New York*, U.S. War Dept. Corps of Engineers and U.S. Shipping Board Port Ser. No. 20, Part I, Washington, 1926. In many instances the Buffalo and Pittsburgh regions are included with this Atlantic group. They definitely lie within the Atlantic-port hinterlands.

The western section of the manufacturing belt is served mainly by the metropolitan region of Chicago (Fig. 47), which also embraces the greater part of the corn-and-hog belt. This city, the economic epitome of the Middle West, combines the functions of both primary and central wholesale markets. It is a vast live-stock and grain market. It has varied industries that are typically concerned with the needs of its hinterland. The chief are slaughtering and meat-packing and the manufacture of iron and steel and machine products. But in addition there are a multitude of miscellaneous industries of which the most important are clothing, electrical supplies, chemicals, and printing

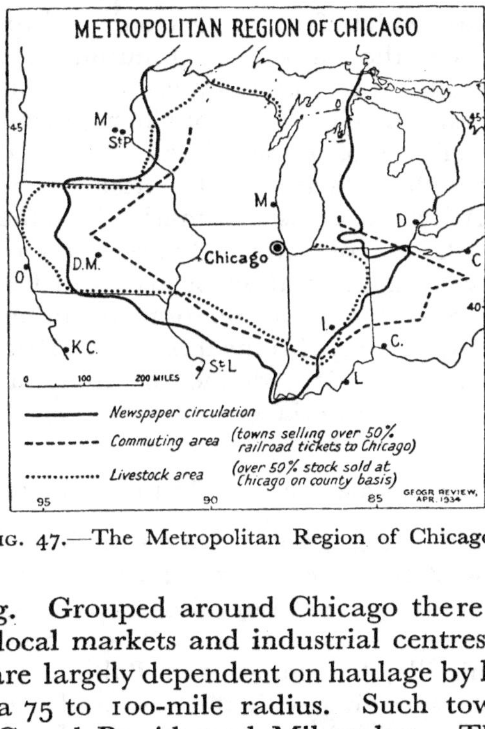

FIG. 47.—The Metropolitan Region of Chicago.

and publishing. Grouped around Chicago there are a number of specialized local markets and industrial centres. The former in particular are largely dependent on haulage by lorry and draw supplies from a 75 to 100-mile radius. Such towns are Peoria, Indianapolis, Grand Rapids and Milwaukee. The major relations of Chicago with its trade area have been examined on a statistical basis (Fig. 47). Statistics are available, for example, of the movement of livestock by wagon-loads from each county to the principal live-stock market centres, and of grain to grain elevators. 70 per cent. of the cattle handled in the Chicago market, 85 per cent. of the pigs and 35 per cent. of the sheep are drawn from the States of Illinois, Iowa, Indiana, southern

Minnesota and Wisconsin. The boundaries of each area from which over 50 per cent. of each class of live-stock transported by rail from each county goes to Chicago closely correspond, so that one can speak of a composite live-stock market area of Chicago. As a grain market Chicago is the main centre for the west of the corn belt, including the States of Illinois, Iowa, northern Indiana, southern Minnesota, with two main areas of outward consignments, northern Illinois and the north-centre of Iowa. The limits of the Federal Reserve Banking District give a broad indication of the sphere of economic influence of Chicago. This includes southern Wisconsin, Iowa, Illinois, Indiana (except their southern parts) and western Michigan (except the eastern part of the State which falls to Detroit). Two other criteria of orientation are the railway commuting zone, which has been defined on the basis of numbers of season tickets issued from stations ; and the circulation of Chicago newspapers, which have been the subject of a special statistical study as the most delicate and reliable indicator of the influence of the city upon its environs. All these areas are marked on the diagram and there is a remarkable coincidence between them. Their limits are fixed partly by consideration of access to Chicago in terms of both mileage and cost in relation to other metropolitan cities such as St. Louis, Minneapolis, St. Paul and Detroit. One can, without hesitation, define the hinterland of Chicago on the basis of these criteria.

7. Centres of the Middle South

To the south of this group of metropolitan regions there is a second east-west tier, the centres of which are in part related to the manufacturing belt but also have relations with, and primarily serve, the Middle South. These are the regions of St. Louis, Louisville, Cincinnati and Baltimore, which form a belt closely corresponding with the corn-and-winter-wheat belt, with tobacco as its chief money-crop. St. Louis, indeed, is at the junction of the hog, cattle, cotton and wheat belts—a favourable location, reflected in its marketing activities and its varied industries. Its historic relationships have been principally with the south-west, where it previously shared with Kansas City an undisputed monopoly of the distributive trades. Indeed, this in some measure still holds good, for it handles almost 40 per cent. of the hardware distribution in the south-western states.[1] Owing to the development of the Gulf Southwest, however, the principal sphere of influence of St. Louis now lies in the middle Mississippi basin.

The Louisville and Cincinnati regions serve the Ohio basin, which includes one of the major areas of tobacco production. Approximately one-fifth of the wholesale trade of Louisville is in farm products, mainly tobacco. To the east, Richmond and Baltimore have similar characteristics. The manufacture of tobacco products is the chief industry of Richmond, situated in the middle of a second major area of tobacco production. Tobacco products account for 20 per cent. of its wholesale trade and 30 per cent. of its occupied workers. Baltimore is the focus of the small Atlantic winter-wheat area. It is the only seaboard city of the Middle Atlantic Group with a grain consignment trade, and it leads in grain-elevator capacity. It has large industries, owing to its seaboard location, and a large distributing trade in Maryland and the Virginias. It serves, with Norfolk and Newport News, the West Virginian coalfield. Thus, like St. Louis and Cincinnati, it is closely linked with the manufacturing belt, while metropolitan functions for a large region to the south are shared with Richmond.

8. Centres of the Cotton Belt (The South)

The southern tier of metropolitan regions closely corresponds with the cotton belt.[2] Freight-rate structure has had an effect on urban development here very different from that of the northern states. In the latter, with the westward progress of the railway, the existing cities that had large distributing trades were able to attract the railways and become " basing points ", with special through rates, but from which, within a definite area, cumulative freight rates are operative. This rate structure has had the effect of assisting the concentration of wholesaling. In the South, however, a number of small towns became local basing points, with special through rates to the north. This has permitted the growth of small wholesaling and jobbing centres, and in addition to the special economic character of the South, has tended to prevent the dominance of any one centre.

[1] W. A. Bowers and W. L. Mitchell, Jr., *Hardware Distribution in the Gulf Southwest*, U.S. Bur. of Foreign and Domestic Commerce, Domestic Commerce Ser. No. 52, 1932.

[2] J. M. Hager, *Commercial Survey of the Southeast*, U.S. Bur. of Foreign and Domestic Commerce, Domestic Commerce Ser. No. 19, 1927.

In the South two rapidly evolving centres of metropolitan integration are Atlanta and Dallas–Fort Worth. This is indicated by their large population increase—40 per cent. for Atlanta and 60 per cent. for Dallas in the 1920–30 decade. " Regional headquarters for national distributors, central banking institutions, and large merchandizing concerns have given Atlanta a metropolitan character and a direct dependence upon the prosperity and progress of every mining, lumbering, manufacturing, and agricultural enterprise in the area." [1] Dallas also claims to have the south-western headquarters of more than 2,000 concerns of sectional or national importance, and 100 of its 750 manufacturing plants are branch plants.[2] Farm supplies (live stock and grain) form over two-thirds of the wholesale trade of Fort Worth and 45 per cent. of Dallas's (cotton). These two cities, however, share the metropolitan functions of the South-west with Houston–Galveston. Similarly, Birmingham and Jacksonville are subsidiary to Atlanta in the South-east.

Between these two regions is that of Memphis, again primarily a cotton market (55 per cent. wholesale sales), but with well-diversified activities. It is to be considered, however, as subordinate to St. Louis, for the live-stock marketing of the whole area is focused on that city, and much of its distributive trade and other economic relations is directed therefrom.

Finally, New Orleans is essentially a through-commodity port and not a centre of metropolitan integration. It has an extensive hinterland, serving the whole of the Mississippi basin. When the Federal Reserve Bank centres were being selected, New Orleans claimed as its hinterland the whole of the South, but the banks that expressed a desire to be associated with New Orleans were, however, located almost entirely in Louisiana and Mississippi. Dallas and Atlanta were therefore selected as more suitable centres.

9. Centres of the West-Centre

The Twin Cities region closely corresponds with the spring-wheat belt. In the marketing of grain the Twin Cities are partnered by Duluth. The large stock-market of St. Paul is a reflection of the size of its supply area and not of intensity of production. The region extends west to the " economic divide ",

[1] Hager, op. cit., p. 142.

[2] C. V. Wallis, *The Southwest Market and Dallas at its Geographic and Economic Center*, Dallas, Tex., 1930, pp. 43–5.

though in southern and western Montana live-stock shipments, as we have seen, are principally directed to Chicago. The distinctiveness of the Twin Cities is further reflected in the great importance of their flour milling, distributive trades, and varied light manufactures and assembly plants.

One of the first studies of the sphere of influence of a metropolitan city was Mildred Hartsough's examination of the Twin Cities.[1] Areas were plotted which were tributary to the cities in respect of grain trade, live-stock trade, wholesale (jobbing) trade and the Federal Reserve banking area. The smaller commercial and industrial centres tributary to Minneapolis and St. Paul were also indicated. These areas were shown only generally, and they have since been studied more exactly from statistical data—notably for live-stock and grain trade, the two primary functions of the Twin Cities, and newspaper circulation. While these areas do not exactly coincide, the composite map shows that the Twin Cities dominate Minnesota, north-western Wisconsin, the Dakotas and part of Montana, an area, be it noted, that has its nucleus in the great spring-wheat growing area with the addition of the semi-arid stock-raising areas on its western border towards the foot of the Rockies, and the wooded dairying area of north-western Wisconsin.

The function of the Twin Cities as a metropolitan focus is reflected in their occupational structure. Functions developed simply to serve the residents in the city itself amounted in 1919 to more than half of the total male workers. Of the remainder, the railroads employed more than a fourth—by far the largest in the group. The remaining three-fourths were employed in activities commonly found concentrated in regional rail centres—flour mills, grain elevators, stockyards and packing plants, warehouses, and jobbing houses, foundries, etc.[2]

To the south there is the Omaha–Kansas City region, closely corresponding with the western—and principal—section of the winter-wheat belt and the beef-cattle area. These two cities, with Sioux City and St. Joseph, are all located at the western extremity of the corn and hog belt. This is reflected in the overwhelming importance, in their wholesale trade, of farm products (live-stock and grain) and, in their manufactures, of slaughtering and meat-packing. Some two-thirds of the wholesale sales of Omaha are

[1] M. Hartsough, *The Twin Cities as a Metropolitan Market*, Studies in Social Sciences, Research Publications of the University of Minnesota, No. 18, 1925.

[2] See R. Hartshorne, "The Twin City District. A Unique Form of Urban Landscape", *Geographical Review*, Vol. XXII, 1932, pp. 431–42.

farm products. Kansas City, with 45 per cent. of its trade in this group, is a larger centre, with a much greater diversity of commercial and industrial activities. It is definitely metropolitan on all criteria ; hence it is selected as the principal centre.

10. Centres of the West

The western states, including the intermontane belt and the Pacific seaboard, are largely dependent on the East for manufactures. San Francisco is still the chief centre of distribution—a fact reflected in its many sales-territory organizations and in the extent of the western Federal Reserve district, for which it is the headquarters. The Panama Canal has brought the intermontane belt considerably nearer to the Pacific ports, which now compete more effectively with distributors and manufacturers from the Middle West. The historic trade monopoly of San Francisco has, however, been broken in this century by the rapid metropolitan development of Los Angeles and of the ports of the Pacific North-west.

The Pacific North-west [1] forms a distinctive geographical unit, focused on the Puget Sound area, with Seattle as the metropolitan centre but sharing its functions with Portland. Some two-thirds of the total wheat production of the whole region (Washington, Oregon, Idaho, north-western Montana) move through these two ports.[2] An increasing proportion of its stock is also being diverted to these centres from the primary markets in the Middle West. Regional distribution is effected by importing goods from San Francisco by water, or by local manufactures, which are increasing rapidly. To the east the region includes the Spokane area and western Montana, from which the movement of grain and stock is directed mainly to the Pacific North-west, while general distribution from Seattle and Portland in competition with the Middle West is rendered possible on an equal freight-rate basis.

Los Angeles did not arise in response to a demand for metropolitan integration : this came subsequently. The city has, however, rapidly acquired metropolitan characteristics and for that reason may be regarded as serving a separate region, though it is closely related to San Francisco.

[1] Edwin Bates, *Commercial Survey of the Pacific Northwest,* U.S. Bur. of Foreign and Domestic Commerce, Domestic Commerce Ser. No. 51, 1932.

[2] Ibid., p. 6, and see the volumes on Seattle, Tacoma, Portland, and other ports in the *Port Series* of the U.S. War Department Corps of Engineers.

All the Pacific ports have local hinterlands, which, when defined on a competitive freight-rate basis, closely correspond with the regions shown on the map. None of the cities has highly specialized wholesaling activities, and none is yet high in manufactures (San Francisco is the highest with $750 *per capita*). Seattle and Portland handle large quantities of lumber and wheat, while petroleum dominates the shipment trade of Los Angeles. In all cases manufactured goods brought from the East make up the bulk of the port receipts, and San Francisco has important coastwise shipments to the Pacific North-west.

Spokane, Salt Lake City–Ogden, and Denver are the metropolitan centres for the intermontane belt.[1] They are the chief centres of collection and distribution of commodities, which are purchased from Pacific or mid-western manufacturers. Manufactures are low but have increased in recent years through both the growth of local concerns and the establishment of branch plants. Each is the centre of a distinct region, and, though small cities (Spokane, 115,000, and Salt Lake City, 150,000), they certainly have metropolitan characteristics of the second order. Spokane is the undisputed focus of a region of agriculture and lumbering, the so-called " Inland Empire " ; one-third of its wholesale trade is in lumbering materials. Denver and Salt Lake City have mining interests and are also large store-stock markets and general distributing centres. Spokane is tributary to Seattle, and Salt Lake City to San Francisco. Denver is considerably larger than the other two and higher in metropolitan status. But it is in some respects subordinate to Kansas City, and its selection as a " first order " city is doubtful, though its region is very definite. The " zone of transition " [2] in Montana and Idaho has several small mining towns and draws supplies from and sends products to, both East and West.

[1] See articles on population, market area, and analysis of wholesaling and manufacturing in the chief cities of the western states in the *University of Denver Business Review*, Denver, Colorado, Vol. VII, 1931.

[2] P. 194.

CHAPTER 8

THE CITY-REGION IN ENGLAND AND WALES

1. The Growth of Urban Population

Comparative study of the regional relations of the cities of Britain must begin with particular reference to their extent and expansion as revealed by population data.

The census of 1931 revealed that 80 per cent. of the population of England and Wales and 70 per cent. of the population of Scotland lived in urban districts. There were 113 urban administrative areas having populations exceeding 50,000. Twenty-eight of these recorded increases of more than 10 per cent. Of these fifteen are in Greater London, one is a suburb of Manchester, and three are holiday resorts in the south-east of England. The remainder are small towns with 50,000–100,000 inhabitants scattered over the land—Oxford, Doncaster, Coventry, Luton, Cambridge, Wakefield, Exeter and Ipswich. Most of these, with several others where the increase has now slackened, record high increases since 1901.

Twenty-five towns decreased during the 1921–31 decade. Nine of these are in Lancashire, and in six of them the decline is a continuation of a loss in the previous decade. With the exception of Leeds and Wakefield, the Yorkshire textile towns have had almost stationary populations in the last three decades. Finally, since 1901, the cities with more than 250,000 inhabitants record increases less than the average for England and Wales, with the exception of Birmingham and Liverpool (1911–31) and Hull (1921–31). In the aggregate they increased only by 1·6 per cent. in the 1920–30 decade.

A half of the total urban population of Britain, and two-fifths of the total population, are contained in the seven great agglomerations or "conurbations".[1] Further, the 37 large towns with more than 100,000 inhabitants each, contain more than half of the total population (53·7 per cent.), and 60 per cent. of the population lived in the sixty-five urban areas with more than 50,000

[1] C. B. Fawcett, " British Conurbations in 1921 ", *Sociological Review*, Vol. XIV, 1922, pp. 111–22, and " Distribution of the Urban Population in Great Britain, 1931 ", *Geographical Journal*, Vol. LXXIX, 1932, pp. 100–16. See maps and statistical analysis accompanying the latter. For definition and criticism see above, pp. 168–9.

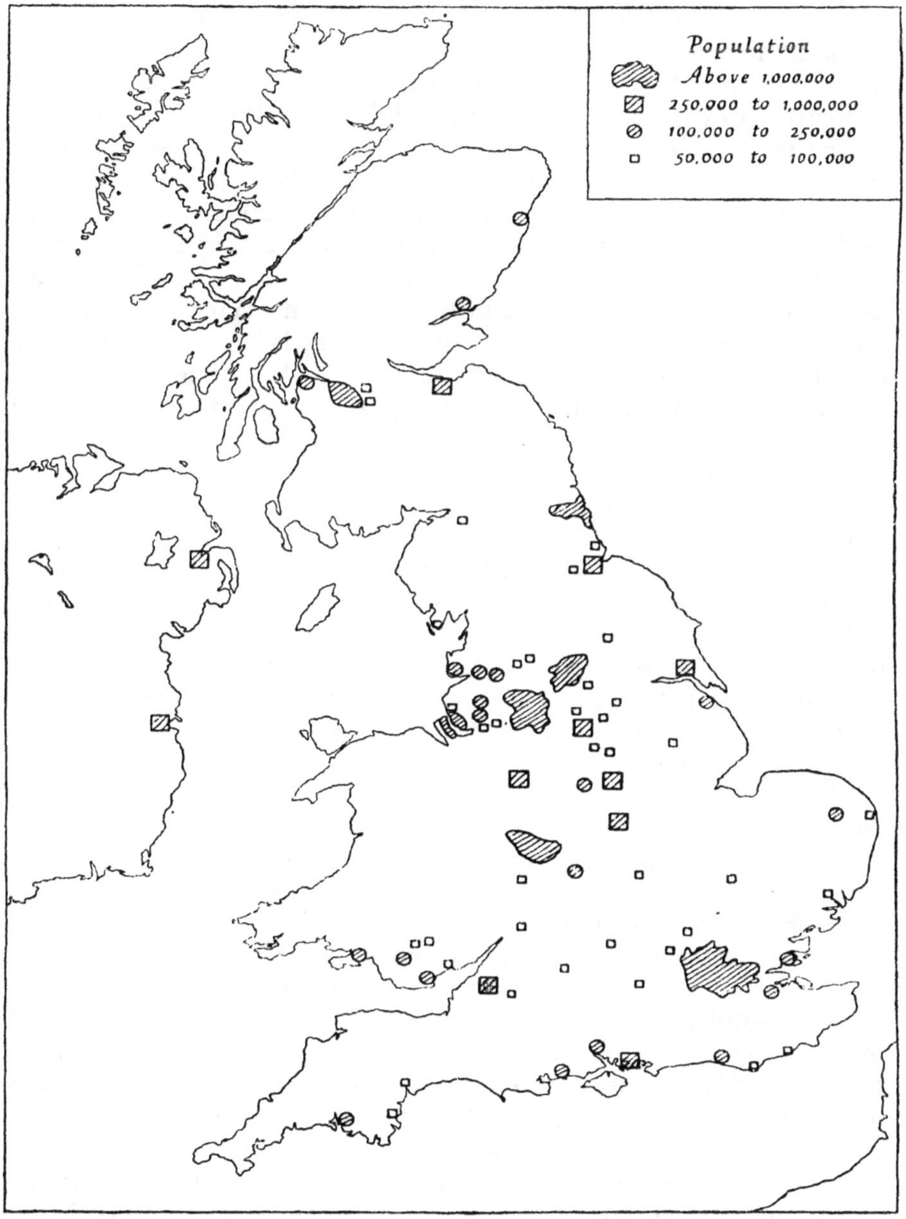

FIG. 48.—Distribution of British Conurbations, 1931 (after C. B. Fawcett).

inhabitants. Great Britain is the most fully urbanized land in the world. The seven great conurbations as defined by Fawcett are Greater London (as defined by the Census Report of 1931, 8,200,000 inhabitants), Manchester–Salford, with its ring of industrial towns (2,427,000), Birmingham or the Black Country (1,861,000), with over half the total population in the central city ; West Yorkshire, consisting of the towns of Leeds, Bradford, Halifax, and Huddersfield, the woollen textile area, with 1,433,000 inhabitants ; Glasgow (1,312,000) ; Merseyside (1,287,000), including Liverpool and adjacent areas, Birkenhead and adjacent (mainly residential) areas in the Wirral Peninsula ; and Tyneside (1,083,000), strung along both banks of the Tyne and the Wear, with its capital in Newcastle. There is a very marked drop before the next group of towns is reached, nine having over 250,000 inhabitants, though the largest, Sheffield, had only 512,000.[1]

The general characteristics of these urban tracts are as follows :

Greater London as defined for Census purposes is roughly co-extensive with the City of London and the Metropolitan Police District, although the built-up urban area extends in parts beyond these limits. The Census area and these latter districts form the Greater London conurbation, with a population of over nine millions, while, if nearby urban areas closely associated with London be included, the total population is about 10 millions.

Greater Manchester consists of a central compact built-up area formed by Manchester and Salford and six smaller urban districts with a population of over one million. Around this Inner Area, within a distance of 8 or 10 miles of Manchester Town Hall, there is a ring of urban districts shaped like a horseshoe open to the west. Built-up strips stretch along the radiating main roads linking up the Inner Area with the outlying urban districts—Bolton, Bury, Rochdale, Oldham, Stockport, and Stalybridge, etc. The whole area occupies the upper basin of the rivers Mersey and Irwell, which is encircled by hills to the north and east with its geographical focus in Manchester–Salford. The total population is about 2½ millions.

The *Birmingham* conurbation has more than half of its population in the county borough of Birmingham. It includes

[1] Sheffield (512,000), Edinburgh (439,000), Bristol (421,000), Stoke-on-Trent (340,000), Nottingham (339,000), Hull (326,000), Teesmouth (297,000), Portsmouth (296,000) and Leicester (259,000). Figures are for 1931.

Wolverhampton, Sutton Coldfield and Stourbridge. It has a total population of nearly 2 millions on about 200 square miles. It lies on the southern edge of the Staffordshire coalfield and is on relatively high ground which is drained in all directions by rivers which are linked by canals passing through Birmingham. This city is peculiar in that it does not lie on a river in a natural basin.

The *West Yorkshire* conurbation lies in the valleys of the Aire and Calder rivers and spreads across the moorland which separates them. It also lies at the northern extremity of the Yorkshire coalfield. The tract is oval-shaped, measuring 20 miles from north-east to south-west, and 12 miles from north-west to south-east. Its two chief cities lie in the Aire valley, Bradford situated in a southern cul-de-sac surrounded by high land, much of which is moorland, and Leeds at the place where the Aire Valley opens out to the plain of York. In the Calder and its tributary valleys are Huddersfield, Halifax and Dewsbury. The intervening high land is not closely built-up, but there are many smaller centres—village nuclei with factories, some bearing town character—with buildings along the main roads. The total population is about 1½ millions.

Greater Glasgow is situated in the lower part of the Clyde valley. Glasgow city itself has a million inhabitants and includes most of the urban area. There are only six other burghs adjacent to Glasgow, bringing the total population to 1·3 millions.

Merseyside lies on both sides of the Mersey estuary, that is, in south-west Lancashire and the Wirral peninsula in Cheshire. Liverpool contains over two-thirds of its population and around it lies compactly the remainder of the urban area in south-west Lancashire, of which the chief town is Bootle. The Wirral Peninsula has its main area in Birkenhead and beyond it along the estuary and behind it stretch residential areas. The total population is about 1·3 millions—almost identical with that of Greater Glasgow.

Tyneside, though the smallest conurbation, is very distinctive in every way. It lies along the lower river Tyne, below Newcastle and Gateshead, stretching to Jarrow and Shields at the mouth of the river. It may also be considered to include the Sunderland district at the mouth of the river Wear, which is only five miles distant. Out of a total of just over one million inhabitants, one-quarter is in the capital of Newcastle.

During the 1921–31 period, the population of Great Britain increased by 4·7 per cent. The seven conurbations increased by

6·5 per cent., and the total population of all the other 30 towns with more than 100,000 people by 4·4 per cent.; that of the 38 towns with 50,000 to 100,000, by 2·6 per cent. In the same period, the rural areas increased by only 1·9 per cent. The greater part of the increase was thus in the conurbations, which absorbed over half of the total increase in the population of the country. But of the total increase, Greater London accounted for over a half and Birmingham for more than a sixth. Manchester, West Yorkshire, Glasgow and Tyneside had increases well below the average for the country, while Merseyside and London increased by nearly 10 per cent.

There has, however, been taking place during the last forty years, and especially in the last two decades, a remarkable redistribution of the population outwards from the central congested city to the suburban fringe, a great part of which is included in the definition of the conurbation, though residential settlement extends well beyond the latter. This trend is evident in the impact of urban land uses and the urban way of living on the countryside. Nearly 4 million houses have been built on rural, that is, farm land, in the period 1919–39, and the great majority of these houses lie on the suburban fringes of the conurbations and other urban areas. This fact is clearly demonstrated by a map in the report of the Scott Committee showing the changes in population from 1931 to 1938.[1] In many cases, the large cities have actually decreased in population or have an increase well below the average for the country as a whole. This phenomenon has appeared in previous census reports over the last fifty years and is due to the fact that these cities have reached " building saturation point " and are now expanding rapidly outwards. The outer suburban areas, on the other hand, have continued to increase rapidly.

" In any one of our great industrial cities of the central belt there is a *decrease* of population in the centre where commercial buildings replace dwelling houses, and a huge *increase* in the surrounding fringes, where country is replaced by suburb and town." [2]

It is not generally realized that since 1901 the *natural increase* in each of the conurbations, in spite of a falling birth-rate since the 'seventies, has exceeded the actual growth of population. In other words, there has been *a net loss* by migration, the reverse

[1] Scott Report, p. 7. See also *Ground Plan of Britain*, The 1940 Council, 1942, p. 27.
[2] Scott Report, p. vi.

of the conditions before 1901 when each conurbation gained by migration. This outward movement of population is due to the expansion of residential areas and the decentralization of industry. It is directed to the peripheries of the conurbations and to small, easily accessible towns; as also to areas in the country that afford better opportunities for employment. The extent of the residential or "commuting" areas in relation to the nuclei of the conurbations, which are the principal foci of the daily ebb and flow of workers, may be obtained from the Census of Workplaces for 1921, an exceedingly valuable addition to the census of England and Wales (though not repeated in 1931). A map showing the distribution of persons working in the large cities but living elsewhere shows that the greater part of the movement is from the conurbations and contiguous rural areas within an hour's journey of the place of work, while farther afield, accessible within two hours, are the inland and coastal residential towns. It is worthy of note that in all cases, with the exception of Birmingham, the commuting areas include sea-coast towns. For example, the towns of North Wales and the coast of Lancashire are both pleasure resorts and residential centres for the well-to-do and retired Lancashire business men; and the coastal towns of south-western England bear the same relation to London.

PERCENTAGE RATE OF GROWTH OF THE SIX PRINCIPAL ENGLISH CONURBATIONS AND CHIEF CITIES[1]

Conurbations and Chief Cities	1891–1901	1901–11	1911–21	1921–31	Population in 1931 in 000's
West Yorkshire	8·2	4·1	0·7	*3·5*	1,433
Leeds	16·7	4·1	0·9	4·2	
Bradford . . .	5·3	3·1	− 0·9	2·4	
Greater Manchester	9·8	10·0	1·3	*3·2*	2,427
Manchester . . .	7·6	10·8	2·2	4·2	
Greater Birmingham	11·8	10·2	8·5	*9·9*	1,861
Birmingham . .	9·2	10·7	9·4	8·7	
Tyneside	23·5	13·7	6·4	*1·1*	1,083
Newcastle . . .	15·6	7·9	3·2	3·0	
Merseyside . . .	13·5	11·4	8·5	*5·9*	1,287
Liverpool . . .	8·8	5·9	6·6	6·3	
Greater London . .	16·8	10·2	3·2	*9·7*	8,200
London County .	7·3	− 0·3	− 0·8	− 2·0	
England and Wales .	12·2	10·9	5·0	*5·4*	40,000

[1] R. E. Dickinson, "Some Changing Features of the Distribution of Population in England and Wales", *Geographical Review*, Vol. XXI, 1931, pp. 446–65.

During the last thirty years there has been an increasing tendency for new industries of a miscellaneous kind and certain old-established industries originally located in the large cities to shift to the semi-rural areas, on and beyond the margins of the conurbations. The development has proceeded along with the concentration of commercial functions in the chief cities of the conurbations and is a reaction from the extreme urban concentration of the last century, with its inevitable burden of high land values and rates and restrictions imposed upon factory extension. It did not appear as a notable tendency until the opening of this century, and the new positive factors permitting it as opposed to the negative factors associated with the old and cramped surroundings of the cities are mainly " the increasing use of electrical power which can be transmitted over long distances with comparative economy, the recent development of motor transport, and the necessity of providing workers with more adequate housing and healthier environment." [1] As a consequence, owing to the wide distribution of the two vital necessities in the localization of industry, i.e. power and transport, given a convenient situation with regard to land, materials and markets, there is a much larger choice of site for the establishment of a new factory. For miscellaneous light industries the balance of these factors is in favour of the semi-rural areas surrounding the densely populated areas. The change is naturally most evident in Greater London, factories being established on its outskirts along the routes of the new roads and the railways, but it is also important around the large provincial cities, particularly Birmingham and Manchester. [2]

Recent industrial development, however, is not merely restricted to the areas contiguous with the conurbations. The same factors permit the location of factories in small and often historic towns settled in rural surroundings. These towns are scattered over the country, but most of them are located in southern England, south of a line from the mouth of the Severn to the Wash, and in the Midlands. Hence the increased rate of growth of many towns with 50,000–100,000 inhabitants.

[1] Scott Report, p. 25. See also *Eighth Annual Report of the Electricity Commissioners*, 1st April, 1927, to 31st of March, 1928, H.M. Stationery Office, London, 1929, p. 8.

[2] See *Annual Reports of the Chief Inspector of Factories and Workshops* from 1901 when the tendency was first noted. The Report for 1925 says " Special mention may be made of the areas surrounding London which continue to develop individually in a remarkable manner ". The Report for 1927 says " The movement of firms from the centre to the outskirts mentioned in connection with London is taking place also in Birmingham ".

2. The City Region in England and Wales :[1] General Characteristics

The character of the regional functions of towns, and of the metropolitan cities in particular, have already been discussed, with detailed reference to Leeds and Bradford (pp. 171–7). Let us now examine the regional functions of the chief English cities under pre-war conditions from the same point of view. Each of the seven conurbations, which rank among the greatest urban agglomerations in the world, has a dominant central city, which is the main focus of its activities ; but they differ from each other in certain important respects. London, Glasgow and Newcastle have sites with marked nodality, i.e. sites on which there is a convergence of important natural routeways, so that each of them has been a principal centre of activities of a large surrounding tributary area for many centuries. As a bridge-town at the head of navigation of a river, where natural land routes converged, each city has been the nucleus of urban expansion to form the conurbation and a large tributary area around it. Merseyside has a similar unity, since it is grouped around the harbour on both its banks. Birmingham is similarly a dominant unchallenged focus : with respect to its administrative area it is the largest city in Britain. Manchester–Salford, though a great industrial centre, is also a metropolitan centre of the highest order, serving the many smaller industrial towns around it. The West Yorkshire conurbation is unique. It lies astride the moorlands between the Aire and Calder valleys, in which are the towns of Leeds and Bradford, Halifax, Huddersfield and Dewsbury. It has no natural dominant focus, and lacks a tradition of unity, but Bradford is the specialized commercial focus of the woollen and worsted industries, and Leeds is its general business and administrative centre.

The chief city in each of the conurbations, as well as the other larger cities in more rural areas, making a total of about fifteen, possess the following distinctive characteristics.

(i) Each is the principal focus of a densely populated and distinctive industrial region, and has a population much greater than that of surrounding towns. The latter tend to be grouped

[1] R. E. Dickinson, " The Commercial Functions of the Nuclei of the English Conurbations ", *Sociological Review*, Vol. XXI, 1929, pp. 38–49. See also A. E. Smailes, " The Urban Hierarchy in England and Wales ", *Geography*, Vol. XXIX, 1944, pp. 41–51, discussed above in Ch. 2, pp. 47–8.

in a circle as on the rim of a wheel at the hub of which is the metropolitan centre.

(ii) Owing largely to the geographical factors of site and location, through the marketing organization of each is effected the world-wide distribution of what may be termed the primary specialized or basic industries of its surrounding area ; and the collection and distribution of vital supplies. A metropolis, however, is not necessarily an independent centre of trade, for its pre-eminence in this capacity is dependent upon adequate transport facilities, and in the case of Leeds, Bradford, Sheffield and Birmingham the trade of the surrounding area is partly oriented towards competitive metropolitan cities, or finds its outlet in a port which is complementary to and serves the same hinterland as the inland city.

(iii) Each city possesses a great number and variety of miscellaneous industries which owe their origin to the demands of the regional market and their localization in the city to its reservoir of labour and its excellent transport facilities. These are the regional secondary industries, as their products are chiefly distributed throughout the surrounding area, though business transactions are by no means restricted to it, for the growth of an industry in production and reputation results in an ever-widening area of distribution. Examples are the printing and brewing industries and the manufacture of furniture and provisions.

(iv) They are all endowed with a varying degree of financial individuality, since they have served in the past as regional centres of banking and insurance. This independence, particularly with regard to banking, they have failed to maintain, with one conspicuous exception, owing to the overwhelming financial dominance of London and the advantage acquired through representation in London.

(v) In virtue of cheap and frequent travelling facilities, each is an outstanding shopping centre—an indirect form of areal commodity distribution for an extensive surrounding area. The maximum extent of the shopping area is roughly co-extensive with the area which is within two hours' journey of the centre (Fig. 49).

It is important to emphasize here a major contrast between the metropolitan regions of the United States and Great Britain which arises from the difference in their size. Britain is small and highly urbanized, comparable in both these respects with New England in the north-eastern corner of the United States.

Its big cities are within 50 or 60 miles of each other. London, it is true, completely dominates southern England—metropolitan England as Mackinder appropriately called it—but even here the historical capitals have maintained their sway even though small in size over areas which are very distinct historical units—East Anglia with Norwich, Wessex with Salisbury, Winchester and

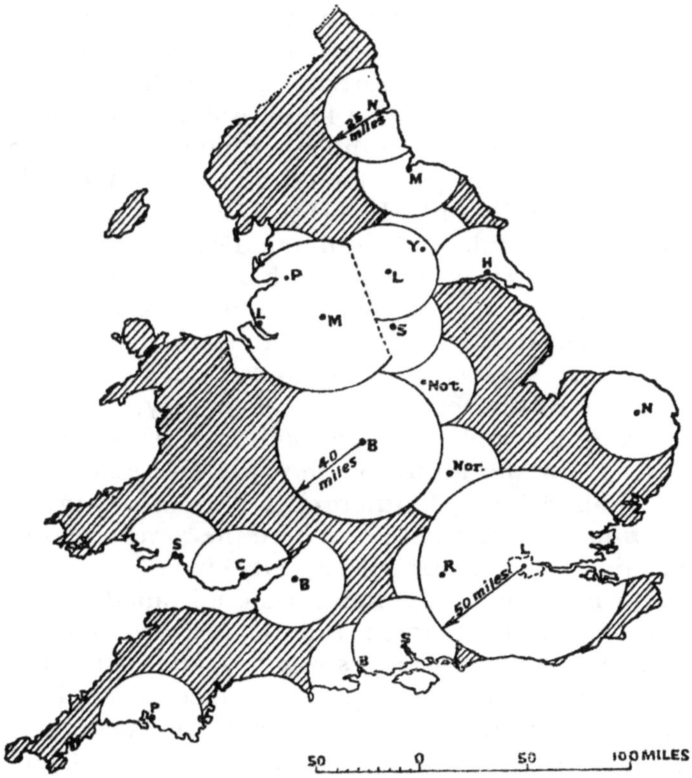

Fig. 49.—England and Wales. Distances from Great Cities (after E. G. R. Taylor).

Radii vary according to the size of the city—London 50 miles ; cities over 1 million, 40 miles ; cities over 100,000, 25 miles. Shaded areas are remote from great cities.

the modern port of Southampton, the Oxford area, the West Country centred on Bristol, the South-West with its centres in Plymouth and Exeter. In the north of England and central Scotland, however, the great urban agglomerations are close together and frequently are contiguous. Moreover, there is a deep-rooted feeling of civic independence as between one urban area and another and as between one smaller cluster and another.

These facts have two results. First, it is not possible to conceive of a clearly defined hinterland for a port or of a clearly defined marketing territory in which one city, because of its great geographical distance from competitors, holds undisputed sway. The threads of trade of any town or factory or port spread over a large part or all of the country. Secondly, the big urban agglomerations and their fringes often merge into each other, so that in theory, the midway line between two centres will be at once the limit of their " region " and of their " hinterland ". This will be the divide of suburban communities, and probably of urban associations on the one hand, and of trade associations and the movement of goods on the other hand. The Manchester region merges, in this respect, for example, into West Yorkshire (Leeds and Bradford) and Sheffield, and while on the west it merges with Merseyside. Again, the Potteries is an old and independent economic unit on the border of the influence of Manchester and Birmingham and has close relations with both.[1]

In the United States, on the other hand, distances are very great—we must reckon in hundreds, not tens of miles geographical contrasts between types of country, their people and products are great, and the metropolitan cities are very widely spaced, except for the group on the Atlantic seaboard (Boston, New York, Philadelphia, Baltimore). These form, in effect, one almost continuous urban belt, and here the same conditions hold as in Britain. The metropolitan city in the United States therefore, far more than in Britain, has undisputed sway over the market around it for a large radius—a sway enjoyed by its wholesalers, its manufacturers, its retailers, and its businesss men. (This is a most important geographical fact that is helped also by the system of freight rates which tends to concentrate manufacturing and wholesaling in selected cities.) There is in consequence a much clearer defined trade area around the metropolitan city in the United States than in Britain and, further, the trade area is much more extensive than in Britain, so that there is a wide geographical difference in the United States between the extensive trade area and the limited city settlement area, which latter is not appreciably different in structure and extent from the British city-region, since it is conditioned by precisely the same forces. The fact of metropolitan orientation in America is something of which the country farmer or townsman is conscious—although he will not know it by this name—just as

[1] See footnote 1 on p. 242.

he recognizes a farming area such as the corn belt, cotton belt, or spring wheat belt.

The metropolitan region, also, in respect of the capital and its tributary area, has more clear-cut traits in France and Germany than in Britain, and this is due also in part to the greater distances, and the concentration of population in a relatively few historical capitals that have grown to modern greatness as seats of industry and commerce. Breslau and Hanover, Frankfurt and Munich, Toulouse and Limoges, Grenoble and Lille are undisputed metropolitan capitals, that for centuries have been leaders of provincial life and even capitals of states, and the development of modern communications, commerce and industry has added to their leadership and has enabled them to establish a firmer grip on their surroundings and, through competition with their neighbours, to bring into stronger relief the framework of their tributary regions.

Let us now examine some of the regional functions of the British cities, remembering that our information is for the inter-war period. The metropolitan centres of the conurbations and, to a lesser degree, of the other large cities, are the principal nodal points for the distribution of produce. This is effected through three main channels, the wholesale markets, wholesale merchants, and branch depots.

3. Meat Distribution

There are about twenty wholesale dead-meat markets in England and Wales, located in those towns which contain, or serve areas which contain, a population exceeding about 500,000.[1] Those with the largest annual turnover are naturally situated in the chief cities.

The trade in imported meat, both fresh and refrigerated, is dominated by London to the extent of about 40 per cent. of the total for the United Kingdom. The meat imports of Liverpool and other ports have increased, however, in the inter-war years at the expense of London, for if an importer requires meat at a provincial market, he arranges for direct consignment, in order to save the extra rail and handling charges entailed by redistribution from London. Further, apart from London, imported meat is not stored at the ports but is delivered direct to the consuming centres,

[1] Ministry of Agriculture and Fisheries, Economic Series, No. 14, Reports on Markets and Fairs in England and Wales. Part II, *Midland Markets*, 1927, p. 16. F. J. Prewett, *The Marketing of Farm Produce*, Part I, *Livestock* (1926), Part II, *Milk* (1927), Oxford.

where cold storage accommodation is provided at convenient centres for distribution. In 1926 Smithfield Market handled over 40 per cent. of our meat imports, and it has been estimated that over 90 per cent. of its supplies are consumed in the Metropolitan and Home Counties Area.[1]

Though London and Liverpool are the giants of the imported meat trade, the smaller importers tend to serve distinctive areas.

Newcastle serves the coal and steel areas of the N. Riding of Yorkshire, Northumberland and Durham, and Cumberland ; Hull is well placed for the W. Riding and the NE. Midlands ; Manchester and Liverpool for Lancashire, Cheshire, W. Riding, Midlands and North Wales ; the Bristol Channel ports for South Wales and the Lower Severn Basin ; Southampton for the Southern Counties ; and London for the Home and Eastern Counties and the South Midlands.[2]

Meat is distributed from the ports within these general limits, but there is much overlapping.[3] From the wholesale meat markets home-produced, and to a smaller extent imported, meat is regularly distributed to the principal towns within a radius of 20–30 miles, where, by means of rapid motor transport, fresh supplies can be delivered and offered for sale by retailers on the same morning as they are purchased.

4. Fruit and Vegetable Distribution

Up to fifty years ago the provincial wholesale markets dealing in fruit and vegetables[4] were concerned with trade in local produce, obtaining all other English, continental and imported produce from Covent Garden, and to a lesser extent from the port auctions. Now they receive home-grown and continental supplies direct from the centres of production, and are rapidly increasing in importance as distributors, independent of Covent Garden. The re-consignment trade of the latter is now mainly restricted to imported fruits and expensive luxury out-of-season produce.

The wholesale trade of Covent Garden, the principal market

[1] Report of the Linlithgow Committee on the Distribution and Prices of Meat, Poultry and Eggs, Cmd. 1927 (1923), p. 84.

[2] Ministry of Agriculture and Fisheries, Economic Series, No. 6, *Report on the Trade in Refrigerated Beef, Mutton and Lamb*, 1925, p. 35.

[3] Similar factors control the distribution of imported dairy produce. London imports 50 per cent. and 70 per cent. of butter and cheese imports respectively. See Report of Imperial Economic Committee on Dairy Produce : also recommendations contained therein for the establishment of provincial markets at the ports of Liverpool or Manchester, Avonmouth and Hull.

[4] Ministry of Agriculture and Fisheries, Economic Series, No. 15, *Report on Fruit Marketing in England and Wales*, 1927. Report of the Linlithgow Committee on the Marketing of Fruit and Vegetables (1924).

in southern England, may be divided into four categories. For the distribution of expensive luxury produce there is a national market. Produce is distributed to the large provincial markets, although this trade has been decreasing *pari passu* with decentralization and provincial competition. Supplies also reach retailers and small wholesale merchants who reside within carting distance and are able to visit the markets in person. With rapid motor transport, the area served by this means has a radius of about 50 miles. In the country south of a line from the Wash to the Severn, districts with a small population are able to supply most of their own requirements locally. They only require imported and continental fruits, and these can be obtained cheaper, with better choice and often better transport facilities, from Covent Garden than from local wholesale markets. Gloucester, West Wiltshire, Dorset, and the South-West find their chief distributing centres in Bristol and Plymouth, although large quantities of imported and luxury produce are preferably obtained by local merchants and retailers from London.

In northern England, south to a line from Louth through Chesterfield and Crewe to the Welsh Coast, the chief markets with their areas of distribution are as follows :

Market	*Distributing Area*
Manchester–Liverpool :	Lancashire, part of Derby, North Staffs, Cheshire, North Wales and the Isle of Man.
Newcastle :	Northumberland, Durham and Cumberland.
Leeds :	Radius of 40 miles to North and North-West : East Riding shared with Hull. To West and South includes West Yorkshire towns, though the largest of these are local competitors.
Sheffield :	Radius of 20 miles, including Rotherham, Barnsley, Doncaster and Chesterfield.
Hull :	East Riding and North Lincoln.

In the Midlands, which area lies south of the Northern area to a line from King's Lynn, south of Northampton to Gloucester, and thence to Aberystwyth, the markets are as follows :

Market	*Distributing Area*
Birmingham :	Four West-Midland Counties : Central Wales with centres at Shrewsbury and Aberystwyth.
Nottingham :	South to Loughborough ; West to Long Eaton ; North to Chesterfield and Retford ; and East to the Coast.
Derby :	A radius of 20 miles.
Leicester :	A radius of 15 miles.

In South Wales the main distributing foci are Swansea and Cardiff. The former is of lesser importance and supplies the thinly populated western portion of South Wales ; the latter, a growing importer of overseas fruits, serves the densely populated valleys of South Wales from Newport to Swansea.

5. Milk Supplies to the Cities [1]

The conurbations are the chief consuming areas of milk, and owing to its perishable character their supplies are drawn mainly by motor transport from the surrounding rural areas within a 50-mile radius, though Newcastle and London receive large quantities by rail, from sources 150–200 miles distant. The chief consuming areas and the areas from which they draw their supplies are as follows :

Consuming Area	*Producing Area*
NE. England :	Northumberland, Durham, Westmorland, Cumberland and south-west Scotland.
SE. Lancashire and Merseyside :	Lancashire, Cheshire, Flint, Denbigh, North Shropshire, North Staffs, North and West Derbyshire.
West Yorkshire :	Yorkshire Dales, East Riding, North Lancashire, Westmorland.
South Yorkshire :	West Riding and Derbyshire.
Greater Birmingham :	Warwick, Salop, Staffs, Worcester, Gloucester, Derbyshire.
South Wales (Monmouth and Glamorgan) :	Pembroke and Caermarthen.
Greater London :	(i) Wilts, Somerset, Dorset ; (ii) Derby, Leicester, Staffs, Cheshire. (iii) Berks, Oxford, Bucks. (iv) Hants, Sussex, Norfolk, Suffolk.

6. Provision Distribution

The distribution of provisions is effected by independent wholesale merchants, branch depots, and the headquarters and branch warehouses of multiple-shop companies. The existence of port facilities has a profound effect on the nature of the business of the first group of distributors and the extent of the areas served by them. In the port of Manchester, for example, large wholesale importing houses specialize in one line of business, and distribute to the smaller wholesalers of surrounding towns for local re-distribution. On the other hand, Leeds, Sheffield and Birmingham receive supplies for re-distribution from the ports by rail. Many large provision and confectionery firms have adopted

[1] Ministry of Agriculture and Fisheries, Economic Series, No. 16, *The Fluid Milk Market in England and Wales*, by R. B. Forrester. Prewett, op. cit., Part II, *Milk*. Report of Linlithgow Committee on Milk (1924). The marketing of milk differs from other produce since it is drawn towards the consuming areas and not distributed from their centres.

a system whereby distributing depots are located at large towns throughout the country to supply the chief consuming areas. Such depots are, therefore, nearly always situated in the big cities. Motor transport is the chief means of distribution used by these firms. Their areas of regular distribution are conditioned by the fact that, first, the maximum limit of distribution will be that distance which can be conveniently covered by road in one day, an approximate radial distance of 60 miles ; and second, delivery over such distances is not regularly practised because the demand in rural areas and scattered small towns is not sufficient to require full and frequent loads. Consequently, provision distributors generally serve an area within a radius of 30 miles by road, comprising the conurbations and their contiguous suburban and rural areas. The same factors condition the distribution of fruit and vegetables and meat from the wholesale markets, and the extent of the areas drawn upon by road by the conurbations for their milk supplies (see p. 170).

7. Business Organization

A great variety of firms, in addition to those engaged in the provision trades, have also established provincial offices and depots. In all cases, the areas served vary widely according to the nature and amount of the business of each company. Such offices are usually situated in London, Birmingham, Manchester and/or Liverpool, Leeds and or Bradford, Newcastle, Cardiff and Bristol, and occasionally in Sheffield, serving south Yorkshire and part of Derbyshire and Nottinghamshire and north Lincolnshire, Plymouth serving Devon and Cornwall, Southampton for Hampshire, and Nottingham or Leicester for the East Midland counties.

The chief city in each conurbation is the centre of the economic activities of the surrounding industrial towns, and in it there are therefore located the Exchanges, the scenes of the daily activities of the business men of the conurbations and their surrounding towns. Examples are the Manchester Royal Exchange, the principal transaction centre for the cotton industries of south-east Lancashire, the Iron and Steel Exchange at Birmingham, the Bradford Exchange serving the woollen and worsted industries of the west Yorkshire conurbation, the Newcastle Commercial Exchange, primarily a coal and coke market, though with varied engineering and shipping interests, the Sheffield Exchange, the centre for the iron, steel and coal trades of south Yorkshire, Nottinghamshire and north Lincolnshire. It is obvious that the sphere of influence of a city in this capacity is to be obtained from the distribution of members of the exchange, and is to be explained in terms of the distribution of those economic activities for which the city is the business centre. There are also located in the city the registered offices of firms with their factories in the surrounding industrial districts ; and the headquarters or branch offices of industrial and trade associations established in the interests of the principal regional industries, and for the general welfare of those whom they employ. An excellent example of a trade association which has adopted a regional system is the Federation of British Industries with eleven districts in England, excluding the London City Area. The extent of each district of these trade associations is determined by the nature of its business, and one district branch should be concerned as far as possible with one group of related industries, and all parts of each district should be within easy reach of the centre.[1]

8. Banking Organization

The amalgamation movement in English banking,[2] dating from the third decade of the nineteenth century, bears ample testimony to the function of the great cities as regional financial centres. At first, joint-stock companies and private banks established in the cities extended their influence in all directions by absorbing small banks and by opening up new branches in surrounding towns, mainly to cater for the growing financial requirements of their principal industries. The amalgamation movement, however (while enhancing the financial strength and individuality of the principal cities in the middle of the nineteenth century), proceeded so far in the latter half, that to-day only four important independent banks remain, the Lancashire group with headquarters in Manchester and Liverpool. The large provincial cities, however, do possess distinctive functions which tend to offset excessive centralization in London. There are ten branches of the Bank of England doing business in the

[1] The zoning system used during the war has sought to parcel out the whole country into zones, each being served from fixed centres of distribution for certain commodities, so as to eliminate, as far as possible, long and uneconomic hauls. Reference to the zones for particular commodities shows that they often broadly correspond with the normal service areas of the principal cities described in this chapter.

[2] J. Sykes, *The Amalgamation Movement in English Banking*, London, 1926.

chief commercial centres. The organization of absorbed banks has also in some cases been maintained, and their provincial headquarters still retain their original capacity in a modified form ; and district boards have been organized in recent years, with their headquarters in large cities.

9. Insurance Organization

The insurance companies of England are mainly concentrated in London, though a reliable indication of the status of a city as a financial centre is afforded by the number of branch offices and headquarters of insurance firms which it contains. Insurance business, owing to its need for constant personal contacts and to rapidly increasing business connections throughout the country, lends itself to the organization of districts which are administered from main district branch offices. Every important insurance company exhibits this tendency, as we have seen in the case of Leeds and Bradford (p. 172). The extent of the districts controlled from the branch offices of a firm varies widely according to the nature and amount of its business, but the main factors which are generally considered in their delimitation are : (1) the amount of business of a district should not be too great for efficient and prompt attention at one office ; (2) all parts of the district must be easily accessible to its branch office to facilitate efficient control and supervision.

The cities of Newcastle, Leeds, Manchester, Liverpool, Birmingham, Bristol and Cardiff are pre-eminent among provincial cities as centres for the location of district branch offices and the headquarters of provincial companies. All the large insurance companies originally established branches in these cities, although the growth of business has often necessitated a still further step in the process of decentralization, whereby smaller districts with independent branches have been established. Thus, Bristol was originally the centre for the lower Severn and south-west counties, but now the latter have often a separate branch office at Plymouth or Exeter. Leeds was the original centre for all Yorkshire, but it has lost Cleveland to Newcastle, and south Yorkshire to Sheffield. London was, in many cases, the original headquarters for south-east England, but now many firms have district branches in Norwich and Southampton, the London district being restricted to the Home Counties.

10. Conclusion

From the foregoing survey it will be clear that around the chief city of each conurbation there is an extensive tributary region, the extent of which is closely associated with the distribution of the principal related industries for which the city is the commercial centre ; and the degree of accessibility of the centre to the surrounding towns and countryside, both by road and rail. Moreover, the nature and potency of the influence exerted by a metropolitan city are affected by the peculiarities of its location ; the competition of other cities which serve the same region, e.g. Leeds and Bradford, Manchester and Liverpool ; the overwhelming financial dominance of London ; and the slow progress of commercial decentralization.

These regions of city orientation do not exist as water-tight compartments. There is much overlapping although it is possible to define large areas within which the influence of one city is dominant. For example, as we have already noticed, the west Yorkshire and south Yorkshire woollen textile and iron and steel areas have their respective centres in Leeds–Bradford and Sheffield. On the western slopes of the Pennines, the compact industrial area of south-east Lancashire has its focus in the city and port of Manchester, which has attained a much more advanced stage of metropolitan development, so that its influence, along with that of its partner, Liverpool, penetrates into the West Riding. Thus, several banks with their original headquarters in the Yorkshire centres have been absorbed by Manchester firms ; the West Riding lies in the hinterland of the Lancashire ports ; and several Manchester retail firms have branches in Leeds and neighbouring towns.

It may be noted that North Wales lies definitely within the sphere of influence of Manchester–Liverpool, and central Wales has close relations with Birmingham and the Midlands. The Potteries district is very divided in its allegiances to Manchester and Birmingham.[1] In south-eastern England, on the other hand,

[1] Lacking an effective capital of its own, the North Staffordshire Potteries is oriented towards Manchester and Birmingham. " The proximity of Manchester to the north and of Birmingham to the south even suggested the superfluity of such a capital, a point which national administration in its delimitation of local government regions seemed to underline, and north-west Staffordshire would seem to be falling more and more into the position of a satellite to the West Midland capital, Birmingham. Local feelings now have been subordinated to a truer regionalism, though Newcastle and Stoke have still to work out its full significance for themselves, and also jointly, the relation of this regionalism itself to the larger provincial loyalty." A. H. Morgan, " Regional Consciousness in the North Staffordshire Potteries ", *Geography*, Vol. XXVII, 1942, pp. 95–102.

the regional capitals are not independent foci. Metropolitan England, as its name indicates, is dominated by the influence of London. Here again, however, for many commercial purposes, the south-western counties have their separate headquarters in Bristol, and sometimes in Plymouth, and the eastern counties have separate headquarters in Norwich and Cambridge.

PART IV
REGIONALISM AND THE REGION

CHAPTER 9
THE CASE FOR THE REGION

So far we have discussed the role of nucleated settlements village, town and city—as regional centres. We now come to consider the major social groupings or regions within the State and the various ways in which these find expression. The metropolitan or city region is the most potent single force in the formation of these modern groupings, but other conditions—historic, cultural and economic—also contribute to their characterization. Our concern is not with the constitutional intricacies of the regional movement, but with a discussion of the characteristics of the major social groupings inherent in society, for it is upon these that any reorganization of new political divisions must be based. After a general discussion of Regionalism, we shall pass to a brief review of its particular aspects in France, England and Wales, and the United States.

1. The Region as an Intermediate Area between Local Government Unit and State

The prominence accorded by the public to the idea of the Region is the spontaneous expression of an urgent need in the life and organization of modern society. The great mobility and the complex structure of modern society in Europe and America have meant that new areas of organization are needed for all aspects of national life, and that existing local government areas have been outmoded by wider areal organizations, and act as deterrents to the efficient functioning of public services. The idea of the Region has also developed in relation to the movement for the decentralization of authority from the central national government to a limited number of provinces, which would relieve the central government of its too onerous responsibilities, foster the development of local responsibility in the truest democratic tradition, and foster provincial and regional differences of tradition and culture. Regions, in the sense of more or less arbitrary groups of contiguous local government authorities, have been established since 1918 in all the countries with which we are concerned, as a necessary extension of the planning of individual towns. This is probably the widest use of the term Regional Planning in our own country. There is also a demand for the delimitation of major homogeneous divisions of the State, to serve as a framework for the long-term development of national resources, and for physical and economic planning. This is the type of region which seems to be the ultimate goal of Regional Planning as the counterpart of National Planning. It has figured prominently in Germany and the United States in the last ten years, and in Britain especially during the 1939–45 war. In all countries the problem is to demarcate a few areas throughout the country larger than, although built on the foundations of, the existing political divisions. It is obvious that for many purposes the *County* in Britain, the *Land* in Germany, the *Département* in France and the *State* in the United States are, in each case, either too small, or too unrelated to the practical needs of modern life, for large-scale planning organized on a nation-wide scale.

2. Cultural and Political Regionalism

The problem of Regionalism in its broadest European aspect was admirably summed up by Patrick Geddes and Victor Branford [1] at the close of World War I:

> The industry and the politics of the nineteenth century progressed alike through extension and unification. Larger industries, further reaching and swifter transports, wider markets, became naturally also associated over greater areas, and these into larger unities of administration and government. Railways and telegraphs, steam-routes and cables at once enlarged industrial towns to world markets, and aggrandized their metropolitan cities into imperial capitals. Practical life and political endeavour were thus at one ; hence the expansion of England ; the centralization of France, the American War of Union, the unification of Germany, and even of Italy, are now seen as kindred processes.
>
> Exceptions to these processes, even dissents from them, were noticeable. But these were simply explained, in terms of limitation or backwardness, e.g. geographical for Switzerland, linguistic for Hungary, sentimental for Alsace and Lorraine, legendary for Ireland, and so on ; and thus as so many survivals, destined to disappear with

[1] Editors' Introduction to C. B. Fawcett's *The Provinces of England : A Study of Some Geographical Aspects of Devolution*, London, 1919.

progress or education, or at worst as petty self-assertions, to be repressed with such firmness as need be. The sun of Progress shone essentially from the ever-growing capitals over their extending empires, and illuminated the unification of their nationalities. under due predominance of their metropolitan types.

Yet, despite Vienna and Austria, Hungary achieved her equality in empire ; and thus diffused a more European influence and example than she knew. The separation of Norway from Sweden was a more peaceful case of this process, albeit a more extreme one : and now, after the war, we see not only the conversion of the small nations, but the rise of new ones—witness the complete break up of Austria into its units, the reunion of Poland, the disintegration of " all the Russias ", and the growing detachment of German States from Berlin and Prussia. The decentralization of " all the Spains " from Madrid is also under discussion ; and, most significant of all, it is from France, though the earliest and most fully centralized of countries, and most unanimous of all throughout the war, that we have longest been receiving alike the best descriptions of her component regions and the most definite projects of legislation towards their renewal. " Regionalism " was indeed a French word : and this not merely in geography, but also in politics, and long before the war. From Brittany to Provence its studies and policy have long been preparing ; and now still more definitely with the return of Alsace and Lorraine. The United Provinces of France are thus in the remaking.

For most of the older generation, whether industrial and liberal, or of imperialist and financial outlooks, this newer movement has seemed reactionary or perverse, and of course not always without cause. Yet as students of social life and its processes we are learning to recognize that every society is a complex web, with its relatively fixed geographical and historic conditions, its regional warp, as the very basis of its economic and political woof. Economics is thus fundamentally regional, since sources of food, materials, and power, conditions of transport and more, are of Nature's making, which we utilize more than we modify. Hence since politics cannot but follow economic lines, it has to become inter-regional as well ; not simply super-regional—i.e. uni-regional, if not positively irregional—as metropolitan bureaucracies are increasingly felt by their external provinces to be. So far as the war settlement and the League of Nations are recognizing these conditions and dual requirements, the regional and the general, their work may thus be effective, and become stable ; or conversely.

Hence this new movement towards regionalism, and all over Europe, despite the impatiences, or even excesses, with which it may be chargeable, is by no means the mere mental disorder or material revolt so alarming to the metropolitan view-point ; since its inmost purpose is not the disruption of larger ties, so far as vital ones, but the legitimate development of local life ; which has been at best but insufficiently fostered, if not positively repressed, from distant centres substantially unacquainted with it. As the first claims of this regional life are granted, inter-regionalism cannot but be advanced anew :

hence the most discerning, and therefore the most intensive regionalists of to-day are also among the most appreciative of truly comprehensive politics, such as the League of Nations. Though the embers and sparks which actually kindled the war were largely from among the Balkan peoples, in their long ill-centralized and still unadjusted regions, it is the well-adjusted cantons of Switzerland, with their different races, languages and religions, their varied yet mutualized sympathies and interests accordingly, and the United Provinces of Holland and of Belgium, the Scandinavian peoples, and of course the United States, which are leading the Great Powers into that League of which they have each so long been samples and examples, in their various ways.

So indeed it is for the British Empire, for which the crude separatism of one generation, and the crude centralization attempted by the next, have in ours been reconciled through wise and large measures of devolution ; and with increasing moral solidarity accordingly, as the war has so vividly shown. Why not then the like in our own islands ? Ireland's predominant demand is not their only difficulty, nor yet Ulster's more exasperated regionalism. Wales, for matters of church, education, etc., Scotland too for her own concerns, and also North England, the Midlands, and more, are all claiming more understanding, and affirming more urgency, for their own affairs, than an overworked central government can give them.

Hence the need of regional geography ; and for this survey, this necessary description and diagnosis before treatment, England offers one of the best fields, the more so since undivided in language or sentiment. . . .

The regional movement began in France and has assumed prominence more recently in Britain, Germany and the United States. In France, regionalism had an early and energetic start as a means of offsetting the excessive centralization of affairs in Paris and giving greater scope to regional culture, representative self-government and traditions. Numerous schemes have been put forward during the past fifty years for the division of France into regions, and many treatises written on the theory of regionalism and the practical form it should take. In France, proposals have been made for the erection of new Regions to displace the Departments as political units, and although they came to nothing, many *ad hoc* developments have taken place. In Britain, interest in the question was active immediately after 1918. World War II and the need for national planning have again fostered public and government interest in the whole question. In the United States the problem has been tackled, especially on its functional side, by sociologists and economists. Marketing areas, newspaper circulation areas, and zones of influence of metropolitan cities, have been the subject of careful investigations, and the regional treatment of planning and the development of resources have been examined under the auspices of the National Resources Committee, established in 1935. In Germany, Regionalism has received a great deal of attention in the last twenty years, for here there exist, side by side, a political and administrative framework that is a chaotic legacy from the past, and a vast economic and social structure, not much more than half a century old, that has called into being new major units of life and orientation. Moreover, an entirely new political set-up has long been realized by many thinkers to be the only answer to the political supremacy of Prussia, since this State embraces nearly two-thirds of the area and of the population of the Reich. Abortive attempts were made to re-cast the political divisions in accordance with the Weimar Constitution in 1919. In the inter-war years numerous public authorities and scholars published the results of detailed investigations of various aspects of the general problem and of particular regions.[1] In the U.S.S.R., and, on a smaller scale, in Portugal and Yugoslavia, new political units have been devised in such a way as to harmonize with the existing structure and needs of society.

3. Regional Town Planning

In the broadest terms, compulsory town planning on a nation-wide scale did not appear until the beginning of this century. The co-operation of contiguous neighbouring public authorities for the settlement of joint problems began in the years following the 1914–18 war. Such inter-town, or, as they are generally called, "regional" groupings, normally have no legal status; they exist to study problems of planning and to make proposals; action normally depends upon the consent of all constituent authorities. From this point of view,

> a region may be imperfectly described as a rural or urban or rural-urban area having common needs and interests, linking up its economic and social life, and forming a unit, with physical conditions and boundaries appropriate for the purpose of planning its future developments.[2]

It is not a community structure in the sense of being an administrative unit, such as a county or borough. It is a composite of

[1] See the author's *Regions of Germany*, 1945, in this series.

[2] Thomas Adams at the International Town Planning Conference, Amsterdam, 1924, *Proceedings*, Part I, p. 51. See also "The Preliminary Survey of the Region" by P. Abercrombie in the same volume.

different municipal units in juxtaposition to each other and having common or overlapping problems relating to their economic life, means of circulation and land uses.[1]

Such an authority is established to make a fact-finding study and to make recommendations with regard to zoning, communications, parks, and open spaces. The extent of such a region is primarily determined by the area covered by those authorities who join to form the region.

In determining the area of a region (for purposes of inter-town planning) the factors that have to be considered are the degree of physical and economic unity of an area and the common interest which exists between a group of adjacent local authorities. The Manchester and district area, for example, comprises a group of closely related industrial towns surrounding the city of Manchester, covering about 1,000 square miles. The value of considering problems of transport, industrial development, housing and open spaces, for the interest of the city and the satellite communities within these environs at the same time as for the city is obvious.[2]

Local government authorities are jealous of their powers, and in spite of the obvious difficulties in the field of planning, they seek to find local solutions through compromise rather than by the extension of certain large units at their expense. This problem is particularly acute in Great Britain, especially as the very business of changing local government boundaries is a long and complicated procedure. The problem of London and its chaos of local government divisions is common knowledge. On the Continent, the expansion of the city administrative area has been a quicker process. There are cases where large cities have expanded their administrative limits well beyond their present built-up areas to embrace the whole of their " potential settlement areas ", and such cities are able to enforce throughout their administrative area town-planning measures which regional planning authorities cannot do. Thus, Amsterdam, Rome and Prague have their plans. A further variation is the particular case of special planning powers being given to a large region round a city. This has recently been enacted by law for the Paris Region which covers an area within a radius of thirty-five kilometres of the city.

Such regional town-planning units have the merit of being larger units for the purpose of planning than the town, which is

[1] Thomas Adams, *Recent Advances in Town Planning*, 1932, p. 114.

[2] Thomas Adams, op. cit., pp. 115–16. For a more detailed consideration of Manchester, see below, p. 278.

too small independently to carry out the very functions with which it has been legally endowed. But a larger area defined in this way is often no more a unit than the single local government unit. This is evident from the extent of such town planning regions in the United States, Great Britain, and Germany, for they frequently cut across areas that are closely interdependent and are, indeed, in themselves closely tied up with other areas contiguous to them. In other words, a consideration of joint town-planning areas automatically involves a further consideration of planning on a nation-wide scale for which is needed a framework of definite provinces which should incorporate several regional town-planning units. The cases of north-west Germany and central Germany illustrate this point.

The Ruhr Regional Planning Federation, the only regional planning authority of its kind, was established in 1921 with legal powers to control matters of traffic, housing, open spaces and light-railways traffic. It covers the whole of the Ruhr industrial area east as far as Hamm, and extends west of the Rhine as far as the Dutch frontier, an area of 3,000 square miles with a population before 1939 of over four and a half millions. The rapid growth of this area as the greatest coalfield in Europe and the seat of four-fifths of Germany's coal and pig-iron production, the concentration in it of through east-west railways and the disordered pattern of its local communications, the spread of its great cities, the need for adequate houses and open spaces, and the control of its water supply, all raised problems demanding the supervision and guidance of a central authority. But though a unit in these senses since 1921, it is quite impossible for the Ruhr authority to handle its problems of economic development, land planning and the rest without reference to the areas around it. The new brown coal or lignite industrial area south-west of Cologne, the new heavy industries concentrated on the Rhine, the close economic interrelations between Cologne, Düsseldorf and the Ruhr, bind this whole area into one economic unit. Further, there are other planning districts around the Ruhr, such as that of Düsseldorf and Cologne, roughly corresponding with the administrative districts of the *Regierungsbezirke*, and three districts in the Province of Westphalia. These are closely connected, although they were (till 1936), like the British town-planning regions, independent in their organization and powers. The whole of this area of north-west Germany obviously forms one complex whole.

In the middle Elbe basin there has occurred the most remarkable development of new industry in a hitherto rural agricultural area containing some of the most fertile soils in Germany. Beneath these soils there are extensive deposits of lignite or brown coal. During the 1914–18 war and since, vast new quarries have been opened up, and near to them gigantic electricity plants have been established, which serve chemical works, producing explosives, manures and chemical products, brick works and other plants. The whole now forms one of the chief industrial areas of Germany, though formerly it was entirely rural in character. The quarries now cover large areas, necessitating the removal of scores of villages. When the quarry is exhausted of its lignite the land must be restored. Here is the most urgent demand for large-scale, far-seeing planning. But the area concerned overlaps several states—Prussia, Thuringia, Anhalt, Brunswick and Saxony. Three regional planning authorities cover it—Merseburg, Leipzig, and Gera—and they combined in the 'twenties to prepare a thorough survey of the basic facts of land use, communications, population distribution, population movements, location of natural resources and industry, and to make recommendations for the zoning of towns and future economic development.[1] This survey, published in 1932, could not stop short with the industrial area itself, for the area provides the greater part of Germany, from Magdeburg to the Czech frontier, with fuel and electricity, and economically it is intimately associated with the great and densely populated industrial area of Saxony. The whole of this greater area, forming a triangle with its apexes in Magdeburg, Plauen and Dresden, its capital in Leipzig, and with the brown coal industrial area in its centre, is one economic unit, and its economic sectors have far closer relations with each other than with any other areas of the same extent around them, and indeed they have more rail-borne traffic with each other than with the rest of Germany together. This area, known to Germans as *Mitteldeutschland*, is a unit, and for many purposes of economic and administrative organization is treated as such.

These are but two examples, probably not so well known as British and American cases, that illustrate very clearly the need for the planning of land and resources in areas very densely peopled, with a close network of settlements, routes, open spaces and

[1] R. E. Dickinson, " Mitteldeutschland : The Middle Elbe Basin as a Geographical Unit ", in *Geographical Journal*, Vol. CIII, 1944, pp. 211–25.

interspersed farm land—one an old industrial area like our own, the other an entirely new industrial area on a scale which we do not know in this country. It is also obvious that, in their common interests, such areas in turn should be linked in some way with the areas around them with which they are closely related economically, socially and historically.

Planning on such a scale obviously demands national or State supervision of some kind with a framework of provinces covering the whole country, an organization for collaboration, a programme and technique of research, and means of carrying proposals into effect. The mention of provinces on this scale brings us face to face with a wider issue, namely, the case for the reorganization of major administrative units, for the decentralization of administration and for the development of a more lively and responsible system of local government.

4. Practical or *Ad Hoc* Regions [1]

The whole of Britain—like France, Germany and the United States—is divided into districts for a great variety of purposes —for the collection of statistical data, for administration by departments of State, for the organization of numerous trades and professions, and for military and civil defence. These districts, defined as a rule quite independently of each other, vary greatly according to their purpose. Many are based on the existing political divisions ; others adopt quite new boundaries ; some are simply determined by the amount of business which a single office staff can conveniently handle ; others hinge on the distribution of one or more occupations, on questions of accessibility and distribution of population. Each country is divided into districts by numerous private concerns which have regional offices, depots or warehouses, to facilitate nation-wide organization and service. This procedure is adopted by State Departments, by trade and professional organizations, and by business concerns dealing in consumers' goods so as to ensure effective and regular contact with all retail dealers, and by nation-wide health services and water and electricity supply. Practically every aspect of business, commerce and administration is now " regionalized " in this sense, with the services concentrated in the principal cities. It should also be noted that while these *ad hoc* regions differ widely, often necessarily so, from

[1] E. W. Gilbert, " Practical Regionalism in England and Wales ", *Geographical Journal*, Vol. XCIV, 1939, pp. 29–44.

each other, many show a remarkable similarity in their geographical extent, especially around the great cities. They are built in large measure on existing administrative units which are often anachronisms. Regions are also under discussion for such special purposes as medical services, education, and industrial and social organizations, each demanding a set of units and authorities suited to its particular problems.

5. Regions for Resources Planning

The range of problems falling within the purview of the national planning of resources as summed up recently by an American writer, and as particularly applicable to the United States, are : land use, population, resettlement, land improvement, water use—including drainage, irrigation and navigation —transport, reafforestation, preservation and development of wild life and fisheries, use and conservation of power resources, industrial development (including mining), and social and economic improvement. The American nation, it is argued, is too large to be dealt with as a unit for these purposes, and, moreover, there are great differences of sentiment and attitude in different parts of the country.

> One of the major concerns of the planner becomes, therefore, a search for a sphere of jurisdiction—a unit of area which will provide both reasonable physical and economic homogeneity, and approximate unity of public opinion. Regional planning centres should be established in those areas where there is a major clustering of related resources coincident with a marked regional consciousness among the inhabitants.[1]

Problems of planning would be centralised and co-ordinated from the regional headquarters, but " regions " for specific purposes would vary considerably. The outstanding instance of regional resources planning in action is the Tennessee Valley Authority, and there seems little doubt that it will serve as a prototype for large-scale planning in the post-war world.

[1] G. T. Renner in *Our Natural Resources and their Conservation*, edited by A. E. Parkins and J. B. Whitaker, New York, 1936.

CHAPTER 10

REGIONS AND REGIONALISM IN FRANCE

1. The Regional Movement

The regional movement was born and cradled in France. During the nineteenth century the country suffered from the excessive concentration of both national and provincial affairs in the national capital, and that was the main cause of the growth of the regional movement. Many schemes were put forward for the creation of new political divisions in place of the *départements*, including those of Auguste Comte, the philosopher, and Frédéric Le Play, the sociologist. In 1898 the *Union Régionaliste bretonne* was founded, and in 1900 the *Fédération Régionaliste française* came into being with the objects of affording a link between all advocates of regionalism and of providing for the propaganda and defence of regional ideas and interests. Its organ was *L'Action Régionaliste* and its chief exponent M. Charles-Brun.[1]

This movement aims at the decentralization of administrative, economic and social activities from the capital, and the revival and free development of regional cultures, activities, interests, and aspirations. The excessive centralization of affairs in Paris failed to recognize that within the State there exist such distinct regions, whose requirements cannot be satisfied by uniform treatment from a national centre, detached, ignorant, and unsympathetic, and too burdened with its manifold duties of State and Empire to devote adequate attention to vital questions of high regional priority and significance. The movement aimed ultimately, as one of its main objectives, at the creation in place of the *départements* of entirely new provinces with a large measure of democratic self-government. The enthusiasm with which the problem was tackled is evident from the many schemes of regions and outlines for the machinery of regional government which appeared in the years immediately preceding the 1914–18 war.

The regional movement was further fostered by the emergency of the 1914–18 war. The idea was discussed in the Chamber,

[1] J. Charles-Brun, *Le Régionalisme*, Bloud, Paris, 1911.

the Institute, the Academy of Moral and Political Sciences, and the Faculty of Law at Paris. However, proposals for the effective establishment of political regionalism by creating entirely new major regions and abolishing the *départements* have all failed. Jean Hennessy put a proposal before the Chamber on April 29, 1915, which would have suppressed the *départements* and established elected regional assemblies with a large measure of

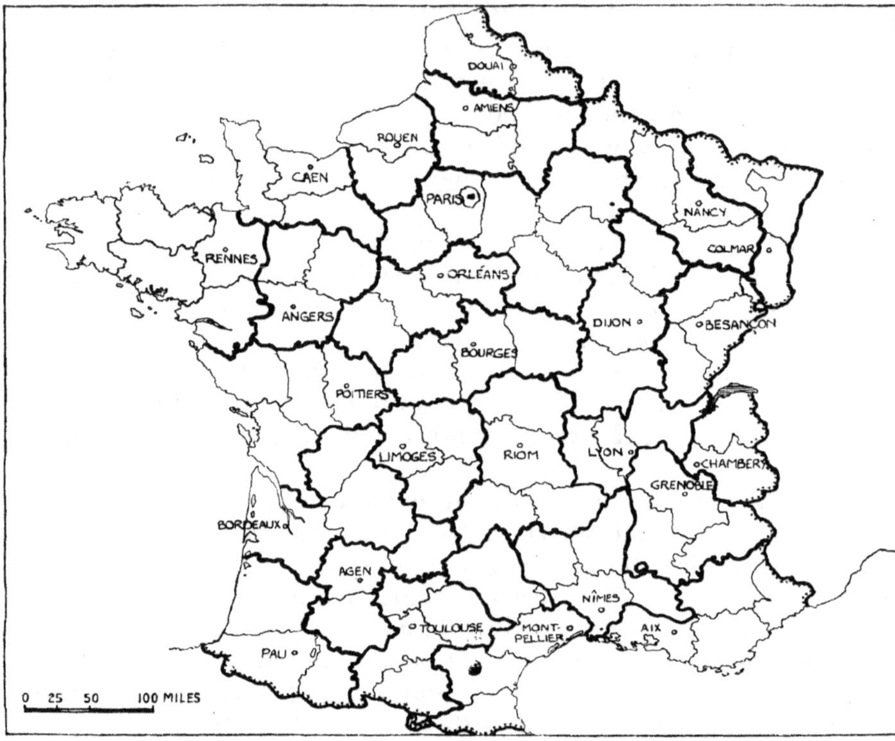

FIG. 50.—France: Regions of Justice (*Atlas de France*).

administrative and legislative decentralization. Similar abortive proposals, were made by Étienne Rognon in 1916 (proposing 34 regions), Victor Peytral and Henri Roy in 1917, and M. Bellet in 1923 (proposing 28 regions). Though in the strict sense political regionalism was not realized, many aspects of political life and organization have, in fact, since been regionalized.

The chief of these are the regions of the Chambers of Commerce.[1] On 22 July, 1915, M. Jean Hennessy, with the aim of "completing military mobilization by economic mobilization",

[1] See F. Prevet, *Le Régionalisme Économique : Conception et Réalisation*, Paris, 1929.

invited the government "not to lose sight of the fact that in a country composed geographically as ours, the component regions differ from each other", and proposed that there should be a technical council on economic matters in each existing military region (as organized in 1914). The decree of October 25 created this organization, the economic council [1] in each military region being charged with "investigation of measures to maintain

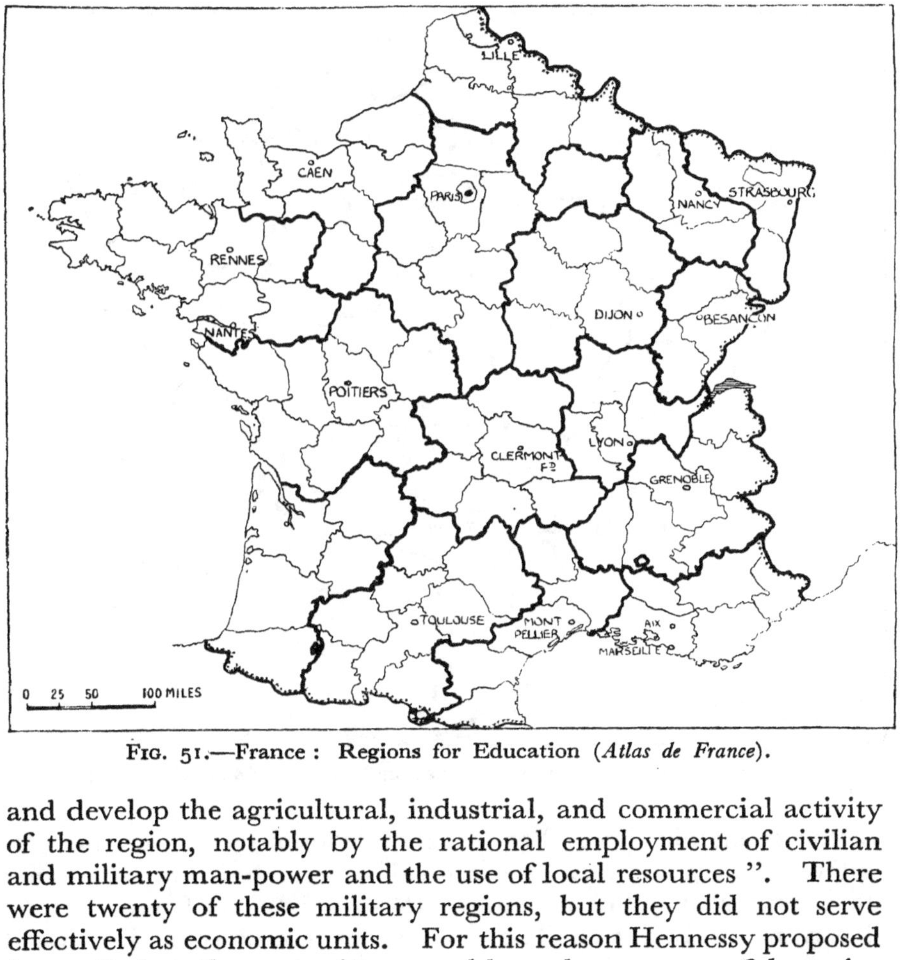

FIG. 51.—France: Regions for Education (*Atlas de France*).

and develop the agricultural, industrial, and commercial activity of the region, notably by the rational employment of civilian and military man-power and the use of local resources". There were twenty of these military regions, but they did not serve effectively as economic units. For this reason Hennessy proposed in 1916 that the committee would render more useful service if the economic units centred in the great cities, as suggested by Vidal de la Blache (see below), were adopted as their spheres of activity instead of the military regions. This proposal was not

[1] *Comité Consultatif d'Action Économique.*

immediately adopted, but in the spring of 1917 M. Clementel, as Minister of Commerce and Industry, undertook an investigation of the economic regions of France under the direction of Fighiera and Hauser. The decree of April 5, 1919, permitted the organization of 136 Chambers of Commerce into seventeen economic regional groups—increased later to twenty. These regions are to-day of fundamental importance in the organization

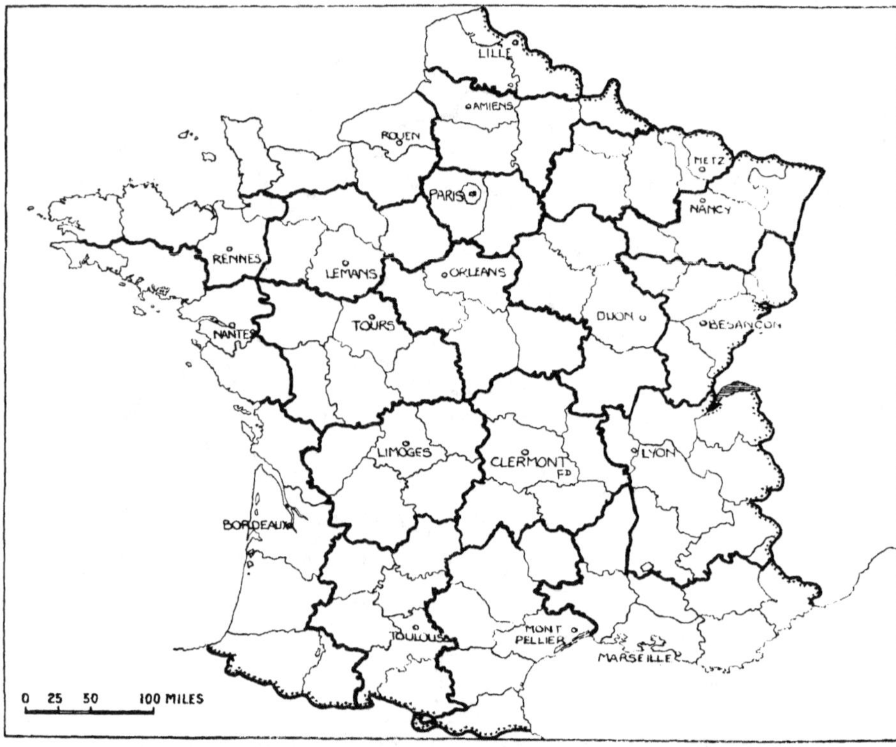

FIG. 52.—France : Regions for National Defence (*Atlas de France*).

of the economic life of France, their committees functioning as intermediaries between the central government and the region they represent. The extent of these regions as fixed in 1939 is given on Fig. 53. Their activities cover such important matters as the development of port facilities, navigation, industrial development, afforestation, irrigation, and general economic surveys.

By 1939 practical regionalism was an established fact in the life of France. There was a regional system for the administration of justice (Fig. 50), for education (Fig. 51),[1] for military organization (Fig. 52), for the associations of the Chambers of Commerce (Fig. 53), and for the agricultural syndicates, to name but the most important purposes. " Spontaneous regionalism " had developed in the cultural activities of the provinces. It also was apparent in the organization of the activities of the country around the cities which, with their roots in the distant

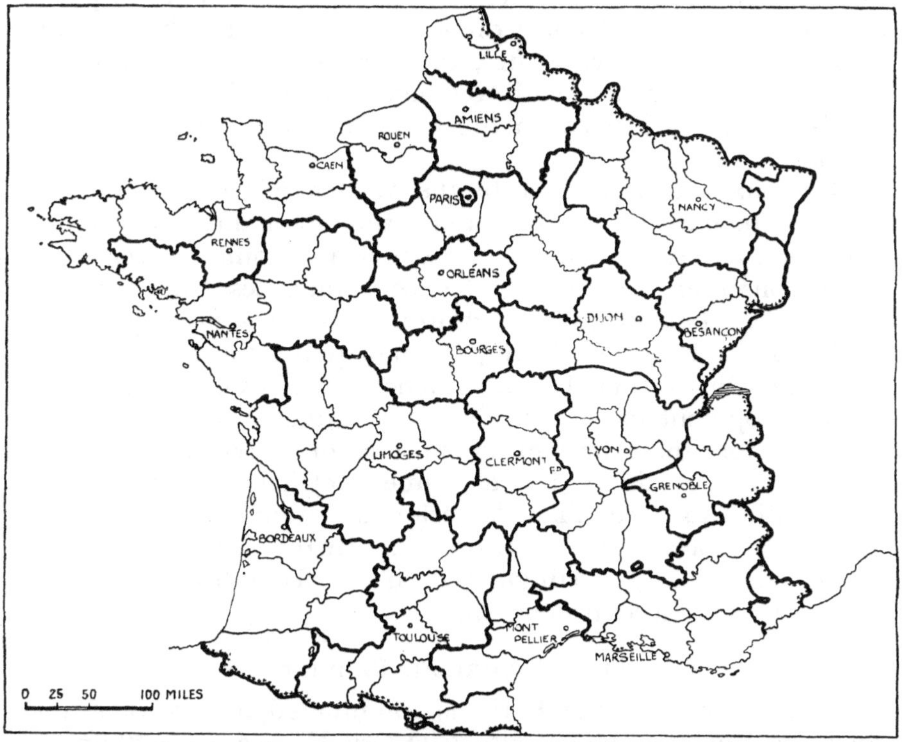

FIG. 53.—France : Economic Regions (*Atlas de France*).

past, have emerged in the last hundred years as outstanding cultural and economic centres of metropolitan character.

Under the Vichy Government emergency regions were established. Pétain encouraged the regionalist idea and sponsored official investigations into the administrative reorganization of France on the basis of new provinces made up of groups of

[1] The country is divided into seventeen educational regions or academies, in which all education is controlled by a Rector who is responsible to the Minister of Education. France has seventeen Universities, as follows : Aix, Besançon, Bordeaux, Caen, Clermont Ferrand, Dijon, Grenoble, Lille, Lyon, Montpellier, Nancy, Paris, Poitiers, Rennes, Strasbourg, Toulouse and, in North Africa, Algiers.

départements. Fourteen industrial regions were organized for supervision by the Inspectors-General of Production. In 1941 Petain decreed the formation of eighteen emergency regions under Prefects who had special administrative, economic and juridical powers. The *Comités d'Organisation Professionnelles* have different *ad hoc* regional divisions to suit the needs of particular industrial occupations. The *Charte de Travail*, effected in 1941 for the settlement of disputes between employer and employee and of personal problems of the latter, provided for the grouping of occupations into syndicates, each group having its own committee (*Comité Social*), with regional unions of the syndicates serving as the link between the *Comité Social* and the *Fédérations* at the centre.

Regionalism was defined by the Académie Française, in 1934, as " une tendance à favoriser tout en maintenant intacte l'unité nationale, le développement particulier autonome des regions et à en conserver les mœurs, les coûtumes, les traditions historiques ". This definition, writes Barathon [1] in a pamphlet dedicated to Marshal Pétain, is inadequate, for regionalism is now a doctrine rather than a tendency ; it aims not at favouring the development of regions but at creating or reviving them ; and has more than the sentimental idea of conserving traditions. It is defined by this writer as " une doctrine politique dont le but est d'instituer, au sein de la nation, des groupements régionaux autonomes dotés d'une vie propre ". In this sense, regionalism should be clearly distinguished from federalism and from administrative decentralization.

2. The Historical Provinces [2]

While it is true that Paris has dominated the life of France since the Revolution, it is also true that France, more than any other country, reveals with the greatest clarity the arrangement of town and countryside in small units of human life and organization with a popular consciousness of that unity. This applies both to the very small human units called *pays* and to the larger pre-1789 units called *provinces*. These are alike in the sense that they have been for centuries units of social life and their names are ancient in origin and popular in their usage.

Under the *ancien régime* France was divided into ecclesiastical,

[1] Claude Barathon, *Le Régionalisme d'hier et de demain*, Les Œuvres Françaises, Paris, 1942.

[2] See Jean Brunhes, *Géographie Humaine de la France*, in G. Hanotaux's *Histoire de la Nation Française*, Tome I, Vol. I, 1920, pp. 337–410.

judicial, financial, military and administrative divisions. The judicial divisions were the *bailliages*, of which there were about two hundred and fifty. For financial matters there were the

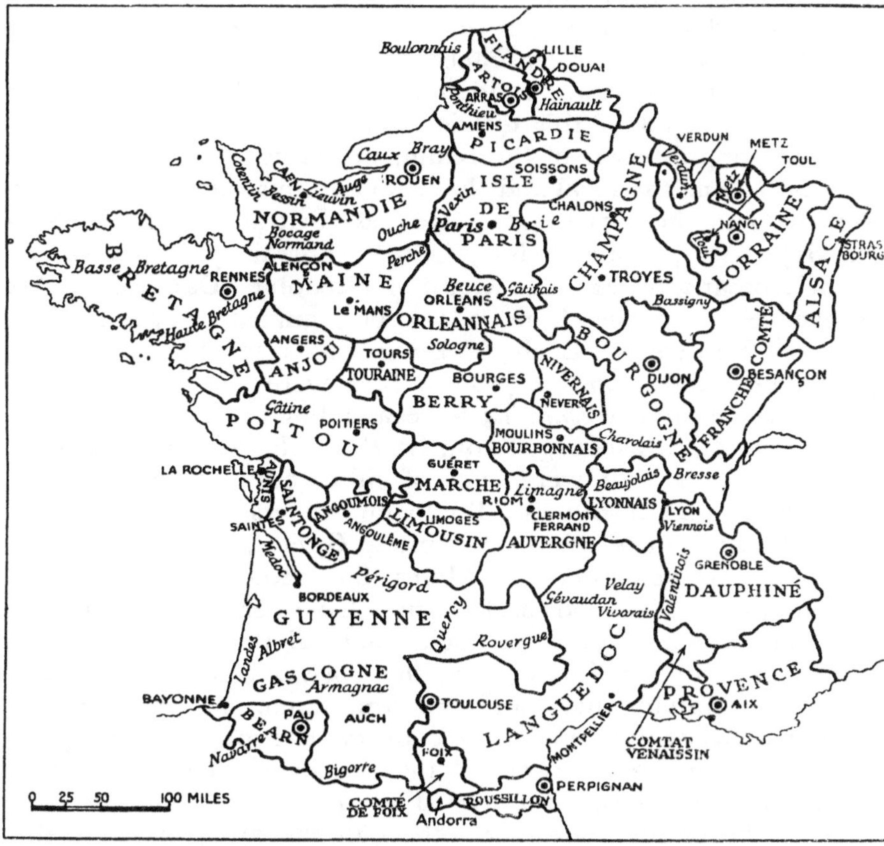

FIG. 54.—France : The Historical Provinces.

The thirty-three political units of the *gouvernements* and their capitals, as shown in this map, give the clearest indication of the approximate extent of the historical provinces. The *pays* are shown in italics. The Comtat Venaissin (1791) and the Comtés of Nice and Savoie (1860), when absorbed by France, became Departments. The cities shown were capitals of the *gouvernements*, and most of these were also capitals of the *généralités*, the most important administrative divisions at that time. Ten cities, however, were the headquarters of *généralités* only. Cities shown with a dot and circle were also the seats of *Parlements* ; they were the outstanding provincial capitals. (From F. Schrader, *Atlas de Géographie Moderne*, 1907, Carte 13. See also W. R. Shepherd, *Historical Atlas*, University of London Press, 1922, pp. 146–7.) (For *Beuce* read *Beauce*.)

pays d'état and the *pays d'élections*. In the former there were provincial assemblies for voting taxation, while in the latter, covering three-quarters of the area of the country, the representative of the king ruled as a virtual viceroy. The *gouvernements* were military districts for the levying of troops under the supervision of a Lieutenant-General. The *généralités* were by far the most important divisions in France. There were thirty-three of these divisions, which sometimes embraced several provinces, and sometimes several were embraced in one province as in the case of Normandy, which was divided into the *généralités* of Rouen and Caen. The head of this district was the Intendant, who was the real administrator of the realm, presiding over the *Parlement* and controlling the legal and financial administration. *Parlements* were held in the *pays d'état* at Rennes (Brittany), Rouen (Normandy), Arras, Douai, Metz, Nancy (Lorraine), Dijon (Bourgogne), Besançon (Franche Comté), Grenoble (Dauphiné), Aix (Provence), Toulouse (Languedoc), Perpignan (Roussillon), Pau (Béarn), Bordeaux. Each of these was also the seat of its *gouvernement* and *généralité*. Cities that were the seats of both a *gouvernement* and a *généralité* were as follows : Tours, Orleans, Lille, Amiens, Strasbourg, Lyon, La Rochelle, Moulins, Bourges, Poitiers. Cities that were seats of either one or the other but not of both were as follows : Boulogne, Angers, Troyes, Chalons, Toul, Valenciennes, Nevers, Saintes, Guéret, Riom, Limoges, Trévoux, Soissons, Alençon, Montpellier, Auch, Montauban.[1]

The names of the *pays* and *provinces* have persisted to this day. They are attached to a particular part of the country with a distinct human individuality—as the Cotswolds in England have a physical individuality—and the same name has been given to various political divisions of the country that roughly correspond with the same area. Some of the provinces have their origin in the Roman *civitas*, such as Touraine (*civitas Turonum*, the Gallic tribe of the *Turones*) ; others in the areas of the Gallic tribal groups such as Poitou (the Gallic tribe of the *Pictones*) ; while the majority emerged as political groupings in the Middle

[1] The initial divisions of France were those of the Gallic tribal groups (French *nations*), and the Roman *civitates* (French *cités*) which often broadly corresponded in extent with the areas of the Gallic tribal groups. The ecclesiastical dioceses followed broadly the outline of the *civitates* and in the great majority of cases the bishoprics were established in the Roman *castra*, the capitals of the *civitates*. Then the Frankish period witnessed the formation of the medieval territorial pattern. The *civitas* was broken up into two or more smaller parts called *pagi* (French *pays*), though some retained the outline of the *civitas*. The medieval unit of the *comté* broadly corresponded with the smaller *pagus* or *pays* and had its capital in the central town. The *généralités* in 1789 were Aix, Alençon, Amiens, Auch-Pau, Besançon, Bordeaux, Bourges, Caen, Châlons, Dijon, Grenoble, Ile de Corse, La Rochelle, Lille, Limoges, Lyon, Metz, Montauban, Montpellier, Moulins, Nancy, Nantes, Orleans, Paris, Perpignan, Poitiers, Rennes, Riom, Rouen, Soissons, Toulouse, Tours, Valenciennes.

Ages, such as Champagne, Languedoc and Aquitaine. Some of the smaller provinces emerged from *pagi*, as Aunis from the *pagus Alionensis*, Senonais from the *pagus Senonicus*. The names of the provinces were in popular usage on the eve of the French Revolution, and although some of the political divisions carried the name of a province, these divisions varied in extent. Thus it is impossible to define the province exactly, any more than one can precisely define, for example, the limits of the Cotswolds or the Weald. " Malgré les fluctuations historiques, malgré les vicissitudes des rattachements ou des sectionnements politiques, il est un certain nombre d'ensembles provinciaux majeurs qui ont conservé ce que nous pourrions appeler une certaine continuité de personnalité, et cela jusqu'à notre siècle même." [1] This unity is often reflected in " un esprit provincial, un art provincial, une littérature provinciale ",[2] and indeed the very mention of the name of a province, writes Brunhes, " éveille et réveille d'un seul coup des ensembles de souvenir, de pensées, de coûtumes, de passions, et d'images correspondant à des séries seculaires de connexions humaines dont la synthèse est encore un fait social, historique et geographique tout actuel ".[3]

The *pays* is usually smaller than the *province*, but, like the latter, it is essentially a social unit. This unity it owes to the distinctive mode of life, common interests and traditions of its inhabitants. Though the *pays* sometimes corresponds with a distinctive physical unit, it is far more characteristically, like the *province*, an amalgam of two or more distinct types of country, whose people are interdependent by reason of the exchange of goods and ideas through the medium of a central capital town, from which it often takes its name.[4] Examples are Touraine, capital Tours ; Anjou, capital Angers ; Poitou, capital Poitiers ; Lyonnais, capital Lyon ; Limousin, capital Limoges ; Périgord, capital Périgueux ; Angoumois, capital Angoulême ; Bordelais, capital Bordeaux ; Agennais, capital Agen : Maconnais, capital Macon ; Laonnais, capital Laon ; Soissonnais, capital Soissons.

[1] Brunhes, op. cit., p. 344.

[2] Claude Barathon, *Le Régionalisme d'hier et de demain*, Les Œuvres Françaises, Paris, 1942, p. 13.

[3] Brunhes, op. cit., p. 342.

The essential trait of a *pays* and the area which it covers may vary in history. The *pays* is not a constant. Thus, in the case of Perche in Normandy, Musset has shown that it first applied to the forested land, then later to the stock-raising area as opposed to the grain land of Beauce farther east, and finally to the area that specializes in the commercial output of the Percheron breed of horse. See R. Musset ; " Le Perche, nom de pays ", *Annales de Géographie*, Vol. XXVIII, 1919, p. 359.

There are certain contrasts in the character of the social groups that have developed historically in the north and in the south and west of France.[1] In the north, in the area characterized in the past by the compact village with a three-field system of cultivation worked on a compulsory communal system, space-groupings have been more permanent and are more real than in the west and south, where the isolated farmstead has been dominant. The *pays* of Beauce, Brie, Vexin and Valois are ancient names antedating that of France itself. Their origins date back to the Gallic tribal divisions and the Roman *civitates*, through the *pagus* to the medieval *comté*. The same stability is characteristic of the *commune*, the successor in 1789 of the *parish*, which in effect was the village community area. In the south and west, on the other hand, the parish was not so clearly defined, as the dispersed farmsteads and hamlets were not suitable for the erection of parishes centred on one village. With the formation of the *communes* as civil units in 1789 these had to be imposed on the countryside, since there was no existing village community area. In the north, the *arrondissement* also became a real unit and shows the same cohesion as the *commune*. " C'est une veritable société homogène, consciente, bien ordonnée autour de sa petite ville comme autour d'une capitale." [2] It is often coincident with the ancient *bailliage* and this, in turn, was often based on a seigneurial district or a fief. In the west and south, space-groupings are neither so homogeneous nor so clearly defined. *Commune* and *arrondissement* are somewhat arbitrary units, since there were no clearly defined social units in 1789 on which they could be based. Ancient *noms de pays* of Gallo-Roman origin are rare. Feudalism did not have nearly the same hold as north of the Seine. The family was the primary social unit situated in the centre of its own lands, and above it was the château of the nobility which formed a distinct aristocratic class, frequently in opposition to the peasantry.

Brunhes divides the historical provinces into six groups, according to their geographical location and the character of their historical development.[3] A group in the centre of the Seine basin, the middle Loire, and the basin of the Garonne, served as nuclei of crystallization for distinct provinces. These are the Île de France ; Guyenne and Gascony ; Lyonnais, Forez, Beaujolais ;

[1] See Philippe Ariès, *Les Traditions Sociales dans les pays de France*, Les Éditions de la Nouvelle France, Paris, 1941.

[2] Philippe Ariès, op. cit., p. 23.

[3] Brunhes, op. cit., p. 342 et seq.

Touraine, Maine, and Anjou ; Alsace. A second group lies on routeways astride the great lowlands—Poitou, Champagne, Picardy and Artois, Burgundy (with Nivernais) and Languedoc—each of which had a wide field of influence, intellectual, commercial and political. A third group contains what were formerly important centres but are now of secondary importance. All of them are situated in the centre of France, between Chartres in the north and Saint-Flour in the south. They include Auvergne, Berry, Bourbonnais, Orléannais and Nivernais. These provinces have suffered particularly through the lack of capital cities sufficiently strong to offset the dominating influence of Paris. The fourth group are isolated, thinly peopled provinces in the barren high plateau of central France ; nevertheless, these provinces are distinct, and their names and character are strongly entrenched in popular feeling and usage—Limousin and Marche ; Périgord and Quercy ; Rouergue ; Gévaudan ; Velay and Vivarais. The fifth group are the frontier provinces which have been absorbed, in part or in their entirety, into France during its expansionist phase when attempting to reach its so-called " natural frontiers " on the Rhine and the watersheds of the Alps and Pyrenees in the seventeenth and eighteenth centuries. These provinces are Roussillon and Cerdagne in the eastern Pyrenees ; Foix, Andorra, Quatre-Vallées and Bigorre in the central Pyrenees ; Béarn, Navarre and the Basque country in the western Pyrenees ; the Dauphiné, Briançonnais and Savoy in the Alps ; Flanders, Lorraine, Barrois, and the three bishoprics (Metz, Toul and Verdun) in Lorraine ; and Franche Comté in the east. The sixth group includes the maritime frontier provinces of Provence in the south-east, with Corsica across the water ; Brittany and Normandy in the north-west ; and Aunis and Boulonnais, two small provinces on the western and northern coastlands respectively.

The scheme drawn up by the *Comité de Constitution* in 1789 for the new administrative departments was based upon these historical provinces, although the names of the provinces were dropped and the *départements* named after principal topographic features. The *départements* corresponded usually to one province or a group of two or three departments to one province. The *départements* all have roughly the same area and were so designed that the central city could be reached within one whole day by road. Each *département* was divided into three or four divisions called *arrondissements*, and these again into about ten divisions, called *cantons*. The same principle of approximately equal size with a centrally placed town as administrative centre was observed throughout.[1] A main political problem that has persisted since the French Revolution is that of offsetting the centralization of affairs in Paris by democratic government in units larger than the *départements*. This has taken many forms —separatism, federalism, administrative decentralization and regionalism. Though these solutions differ in their political aspects they have a large measure of common ground in that they recognize the need for new units that are neither the *départements* nor an attempted revival of the provinces, but are based on the real socio-economic groupings of to-day.

Growing partly from the ancient provinces, but in large measure reoriented around the chief regional centres, there have emerged in the last hundred years new provinces or regions. Here have crystallized both the imposed regionalization of nation-wide activities, and the spontaneous development of regional activities, with their founts in the leading cities. The problem of defining new Regions is, in fact, an attempt to define the areas of social and economic association that have emerged in the structure of modern society.[2]

3. Modern Regions and their Capitals

A main problem of political regionalism in France is the actual definition of regions. M. Charles-Brun set out the factors which should be considered and, though old, his enunciation of principles is still of interest.[3] He began with general matters of climate, geology, relief, orientation, natural products, race, customs, history and language. Next, he pointed out that " homogeneity, which is the fundamental feature of the *pays*, should not, in so far as the region is concerned, be exclusive of a certain amount of variety ". The region should " combine

[1] An even distribution of urban settlements permitted the formation of a geometrical pattern of administrative divisions, each with its central capital. " *Chefs-lieux* de départements et arrondissements y sont disposés comme les pièces d'un damier à distances convenables, chacun avec son rayon limité d'action ". Vidal de La Blache, in the symposium *Les Divisions Régionales de la France*, 1913.

[2] The best general work that deals specifically with these problems is a symposium entitled *Les Divisions Régionales de la France*, 1913. It contains articles on the regions of France, the geographical divisions of France in 1789, the development of transport in the nineteenth century, the growth of urban agglomerations, the development of ports, and studies of individual regions—Brittany, Lorraine, and the regions of Nantes and Rouen. All are written by outstanding scholars.

[3] Charles-Brun, *Le Régionalisme*, Paris, 1911.

opposed elements ". Thirdly, he argued that new tendencies, particularly of an economic character, should be taken into account, since professional organizations will be based on a regional system. Under this head he includes population and commercial relations. Finally, he considers size. The new regions should be large enough, not only in area, but also in resources and population, to withstand the influence of Paris, and they should all be roughly of the same size. This last

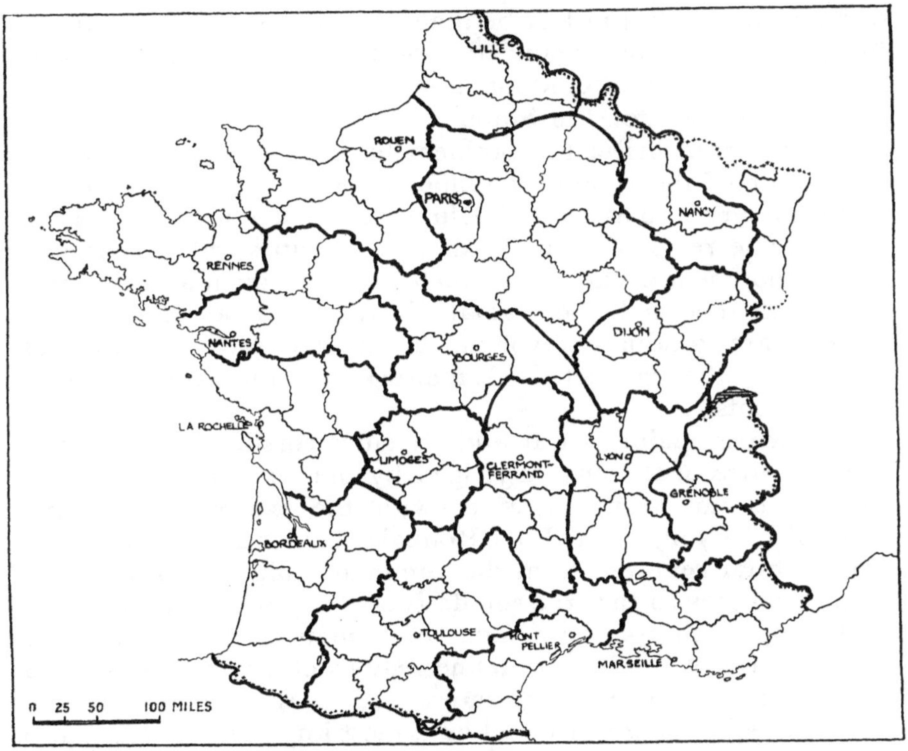

FIG. 55.—Regions of France proposed by Charles-Brun (1911).

requirement, he realized, could not be fulfilled with mathematical exactness, since Nature does not permit absolute equality of treatment, but some rough correspondence should be attained so that all regions might be sufficiently strong and vigorous, absorption by the stronger and bigger would be avoided, and an equilibrium would be maintained. His scheme of regions is shown on Fig. 55.

This heterogeneous collection of criteria does not greatly help in providing for specific needs. It has often been pointed out by French regionalists that the definition of new regions does, in fact, involve a dual problem. The region must function as an effective unit of government, so that every part is easily accessible to the centre ; and it must be a balanced economic unit, that is, it should possess a common basis of activity and interests. Further, the regional capital should be strong enough in virtue of its history and tradition, and of its large population and modern commercial importance, to withstand the influence of Paris, and should in fact be the principal natural (real) centre for the activities and organization of its region. In other words, the regional capital should function for the region as the national capital functions for the State.

French students of the problem have long been agreed that the extent of the sphere of influence of the dominant cities is of paramount importance in defining such homogeneous human units. The regional capitals are " large cities indispensable to the development of the areas which surround them and throughout which their influence radiates ",[1] and around which " new regions are gradually evolving from the ancient provinces, shattering the restricted framework of the administrative Departments ".

France is mainly rural and its cities small. Consequently, the influence of the city, if judged by the amount of brick and mortar, is not very extensive—with the exceptions of Paris, Marseille, Lyon, and Lille–Roubaix–Tourcoing. Very large areas are far removed from the dominant influence of any great urban complex so that the small historic city plays a correspondingly important role as a focus of human activities. For this reason it is not possible to define regions clearly on the basis of service areas of a few large cities. Thus, while many schemes for new regions of government and, most important of all, that of Vidal de la Blache (Fig. 56),[2] include from twelve to seventeen large regions, there is bound to be an arbitrary factor in the definition of such large units. New economic regions, as Brunhes has argued,[3] cannot be resolved simply by the compass or by railway time-tables and bus services, that is, on the basis of

[1] Jean Hennessy, *Régions de France (1911–1916)*, Paris, 1916.
[2] P. Vidal de la Blache, " Les Régions Françaises ", *Révue de Paris*, December 1910.
[3] Jean Brunhes and H. Pierre Deffontaines, *Géographie Politique et Géographie du Travail*, in G. Hanotaux's *Histoire de la Nation Française*, Tome II, Vol. II, 1926, pp. 51–78.

accessibility of the suggested capital city. Moreover, there are many other regional factors to be taken into account, such as types of farming and industry, and regional needs and interests. In order to meet these needs, Brunhes maintains that effective human units must be smaller in size, and, still adhering to the basic idea of the city as the focal point of the region, he suggests about twenty-five to thirty regions, each with a historic and modern "regional metropolis" as its focus. These regions in

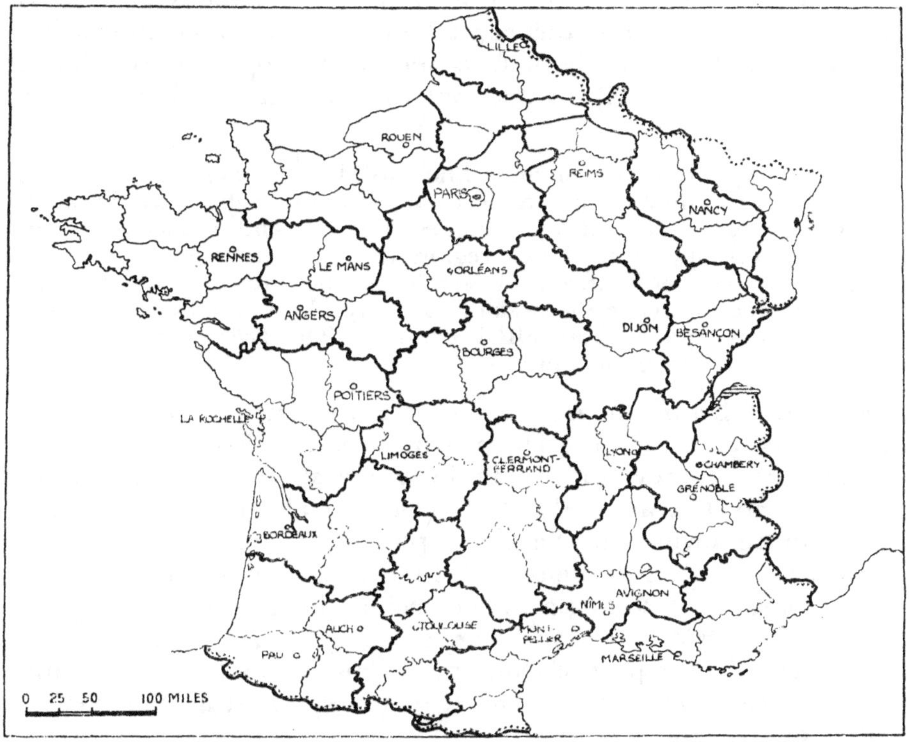

FIG. 56.—Regions of France proposed by Vidal de La Blache (1910).

groups of two or three would form in some cases the major regions. The following is his selection of centres: Rouen and Caen are capitals for eastern and western Normandy respectively; Orleans, Tours, Angers, and Bourges are capitals and serve the middle Loire lands—and to these we would add Poitiers; Rennes and Nantes serve the west, the first for Brittany, the second for that portion of the province of Brittany and the lower Loire lands which is centred on Nantes and its outport of Saint-Nazaire; Bordeaux is the capital for the south-west, and

La Rochelle for the lands between Poitiers and the Garonne, and Limoges and the west coast; Limoges serves the west of the Central Plateau, with its nucleus in Limousin, and Clermont Ferrand the heart of the Central Plateau, with its nucleus in the lowland of the river Loire and the province of Auvergne; Toulouse and Montpellier are pre-eminent historical and cultural centres in the south, with, to name but one trait, Universities dating back to the thirteenth century; Lyon, Marseille and Grenoble are the capitals of the south-east, for the Rhone lands and the Alps. Lyon, the capital of the silk textile industries, is an admirable example of an economic metropolis in the fullest sense of the term, with a regional role as great as that of Paris itself.

> From Dijon to Besançon and almost to Montélimar, from Clermont Ferrand to Chambéry, the greatest part of the industries are directed, controlled and financed by societies, boards, and banks of Lyon. The city also supplies all this region with products for its factories and its wholesale houses and serves as the intermediary for the export of goods outside the region. Finally, through its stations there pass in transit goods from all directions. The Lyon region is thus a remarkable economic unity.[1]

Dijon and Besançon are capitals for the ancient provinces of Burgundy and the frontier province of the Franche Comté; Strasbourg, Nancy and Metz in the east are the capitals of Alsace and Lorraine, the latter of special modern economic importance owing to the development of the iron and steel industry; Reims, Amiens and Lille are capitals in the north of France, the last in particular being the centre of the great textile industrial area and the coalfield which serves as the hinterland for the port of Dunkirk. This gives a total, excluding Paris, of twenty-five cities, and it includes almost all the leading cities of the country.

Two other cities, however, might be added. Their claims to be independent centres of a high order are evident from their insistence on being treated as separate units in the regional organization of the chambers of commerce. They are Saint-Étienne (350,000), and Nice (270,000). The first is an ancient and modern industrial centre complementary to, but in large measure independent of, Lyon. Saint-Étienne is the centre of a small conurbation on a coalfield with about 350,000 inhabitants, and the focus for the industries—engineering, ribbon wear and

[1] R. Blanchard, *Grenoble*, 1935, p. 222.

knitted goods—in the homes and small workshops of the *pays* of Velay and Forez. The Saint-Etienne district long lay outside the organization of economic regions and did not accept till lately absorption into the Lyon region. The second, Nice, is a new city, the centre of a horticultural and luxury residential area, which has become the capital for the whole of the Maritime Alps north to the Var river ; it may be noted that in the organization of the economic regions, it preferred union with Grenoble rather than with Marseille.

In the schemes for fewer larger regions on the lines of Vidal's plan, the following cities appear almost always as the suggested capitals, and may, therefore, be regarded in fact as the outstanding cities with real metropolitan character Clermont Ferrand, Limoges, Bourges, Dijon, Rennes, Grenoble, Lille, Toulouse, Nancy, Lyon, Marseille, Rouen, and Nantes. These are the cities of the first order in France ; the remainder, noted above, are of the second order.

The sequence of the historical development of these natural capitals of France is much the same in all cases. Each began as the centre of a Gallic tribe, usually on a hill-top. This was followed by the location of the Roman settlement on the flatter land by the river-side commanding the river crossing, the centre being the headquarters of a *civitas* that in turn was based on the area of the Gallic tribe. The same centres and areas were used in the Middle Ages for the siting of bishops' seats and their dioceses. In the same period they became outstanding commercial and industrial centres for their surroundings. Under the *ancien régime* they were capitals of Provinces, several had their own *Parlements* and were seats of the nobility. They also had Universities, thus enjoying the fullest status of a great city. Deprived of many of their functions in 1789, they obtained a new lease of life with the coming of the railway and the growth of industry and commerce, and have in general regained their historical status, in modified character and degree, as metropolitan centres, and regional seats of commerce, industry, culture, and administration.

APPENDIX TO CHAPTER 10

The new emergency regions of the Regional Prefects as established under the Vichy regime were as follows (February, 1942).

Unoccupied Zone

Region	*Departments*
Lyon	Rhône, Loire, Ain, Saône-et-Loire, Jura, Ardèche, Drôme, Savoie, Haute-Savoie, Isère.
Marseille . .	Bouches-du-Rhône, Var, Alpes Maritimes, Basses-Alpes, Vaucluse, Gard, Hautes-Alpes, Corse.
Montpellier . .	Hérault, Lozère, Aveyron, Aude, Pyrénées-Orientales.
Clermont Ferrand.	Puy-de-Dôme, Cantal, Allier, Haute-Loire.
Limoges . . .	Haute-Vienne, Corrèze, Creuse, Dordogne, Charente, Vienne, Indre, Cher, Indre-et-Loire.
Toulouse . . .	Lot, Haute-Garonne, Tarn, Tarn-et-Garonne, Lot-et-Garonne, Ariège, Gers, Hautes-Pyrénées, Basses-Pyrénées, Landes, Gironde.

Occupied Zone

Bordeaux . .	Gironde, Landes, Basses-Pyrénées.
Poitiers . . .	Vienne, Deux-Sèvres, Charente, Vendée, Charente-Maritime.
Angers	Loire Inférieure, Maine-et-Loire, Indre-et-Loire, Loir-et-Cher, Cher.
Orleans. . . .	Loiret, Eure-et-Loir, Loir-et-Cher, Cher.
Rennes	Ille-et-Vilaine, Côtes-du-Nord, Morbihan, Finistère.
Rouen	Manche, Calvados, Orne, Eure, Seine-Inférieure.
Nancy	Meuse, Meurthe-et-Moselle, Vosges.
Dijon	Côte-d'Or, Haute-Saône, Doubs, Jura, Saône-et-Loire, Nièvre, Yonne.
Saint-Quentin .	Somme, Aisne, Ardennes.
Chalons . . .	Marne, Aube, Haute-Marne.
Lille	Nord, Pas-de-Calais.
Strasbourg .	Bas-Rhin, Moselle, Haut-Rhin.

CHAPTER 11

REGIONS AND REGIONALISM IN ENGLAND AND WALES

1. The Demand for New Regions

Regionalism has never been as popular a movement in Britain as in France. It has, in fact, been confined to a few scholars—geographers, economists and political philosophers—and it is only in the last few years that it has become a subject of wider interest and concern. This interest centres on the administrative aspect, that is, the question of reorganization of local government districts and the devolution of administrative authority from Whitehall. Indeed, Regionalism is usually considered in this country as synonymous with administrative devolution in a framework of new local government units.

The civil administration of England is carried out on a triple system—the County (with its urban counterpart in the County Borough), the Rural District (with its counterpart in the Urban District, which in country districts is usually a small town near the geographical centre of the Rural District) and the Civil Parish. The Rural and Urban Districts, formed in 1894 from the sanitary districts of 1872, were based upon the Poor Law Union districts which were created under the Poor Law Amendment Act of 1834. Before this date, the triple system consisted of the County, the Hundred, and the Parish, and it can be traced back to the Norman Conquest.

The County or Shire is the major statutory administrative unit in Britain. It was already in existence before the Norman Conquest. The counties south of the Thames and in eastern England were recognised as " shires " in the time of King Alfred. Some of them correspond with the initial areas of settlement of the Anglo-Saxon tribes, so that each had a nucleus of open, settled land and a periphery of marsh and forest on the lower land. In the densely wooded and thinly peopled lands of central England, which was then the kingdom of Mercia, the shires came into being during the tenth century. Each was defined as a group of Hundreds conveniently accessible in one day to a fortified administrative centre, which became the county town. These midland counties, each with the suffix " shire ", are all approximately of the same size, and the shire bears the same name as its capital. The counties in the north and south-west of England are the largest ; they were not organized as definite county administrative areas until the twelfth century, and were not effectively absorbed into England until after the time of the Tudors. In Wales [1] the political divisions that existed before the English conquest were based on the ancient tribal groupings of pastoral communities who lived on the lower slopes of the hills below the moorlands and above the forested river valleys. The hierarchy of tribal space-groupings culminated in nine major areas each under the rule of an overlord. By the Statute of Rhuddlan in 1284, Edward created the Principality in west and north Wales with five new counties based on the tribal overlordships. The March remained under the disputed control of about 150 marcher lords until 1536, when by the Act of Union five more new counties were created, the boundaries of which were based on those of the lordships and not upon physical divides.

The main fact we would emphasize in connection with these counties is that they came into being before the Norman Conquest and already in the Middle Ages their boundaries had often become areas of close settlement. This has been greatly emphasized in the last hundred years, so that to-day there is often little relation between the county boundaries and the present distribution and movements of population.

Regionalism as a popular cultural movement (though not bearing this name) has been mainly associated with the linguistic revivals in Scotland, Wales, and Ireland, each of which has expressed itself, in more or less degree, in a Home Rule Movement. These movements are analogous to the claims to autonomy of groups inside the west European States, such as Provence, Brittany and Alsace-Lorraine in France, Flanders in Belgium, and Catalonia in Spain, although such movements on the Continent have been regarded as endangering the unity of the State and have been treated with suspicion. In England proper, there are no deep-seated cultural differences. There is, of course, a popular consciousness of regional association, as manifested by a regional literature—Hardy's Wessex novels, for example—and by differences of dialect, popularized in some measure by the wireless. There are also " regional " or " district " social, cultural and trade associations. But in its national and cultural

[1] See E. G. Bowen, M.A., *Wales : A Study in Geography and History*, Cardiff, 1941.

life England is a unit, for reasons of history and of the small size of the country, and perhaps most of all in the last century, because of the overwhelming dominance of the urban way of life. Hitherto, the rural way of life—dialect, customs, temperament and the like—have been swamped by the influence of city ways.

There are big differences between France and Britain in socio-geographical structure. In France the old political divisions and the very names of the provinces were abolished in 1789 and the new and smaller Departments established. In Britain, the counties are historical provinces and have persisted, unchanged, with real administrative significance, to this day. Nevertheless, there have been periods when larger units were required. The chief of these were the military governorships established under Cromwell. Among the few popular names applicable to such larger areas are East Anglia, the Fens, Wessex, and the Weald. Britain, however, is lacking in geographical names that refer to permanent social groupings like the *pays*. Possible examples are the Craven, Hallamshire, Forest of Dean, and Holderness districts, but these are of little significance to-day. This whole question, it may be added in passing, deserves much more careful study than it has yet received. Lastly, with few exceptions, urbanism has not seriously affected the traditional social patterns in France. In Britain, especially in northern England, the Midlands, the Home Counties, and central Scotland, these have been profoundly changed. Yet the county has always been the chief social and political unit in Britain and still is a popular unit for trade, professional and cultural associations, army regiments, football and cricket clubs, and the like. It is most popular in rural areas, but in the vicinity of the great cities the old-time associations have been all but obliterated, and here the need for a reorganization is most urgent and most needed. Any scheme to introduce new divisions and abolish the old must contend with very real and valuable social forces of tradition and conservatism.

But if the movement for the creation of entirely new divisions to replace the counties has never been widespread in Britain, in practice, as we have already seen,[1] regions exist for a great variety of purposes, and in recent years the need for new regions for many aspects of public life has become more obvious and urgent. As far back as 1905 the Fabian Society championed

[1] See above, Chapter 8.

the idea of regionalism with a series of pamphlets called *The New Heptarchy*. The first of the series[1] declared that "the great towns of Liverpool, Manchester, Birmingham, Newcastle, Nottingham, Leeds, etc., must be considered as centres, and not as self-contained units for all local government purposes, particularly in regard to such services as transport, electricity, and water supply". This idea of a new Heptarchy attracted some attention and was worked out in greater detail by C. B. Fawcett, G. D. H. Cole and W. A. Robson. More recently a number of official reports have presented somewhat similar conclusions, as the Report of the Hadow Committee on the training of Local Government Officers (1934), the Report on Greater London Drainage (1935), the Report of the Royal Commission on Local Government in the Tyneside Area (1937), and, still more recently, the Report of the Royal Commissions on the Geographical Distribution of Industrial Population (1940), the Report of the Scott Committee on Land Utilization in Rural Areas (1942), and the Report of the Uthwatt Committee on Compensation and Betterment (1942). Meanwhile the needs of the last war led to the setting up of Civil Defence Regions (Fig. 57). These were administrative regions designed to co-ordinate the functions of the various government departments. Regional divisions of the country have been suggested during World War II in reports and Bills for purposes of education, employment, medical services, and the organization of the coal industry. A basic and comprehensive survey of the major regional divisions has also been put forward in a recent article in *The Times* for purposes of industrial reorganization and development.[2] Thus, while scarcely agreeing that "at last we have established regionalism", as is claimed by a writer on the Civil Defence Regions,[3] it is undoubtedly true that regional treatment is now generally recognized as an urgent need in the solution of our domestic problems.

As each year passed, the obsolete and inefficient character of our local government organization became increasingly obvious. Every extension of the scope of the functions which local authorities were required or permitted to discharge emphasized the hopeless inadequacy of the existing system of areas. Far larger units of administra-

[1] W. Stephen Sanders, *The Municipalization by Provinces*, 1905.

[2] "An Industrial Survey, The Role of Regional Research and Development", *The Times*, April 3, 1945.

[3] *Regional Government*, published by the Fabian Society, Research Series, No. 63, 1942.

tion were needed not only for relatively new services such as town and country planning, housing, electricity supply, road passenger transport, higher and technical education, but also for the older functions such as police, highways and sewage disposal. The expansion in the area of the daily movement of the people for economic, social and political purposes, consequent upon improved methods of transport and communication, has for long necessitated a corresponding extension in the units of local government ; for no principle of political organization is more firmly established than that the areas of public administration should approximate to the areas of diurnal movement. Above all the dichotomy between town and country which is implicit in the rigid separation of county councils and county borough councils became manifestly absurd when vast numbers of persons who work and earn their living in cities were enabled to live in semi-rural or suburban dormitories situated at considerable distances outside their boundaries. An enlargement of local government areas on the one hand, and an integration of town and country authorities on the other, thus became insistent needs which no amount of obtuseness or resistance to reform on the part of local councils or their associations could overcome. The facts were too eloquent to be silenced.[1]

2. The Town Planning Region

Town planning in Britain has proceeded, in the phrase of Sir Patrick Abercrombie, from the particular to the general. During the nineteenth century, as the result of private enterprise, much pioneer work was done in estate and village planning. Well-known villages attached to factories, the co-partnership tenants' estates, and the great achievement of Hampstead Garden Suburb, all aimed at the reform of the domestic environment. The need to extend those principles to the town led to the first Town Planning Act of 1909. In its somewhat grudging treatment of old built-up areas this act showed a strong bias towards suburban if not estate planning, and in due course it was found inadequate. The next Act of 1919 led to the treatment of towns in their relationship to neighbouring towns and to the wider setting of the town in the countryside. Thus, regional planning was almost unconsciously initiated, and there are now more than a hundred town-planning regions covering all the most populous areas in a belt from the Solway Firth to the Straits of Dover, as well as elsewhere in the country.[2] The disadvantages of such

[1] Ibid.

[2] See a useful summary by C. B. Fawcett on Regional Planning in England and Wales, in the Report of the International Geographical Congress, July, 1928, and a more recent report by G. L. Pepler in *Städtebau und Wohnungswesen der Welt*, published under the auspices of the *Deutsche Verein für Wohnungsreform*, edited by Bruno Schwan, 1935.

regions have already been indicated in general terms. Here it may be added that they frequently do not coincide with the natural geographical region (as defined in this book), and this, in turn, does not coincide with existing administrative boundaries.

The problem of the town-planning region is well illustrated by the case of Manchester,[1] the administrative area of which stretches for twelve miles from north to south but only three and a half miles from east to west. Contiguous to it are closely built-up areas like Salford, Sale and Stretford, and the small towns of north Cheshire, that are its best residential areas. About two million people live within a radius of ten miles and one million within a radius of one mile of the Town Hall. It is the capital for a densely populated industrial area with an aggregate population of over four millions.[2]

The city of Manchester has grown " from the centre outwards by continuous additions on the fringe of the existing built-up area, and as new districts which adjoined the city were developed, they were incorporated by an extension of boundaries ". In consequence, the city falls into four broadly concentric belts that coincide broadly with the concentric belts discussed in Chapter 4. The business centre is an area of one square mile in which much of the property is old and will have to be rebuilt and streets widened. The Slum Belt is a circular zone about half to two miles wide around the business centre with buildings erected before 1890—a mixture of houses, factories, and warehouses. There are 80,000 houses in this belt, out of a total of 180,000 for the whole of Manchester, on an area of 3,000 acres. The Suburban Belt falls into two concentric belts, the inner one containing by-law houses built between 1890 and 1914, that are " dreary, and depressing " but not " unhealthy or unsanitary " ; the outer one containing inter-war housing estates, including the best residential districts to the south of the city and including Wythenshawe south of the Mersey.

The reconstruction of the Slum Belt is the greatest problem of every British city. The scheme put forward by the City Council in Manchester aims at converting the slum belt into " a really fine, healthy, and attractive residential area " by a comprehensive replanning scheme that makes provision for new roads, parks, schools, playing fields and the best lay-out of blocks of flats at a

[1] Sir Ernest D. Simon, *The Rebuilding of Manchester*, London, 1935, and, more recently, in *Rebuilding Britain, A Twenty Year Plan*, London, 1945.

[2] See Simon, *Rebuilding Britain*, 1945, pp. 194–212.

density of forty to the acre, with the relegation of industry to the zones in which it is segregated near the railways and canals. This scheme, however, will permit building on only one-third of the whole area of the Slum Belt of 3,000 acres, housing a total of 40,000 families. This leaves the other half of

FIG. 57.—England and Wales. Civil Defence Regions and twelve other Administrative Divisions.

1. County boundary. 2. Composite boundary varying in width according to the number of Government Departments that use it. 3. Civil Defence headquarters. 4. Civil Defence Region with number. Thin lines are boundaries used by only one Department.[1]

the houses in the belt, plus an outstanding deficit of 20,000 houses, to be built on the periphery of the built-up area. It is proposed to build these as single family houses. There is room for 5,000 new houses in the present city and 25,000 in Wythenshawe, so that about 25,000 houses will be needed outside

[1] Fig. 57 is prepared and reproduced by permission of the Association for Planning and Regional Reconstruction, together with Figs. 58, 59, and 60.

the city area.[1] The demand for the extension of the city's administrative boundaries is obvious. Manchester, Salford, Stretford and north Cheshire ought at least to form one administrative unit. At present the people living in the best residential districts in north Cheshire (Bowdon, Knutsford, Alderley, Wilmslow)

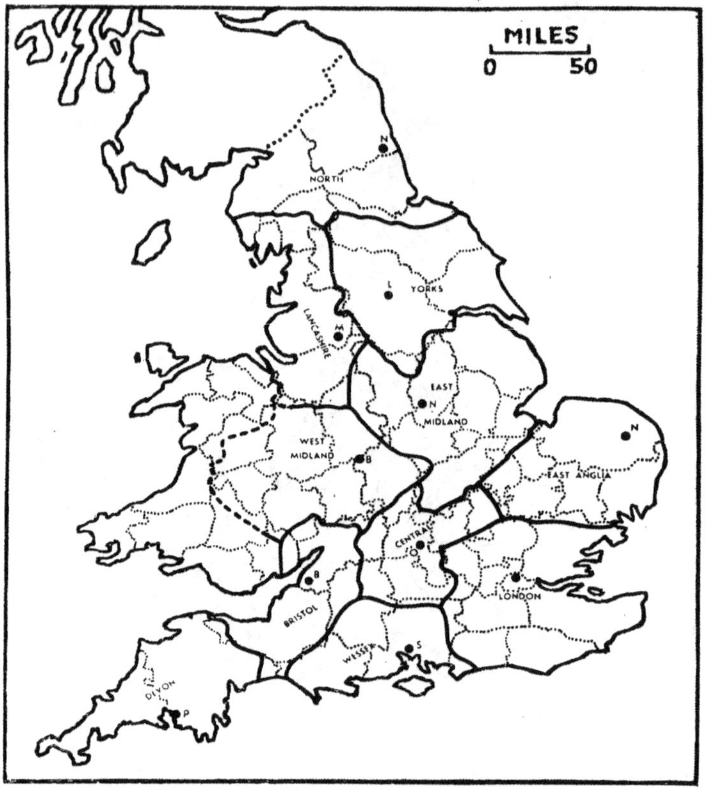

FIG. 58.—England and Wales. Divisions suggested by C. B. Fawcett in 1917, revised in 1942.

for the most part work in Manchester and enjoy its amenities, but escape all financial obligations for its upkeep.[2] Moreover,

[1] The Reconstruction Committee of the City Council have recommended the construction of 75,000 new houses, 50,000 of which should be built *outside* the present area of the city, thus allowing for an overspill of 150,000 persons. See Simon, *Rebuilding Britain*, p. 204.

[2] This is one of the most serious problems raised by the explosion of the big city beyond its administrative limits. The shift of wealthier people from the less desirable residential areas in the city to the pleasanter sites outside it is reflected in the reduction of rateable values in the city by 5 per cent. and a tremendous increase (85 per cent. between 1928 and 1938) in the surrounding districts. There is also a wide disparity in the rates between the two. See Simon, op. cit., p. 209.

the boundary between Manchester and Salford runs through the built-up area along the river Irwell and is almost contiguous with the city centre of Manchester, so that all plans for the reconstruction of Manchester—its core, its slums, its outer fringe, its ring and radial roads and the rest—must be dependent on

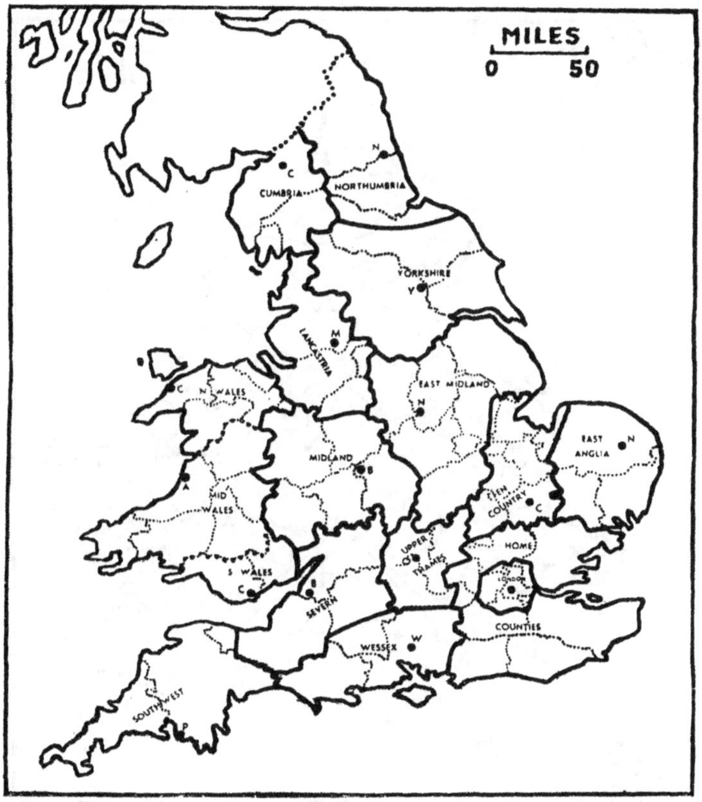

FIG. 59.—England and Wales. Divisions suggested by E. W. Gilbert, 1941.

some sort of adjustment to the needs and plans of the Salford town council. This state of affairs is repeated in practically every British city.

3. Regional Schemes in Theory and Practice

We have seen how nation-wide and town-planning regions have grown up in this country. The purposes for which the former have been created are so varied that a single set of regions cannot possibly serve them all. But we ought to be clear about the purposes and about the principles which are to be observed in the delimitation of new regions. Mr. G. D. H. Cole has asked these questions : Planning for what purpose ? Planning by what machinery ? Planning under whose auspices ? In answer to the first it has already been made clear that widely different con-

FIG. 60.—England and Wales. Divisions suggested by E. G. R. Taylor, 1941.

siderations arise. Industry, electricity supply, banking organization, and so on, require different regions. In the second question we must distinguish between decentralization of the national government, and the reorganization and development, in large units, of local government. For the former, large regions like the Civil Defence Regions may best serve, and those who have primarily considered this need think in terms of such regions. For the latter, small units are required for the effective working of local government. Mr. Cole's " basic conditions " are " that there shall be *some real consciousness of unity* (italics ours) among ordinary citizens extending over that region as a whole, and marking it off from the other regions of the country ". Such a region should be thought of in terms of " a biggish town, or sometimes a group of very close together biggish towns, together

FIG. 61.—England and Wales. Divisions suggested by G. D. H. Cole (1921).

with the surrounding country which is served by that single town or by that conurbation ".[1]

These views closely resemble those of Dr. W. A. Robson, who writes : [2]

If we examine the demands for larger areas put forward by official committees and various disinterested experts, or survey the territories comprised in the most promising developments in electricity, town

[1] " Discussion on Geographical Aspects of Regional Planning ", *Geographical Journal*, Vol. XCIX, 1942, p. 65. See also leading articles in *The Times*, October 5, 6, 7, 1944, quoted on pp. 2–3.

[2] W. A. Robson, *The Development of Local Government*, London, 1931, p. 130.

planning and numerous other services, we shall find that what is required is not a single set of areas, however large or comprehensive, but a whole series of different areas for separate and distinct purposes. There is no division of the country which will suit all municipal functions.

While these areas will and must differ from each other, continues Robson, " the essential principle is that diverse areas should be

FIG. 62.—England and Wales. Local Government Regions proposed by *Housing* (1920).

built up from certain more or less stable units ".[1] Thus both writers accept the principle that two types of region are required. The large region is necessary for decentralization, for national planning in all its aspects, and for statistical purposes : the smaller region is necessary for purposes of local government. The latter units do exist, in fact if not in name, and are mainly centred around the metropolitan cities of the country. Such a

[1] Op. cit., p. 140.

region is that centred on Manchester, which we have already discussed with particular reference to town planning. It remains to examine the schemes which have been put forward for new regions, and the *ad hoc* regions which have come into existence.

Lord Passfield (Mr. Sidney Webb) has considered the problem[1] but paid no attention to the delimitation of the region and the nature and function of its capital. His main purpose

FIG. 63.—Geographical Counties of England and Wales.

was to discuss the organization of the region in a scheme of administrative devolution based upon the principles of socialism. Professor C. B. Fawcett[2] and Mr. G. D. H. Cole[3] both produced a scheme of regions (Figs. 58, 61). The latter, although claiming not to have based his suggestions "on any one predominant principle, least of all that of geography" (by which he evidently

[1] Sidney and Beatrice Webb, *A Constitution for the Socialist Commonwealth of Great Britain*, 1920.
[2] C. B. Fawcett, *The Provinces of England*, 1919.
[3] G. D. H. Cole, *The Future of Local Government*, 1921.

means hills, valleys and rivers), none the less adheres throughout to considerations of human geography, as understood by geographers, although not suggesting such radical changes in the existing county boundaries as does Professor Fawcett. Cole follows the method of first locating the regional capitals and then

FIG. 64.—England and Wales. Principal Administrative Divisions of the Ministries of National Service, Labour and Munitions (September 1918).

tracing very roughly the boundaries of the regions[1] around them which are sparsely populated. The six "axioms" enunciated by Professor Fawcett, who is a geographer, in the delimitation of his twelve provinces are as follows :

1. The provincial boundaries should be so chosen as to interfere as little as possible with the ordinary movements and activities of the people.
2. There should be in each province a definite capital, which should be the real focus of regional life. This implies,

[1] For Cole's general definition of a region, see Chapter I, p. 8.

further, that the area and communications of the province should be such that the capital is easily accessible from every part of it.

3. The least of the provinces should contain a population sufficiently large to justify self-government.

FIG. 65.—England and Wales. Divisions used by the Registrar-General for the Census, 1931.

4. No one province should be so populous as to be able to dominate the Federation.
5. The provincial boundaries should be drawn near the watersheds rather than across the valleys, and very rarely along streams.
6. The grouping of areas must pay regard to local patriotism and to tradition.

The methods and conclusions of both writers are broadly similar, though Cole, as an administrator, devotes more attention to " the coming of the region " and to the evolution of a system of regional government, than to a detailed study of the character of the regions he suggests. Another scheme was suggested some twenty-five years ago in a series of articles entitled *Towards a National Survey*[1] (Fig. 62). This scheme was drawn up for the development of a national housing policy based upon a regional

FIG. 66.—England and Wales. Divisions of the Central Electricity Board.

plan, and not with any idea of a provincial system of government. Yet the principles upon which the divisions were made strikingly resemble the axioms of Professor Fawcett. Since this scheme is not so generally known as it deserves, we quote in full the principles adopted.

The aim has been in the first place to form areas which have a community of economic interest. Such an aim necessarily cuts across many local government boundaries . . . but it is recognized that it is very undesirable to do this more than is absolutely necessary.

[1] Published by the Ministry of Health in *Housing*, November 1921.

Generally speaking, therefore, the areas of the boroughs, and urban and rural districts have been adhered to, but it has been found necessary in many instances to ignore the county boundaries.

For the first broad division of the country, the principal watersheds have been taken as guiding lines. Following these but without adhering to them too closely where other factors seem to counteract their influences the country has been parcelled out into 15 main divisions (see Fig. 62). These main divisions have been subdivided

FIG. 67.—England and Wales. Administrative Divisions of the Employment Exchange Service of the Ministry of Labour with headquarters office of each Division (1939).

into 59 " regions " which have been in most cases again sub-divided into groups. For the purpose of these sub-divisions a number of factors have been taken into consideration. In some cases, of course, there is practically an industrial identity between adjoining districts. Far more frequently one area acts as a dormitory for another. Communications by road and rail have a very strong influence on the grouping of areas and it is recognized that a river, where its banks are used for industrial purposes . . . should be looked upon as the centre of a unit.

Another type of scheme was prepared by a Committee of geographers appointed by the Standing Committee on Regional Surveys and Local Studies of the Geographical Association, for submission to the Ministry of Reconstruction during the 1914–18 war.[1] Their conception of a province and the function of its capital differ from the others in two respects.

1. "The Provinces should not . . . be uniform in character throughout, but should have diversity within unity . . . In other words, a satisfactory province would be one which contained several *pays* or sub-regions each with its distinctive individuality, but yet so related to one another as to form a natural whole." The average size of a province, based on three regions in Europe which have been distinct provinces throughout recorded history—Wales, Brittany, and Tuscany—is calculated at 5,000,000 acres.

2. The second difference lies in the conception of the function of the Regional Capital. It is suggested that in each region, instead of all its activities being focussed on one capital, "there should be towns of contrasted types in each province; and regional centres representing different kinds of activities, rather than a single metropolis dominating the life of the provinces", since this would "tend somewhat towards decentralization within the province". A suggestion put forward is Wessex, with Winchester as the administrative centre; Southampton as the industrial and commercial capital; Oxford as the intellectual and educational capital, and Salisbury as the ecclesiastical capital.

In recent years, and especially since 1940, with the growing attention given to problems of post-war reconstruction, the question of new regional divisions has been increasingly prominent. Two geographers, Professor Eva G. R. Taylor and Mr. E. W. Gilbert, have made notable contributions.[2] The Association for Regional Planning and Reconstruction is also actively interested in the problem and has issued broadsheets devoted to it. Professor Taylor suggested a division (Fig. 60) which broke away completely from existing county boundaries, while Mr. Gilbert reluctantly, but wisely, uses them as far as possible. The latter course is more practical both from the point of view of expediency and because the county, in spite of all that

[1] H. J. E. Peake, "Devolution: A Regional Movement", *Sociological Review*, Vol. XI, No. 2, 1917.

[2] Professor Taylor's suggested regions are in the *Geographical Journal*, Vol. XCIX, 1942, p. 62, and those of Mr. Gilbert on p. 76 of the same volume. Mr. Gilbert's valuable paper, "Practical Regionalism in England and Wales", *Geographical Journal*, Vol. XCIV, 1939, p. 29, should be consulted for a variety of *ad hoc* regions. Some of these are reproduced in Figs. 61 to 67 with the kind permission of the Editor of the Journal and of Mr. Gilbert. Also A. M. Carr-Saunders and D. Caradog Jones, *A Survey of the Social Structure of England and Wales*, 2nd ed., 1937, pp. 25–32.

can be said against it, is an established unit and often has popular associations. Mr. Gilbert (Fig. 59), unlike most other contributors, makes the fullest use of the Civil Defence Regions.

The Civil Defence Regions have come in for much criticism ; they are too large and too arbitrary to serve as effective political units, and they pay little attention to natural geographical regions. Several of them are no more than groups of counties.

The Civil Defence Regions in England and Wales are shown on Fig. 57, together with the counties and the divisions of eleven Government Departments. The Regions are numbered and their boundaries are shown by black lines. The heavier black lines that run on both sides of the boundaries of the Regions are proportional in width to the number of boundaries of Departmental divisions that are coincident. The Government Departments so considered are as follows :

Number of Divisions in Great Britain

Civil Defence	12
County and County Divisions	84
Ministry of Labour and National Service	11
Ministry of War Transport	12
Ministry of Food	8
Board of Education Inspectorate	13
Factory Inspectorate	12
Ministry of Works Contractors	15
Ministry of Agriculture Advisory Provinces	16
Ministry of Agriculture Statistical Divisions	15
Board of Trade Census of Production, 1935	17
Registrar General for the Census	16
Post Office Regions	8

Scotland and Wales each form one Civil Defence Region, with capitals at Edinburgh and Cardiff. England is divided into ten Regions—Northern (Newcastle), North-eastern (Leeds), North Midland (Nottingham), Eastern (Cambridge), London (Metropolitan Police Area), Southern (Reading), South-western (Bristol), Midland (Birmingham), North-western (Manchester), and South-eastern (Tunbridge Wells). The Regional Commissioners had three war-time tasks—to represent the central government in each Region ; to carry out the government's Civil Defence and A.R.P. measures ; and to take the place of the central government in the event of a breakdown of communications. Their duties and powers were thus somewhat nebulous. But it is especially important to note that many government departments have appointed Divisional Officers in each Region, so that there is a co-ordination of areas and duties within each Region. These departments include the Ministries of Food, Information, Health, Labour and National Service, Supply, Aircraft Production, War Transport, Works and Buildings, Mines Department, Petroleum Department, Post Office and the Assistance Board, and the Ministry of Home Security.

These Civil Defence Regions are thus administrative regions, organized by the central government as a means of devolution. They are not intended as planning regions and they do not carry out any functions with regard to land planning. They are also not local government regions with elected local representatives. If larger regions are to be formed as units of democratic government with elected legislative bodies, and legislative powers over regional affairs, this larger region would probably be the higher of a twofold order of regions, for smaller districts will be required to replace the county and country authorities.[1] The Civil Defence Regions are a long way from satisfying the demands of regionalism, if the principles we have discussed are sound, and substantial adjustments would have to be made to make them into coherent units. For example, there is much to be said for including northern Nottinghamshire with Sheffield. Reference to our study of the city trade areas gives some further indication of changes. More detailed study of the space relations of border areas, like the Potteries, which are divided in their allegiance between two centres—in this case between Birmingham and Manchester (see footnote on p. 242) and yet possess a lively regional consciousness of their own, would assist in a refinement of solutions. The essential fact remains that the Civil Defence Regions have come into being for the first time as co-ordinated administrative units and should serve as a starting-point for the organization of more permanent regions after the war.

It is relevant to conclude this chapter with a note on the necessity for a far more systematic examination of the facts of social and economic distributions and human associations and movements on a regional basis, since these must serve not only as a basis for reorganization of political units but also for a better understanding of the structure of society, and the problems of planning. This country has not yet got a National Atlas,

[1] A plan is outlined in the Fabian Society Publication on *Regional Government* by Regionaliter, Research Series, No. 63, 1942. See also three leaders in *The Times* for October 5, 6 and 7, 1944, on Local Government (p. 2).

although it is understood that one is now in preparation under the auspices of the Ministry of Town and Country Planning. This, however, is but one step. What is just as urgently required is a series of atlases of the major divisions of Britain—divisions of the same order as the Civil Defence Regions or Regions determined by the Ministry of Town and Country Planning. Each atlas should contain an ordered series of maps on standard scales with standard cartography. Similar systematic studies are needed for the urban areas and their peripheries as we have already suggested.

Finally, the question of the delimitation of regions—their number, size and exact boundaries—must largely be the resultant of the judgement of the individual investigator who selects his own criteria. Social units, envisaged as areas with a large measure of common interests and organization, that is, areas of common living, do not have exact geographical limits and any attempt to define them, no matter how well considered, must be in some measure arbitrary. In order to define regions for any special purpose one should know exactly why the regions are needed and what conditions they should satisfy. In recent years, White Papers and Bills have proposed, and Acts of Parliament have in some cases decreed, that many aspects and problems of our national life should be regionalized—the location of industry, housing, post-war unemployment, medical services, education, in addition to the numerous regional divisions adopted by the government departments. It is therefore the more essential that all these *ad hoc* functions should fit, as far as is practicable, into the same framework, in the interests of the co-ordination of administration from the standpoint both of the administrator and of the individual communities which are administered.

CHAPTER 12

REGIONS AND REGIONALISM IN THE UNITED STATES

1. The Demand for New Regions

The United States is of special interest in connection with the study of all aspects of regionalism, owing to its vast area, roughly three-quarters that of the whole of Europe, forming one political and economic unit ; its rapid modern development and transformation from a producer and exporter of primary products, to a dominantly manufacturing country absorbing the great bulk of its own agricultural production ; the great and rapid growth of large metropolitan cities ; the recent development of federal enterprise in place of the " rugged individualism " of the nineteenth century ; and the development of principles of nation-wide planning based on the conception of regional development. Two chief problems in the field of regional planning have received much attention in recent years. First, there is need for the conservation and scientific development of the country's natural resources, and second, changes have been brought about in the social structure by the increasing dominance of city life, made possible, above all, by the advent of the automobile, which in the United States is not a sign of affluence, but a first claim for every citizen. The boundaries of the constituent States of the Union and their divisions—county and township—are entirely arbitrary and there is need for the creation of new units, large and small, more in conformity with conditions of living and organization.[1]

Planning in the United States is based on the existing administrative units. The city is in many ways inadequate as such a unit, and the metropolitan district or the county has been adopted frequently in its place. The metropolitan unit includes the city and its nearest satellite towns and the contiguous urban and rural districts. It is not the same as the metropolitan district of the Census as defined on p. 198, and its boundaries cut across county and even State boundaries. The chief example

[1] *Our Cities : Their Role in the National Economy*, National Resources Committee, 1937, and *Regional Factors in National Planning and Development*, National Resources Committee, Washington, D.C., 1935.

is that of the New York Region. The county with a city as its centre is a second type of unit, and is adopted by many planning organizations. It has the great advantage, as against the metropolitan district, of being a single administrative unit. Two additional areas, larger than the above, are also used for planning, namely, the State itself, and a large area cutting across State boundaries within an arbitrary limit. In many ways the State is a good unit for planning, primarily because it is an existing political unit. It is a good unit for the study of natural resources and conservation problems. The collection of basic data, the formulation of a plan, and the eventual carrying out of the plan are also facilitated by State arrangements, but the planning of the many problems of the State demands that its regional contrasts should be given full recognition. The Tennessee Valley Authority governs the best known area of the second type, the boundaries of the region being the watershed of the Tennessee river.

In January 1937, there were over 1,500 city, town, metropolitan and county planning agencies in the United States. There were just over 500 metropolitan and county agencies as compared with 85 in 1933. Metropolitan plans include those of Boston and New York. Other such regions organized for planning are Philadelphia, Chicago, St. Louis, Washington and Los Angeles. More than 250 counties have been organized for regional planning, and in these cases the county is usually a predominantly rural area with a few small towns. There are also instances of a county plan which in effect is a metropolitan district and has been organized primarily for the purpose of metropolitan planning. State planning organizations are now in existence in all the 48 American States, with the exception of Delaware.[1]

2. State Planning

State planning began in 1933 with the appointment of the National Planning Board, later called the National Resources Planning Board. Before this there had been independent State planning arising from such matters as the conservation of forest resources, the improvement of cut-over land, highway planning, health and park-lands. The Board established State Planning Boards in relation to the federal government's public works programme, as well as for such matters as land use and transport.

[1] G. Galloway and associates, *Planning for America*, New York, 1941, p. 543.

Within two years there were State Planning Boards in all of the States. These Boards are concerned with such matters as land classification, surveys of water resources, forests, minerals and fisheries.

State planning has been defined by the National Resources Planning Board as " the systematic, continuous, far-sighted application of the best intelligence available, to programmes and problems of State development and organization, in order to provide higher standards of living and greater security for the people of the State ". There has been a very rapid development of State planning in recent years, since the State is the largest existing political unit available for planning of resources. The abandonment of hill farms, and the deterioration of the soil in New York State led to a report in 1920 by the New York State Commission of Housing and Regional Planning, which is generally considered to be the first comprehensive planning report in the United States. The problem of deteriorated and tax-delinquent land led to such investigations as that of the Michigan Land Economic Survey and the Land Economic Inventory of the State of Wisconsin.

The National Planning Board aided in the organization of planning boards in all the States, and assigned to them regional planning consultants, so that State planning rapidly gained momentum. State planning reports were drafted under the National Resources Board (1934) on the resources of the individual States, and on the basis of these reports the National Resources Board submitted to the President in December 1934 a most thorough inventory of national resources.[1]

Notwithstanding the great importance of the State as both a planning unit and a planning authority, one must recognize as of vital importance the fact that there are regional variations in type of country and types of resources with distinctive problems calling for special attention.

3. Regional Planning

Regional planning [2] has come into being from " grass-roots ", to use an American term (as opposed to the totalitarian system

[1] The major items dealt with in these planning reports are : land resources, water resources, mineral resources, manufacturing resources, commerce and commercial assets, transport facilities and patterns, urban formations and their problems, population trends, recreational needs, social conditions and institutions, local government, public services, and public works.

[2] See Galloway and associates, *Planning for America*, Chapter 26, " Regional Planning " by Earle S. Draper.

of imposition from above), through educational institutions as seats of research; semi-public commissions, such as the New England Council; Regional Planning Commissions, such as the Pacific Northwest Regional Planning Commission (consisting of the heads of the State planning boards of Washington, Oregon, Idaho and Montana); commissions derived from "Interstate Compacts", such as the Colorado River Compact (1922) between seven States for the equitable division and apportionment of the waters of the Colorado river system; and federal regional development agencies of which the most famous is the Tennessee Valley Authority (T.V.A.), based on hydro-electric development in that river basin but covering a great number of planning problems associated with it.

Of special significance are the Regional Planning Commissions in and around the great metropolitan cities, that were established as voluntary bodies, with the task of making surveys and advising on future developments. The Regional Plan of New York and Its Environs, the outgrowth of an extensive survey—the Regional Survey of New York and Its Environs, is the most comprehensive programme ever formulated in the United States. The survey runs into ten volumes. The boundaries of the environs are described by the planners as follows: "(1) the area within which the population can and does travel in reasonable time from home to place of work—the commuting area; (2) the large outlying recreational areas within easy reach of the metropolitan centre; and (3) the cities and counties at the periphery of these areas".[1] The actual plan, however, is limited to the area immediately contiguous with the city, that is, it applies especially to the Metropolitan District of New York City as defined by the Census.

The Chicago Regional Planning Association has sponsored many particular scientific surveys on a far more modest scale than the New York survey. Much, however, has also been achieved by it in the realm of practical city and regional planning, including the creation of a system of highways outside the city, the development of an outer belt of parks for public recreation, and the reclamation of land along the Lake Shore front.

In 1934 the St. Louis Regional Planning Association, in connection with the National Planning Board, made a detailed survey of the St. Louis area, covering 3,000 square miles within

[1] *The Regional Plan of New York and Its Environs* (two volumes), New York, Vol. I, p. 133, 1929.

a radius of 35 miles of the city centre. This survey is to be the basis for a comprehensive regional and city plan. The Regional Planning Federation of the Philadelphia Tri-State District has been active for some years in the preparation of a regional plan. The Los Angeles County Regional Planning Commission has a plan covering many cities and open areas outside the metropolis but inside the county. The Massachusetts Division of Metropolitan Planning is mainly concerned with the problem of co-ordinating transport in the Boston area. A plan has been initiated for the Washington–Baltimore–Annapolis area. Mention may also be made of the Milwaukee County Regional Planning Department and the Alleghany County Planning Commission for Pittsburgh, both of which, within the county, include the city and its wider environs. The greatest of all "regional" planning programmes, endowed with authority to carry out its proposals (as opposed to the above surveys, where ability to carry out proposals is not known), is the Tennessee Valley Authority (T.V.A.). This covers a large area in the Tennessee river valley and embraces cities, towns and villages and a vast rural area.[1]

4. Geographical Regions [2]

In a country as large as the United States there are many great contrasts in physical conditions, historical development, economic and cultural conditions, cutting across the artificial State boundaries. Such regions have already been the subject of much scientific investigation as ends in themselves and as a means to the solution of particular problems. Physiographic regions have been the special contribution of American physical geographers. Geographers and agricultural specialists are concerned with "agricultural regions" (Fig. 68), determined on the basis of common systems of farming.[3] Sociologists have elaborated the concept of metropolitan or city regions (Figs. 46 and 74),[4] based on the spheres of influence of the principal cities,

[1] For a full discussion of the programme and achievements of this organization, see Julian Huxley, *T.V.A. Adventure in Planning*, The Architectural Press, 1943.

[2] Figs. 68 to 75 inclusive are reproduced from *Regional Factors in National Planning and Development*, 1935, by permission of the U.S. Government Printing Office, Washington, D.C.

[3] See especially the long series of articles in *Economic Geography*, on the Agricultural Regions of North America, by Dr. O. E. Baker, the major contributor to this field, especially Vol. II, 1926, p. 459, where he defines agricultural regions.

[4] See especially R. D. McKenzie, *The Metropolitan Community*, Recent Social Trends Monographs, 1933.

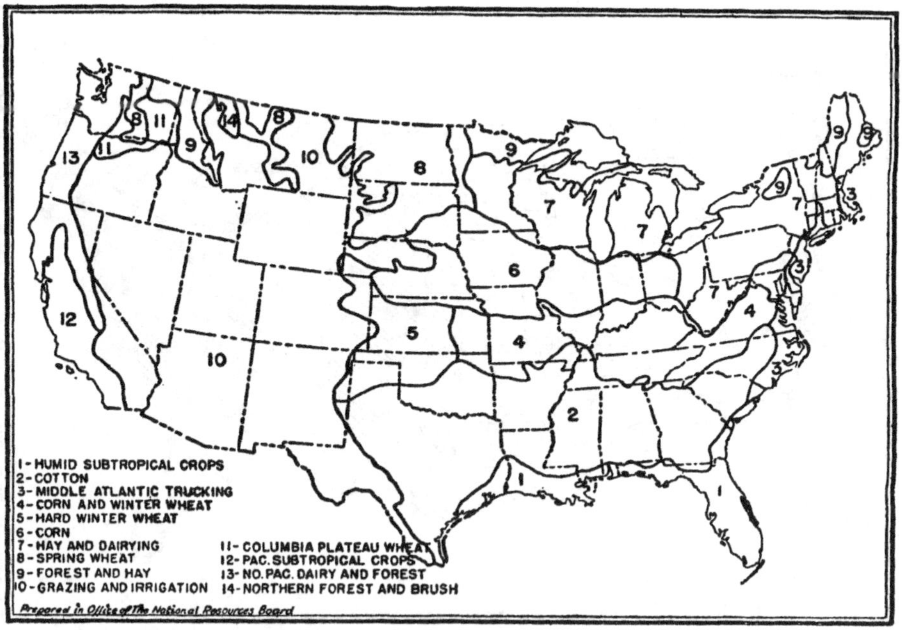

FIG. 68.—United States. Agricultural Regions (after O. E. Baker).

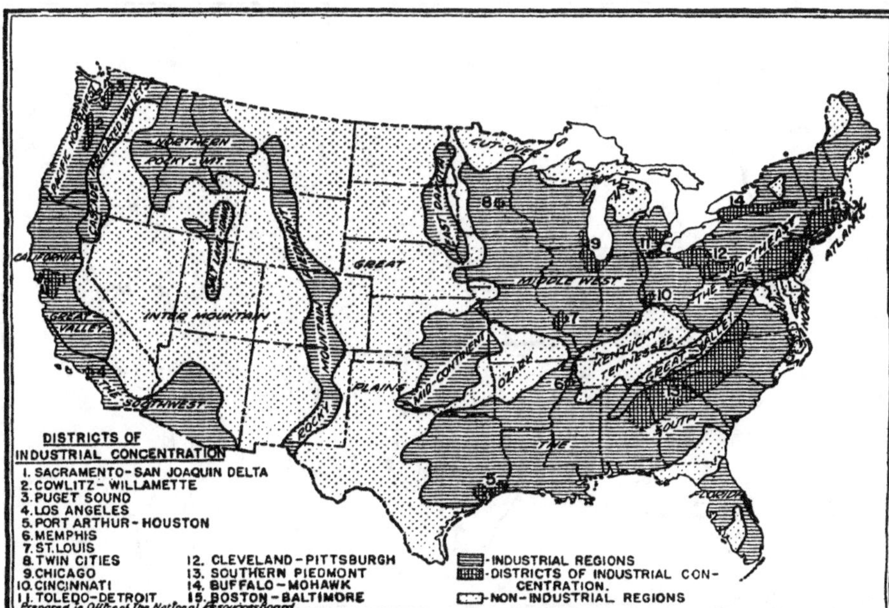

FIG. 69.—United States. Industrial Regions (after Helen M. Strong).

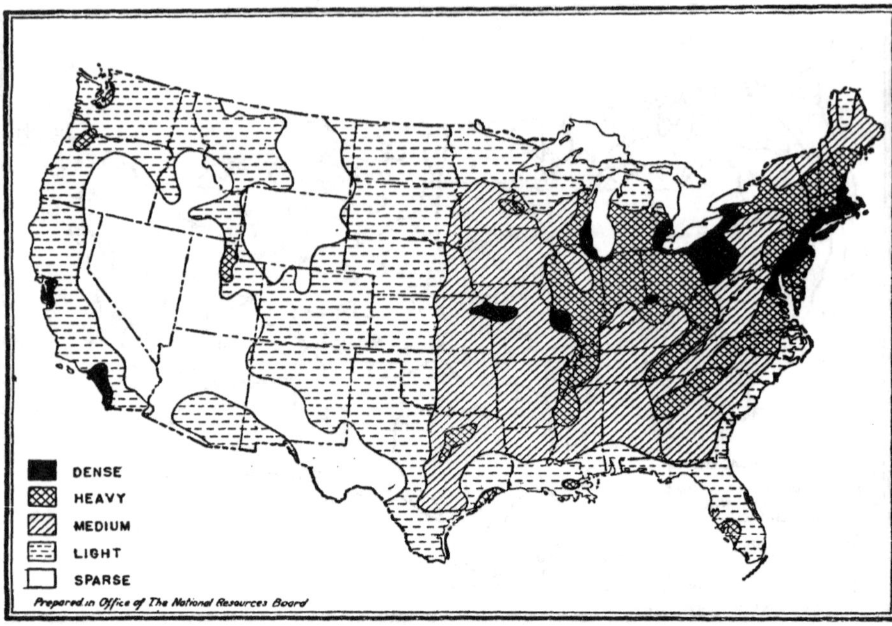

FIG. 70.—United States. Population Regions (after U.S. Census).

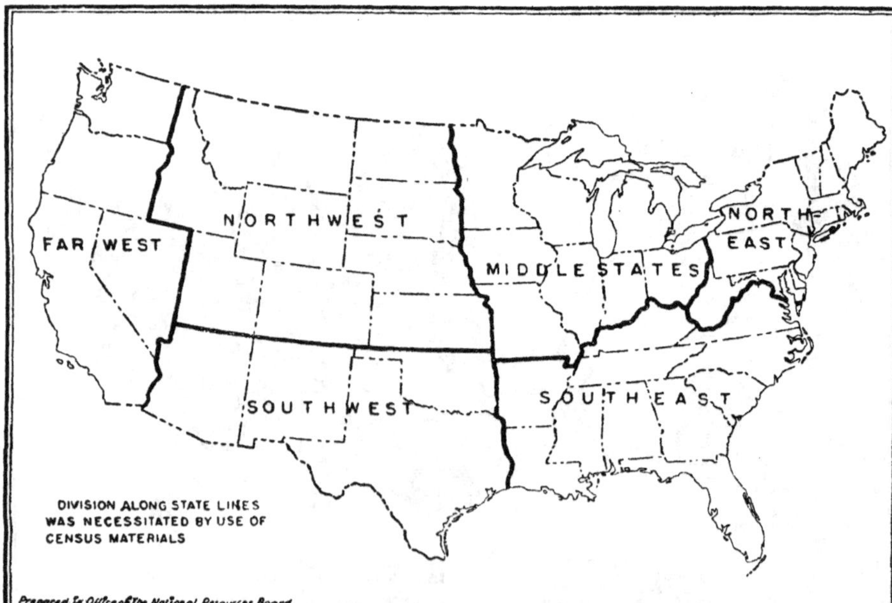

FIG. 71.—United States. Socio-Economic Regions (after Odum and Wooster).

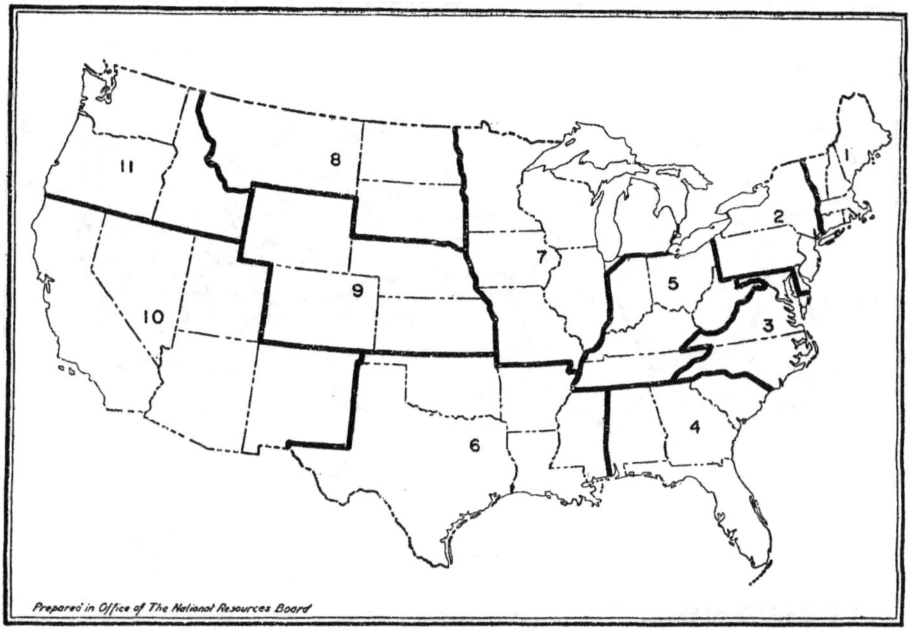

FIG. 72.—United States. National Resources Board Planning Districts, based upon Group-of-States arrangement.

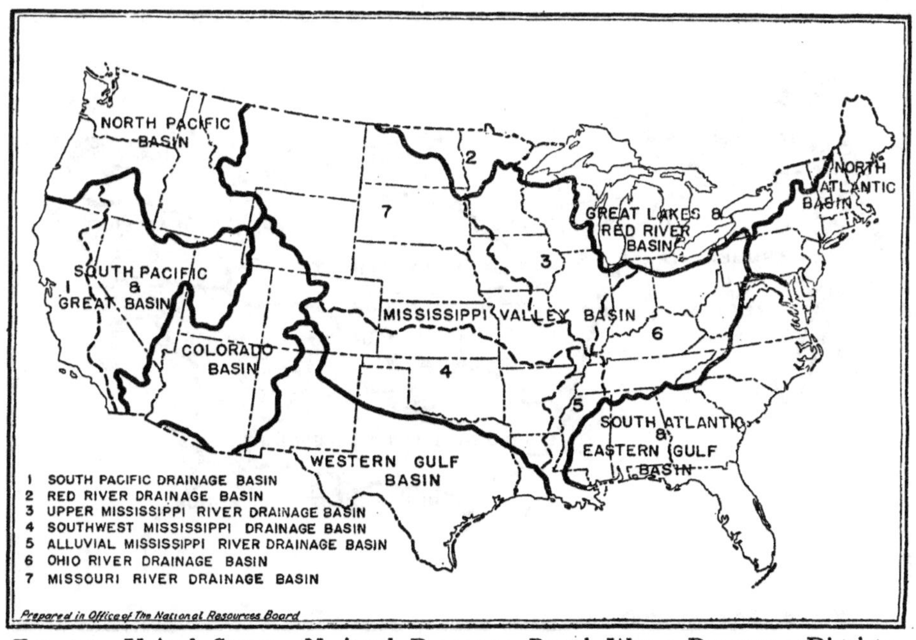

FIG. 73.—United States. National Resources Board Water Resources Districts, based upon a single function.

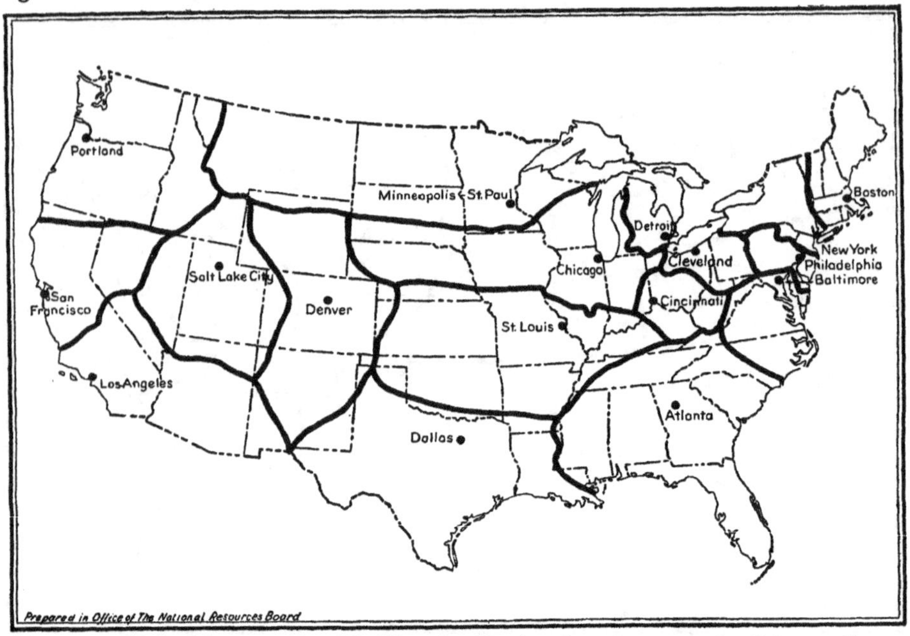

FIG. 74.—United States. Possible Planning Regions based upon major Metropolitan Influence.

FIG. 75.—United States. Possible Planning Regions based upon composite Planning Problems.

as a framework for the understanding of many fundamental social and economic problems of the day that are associated with city life. The National Resources Committee suggests a regional division for planning problems in connection with natural resources, which corresponds closely with the accepted agricultural regions (Fig. 75).[1] The United States Census Bureau uses a threefold division into North, South and West, with subdivisions, bringing the total number of divisions to nine, though these statistical divisions rarely correspond with the other geographical units. An interesting, and what will probably be an extremely useful type of region, is the manufacturing or industrial region to which geographers have recently given attention in America (Fig. 69).[2] Each of these types of homogeneous region is distinguished by a particular range of characteristics and therefore forms a useful basis for the investigation of particular problems.

5. National Planning and National Development Regions [3]

Physical planning is a matter of " designing a pattern of human works and constructions which will bear harmonious relations to the underlying resources ". Such planning cannot be done for the nation as a whole. " The national area is too large and too lacking in homogeneity to be viewed from a single vantage point." Many avenues of approach to planning have been developed for *ad hoc* purposes, in great variety—by groups interested in particular commodities and industries, or particular services (e.g. banking) ; by departments of the federal and State governments (e.g. administration of forest and range land and waterways) ; and by governmental agencies which perform scientific functions (e.g. the United States Geological Survey). " There is, therefore, a fundamental necessity for relating the

[1] *Regional Factors in National Planning and Development*, National Resources Committee, 1935, p. 166.

[2] A. J. Wright, " Manufacturing Districts of the United States ", *Economic Geography*, Vol. XIV, 1938, pp. 195–200 ; C. F. Jones, " Areal Distribution of Manufacturing in the United States ", *Economic Geography*, Vol. XIV, 1938, pp. 217–22 ; Helen M. Strong, " Regions of Manufacturing Intensity in the United States ", *Annals Association of American Geographers*, Vol. XXVII, pp. 23–43 ; R. Hartshorne, " A New Map of the Manufacturing Belt in North America ", *Economic Geography*, Vol. XII, 1936, pp. 45–53.

[3] See chapter by George Renner in *Our Natural Resources and their Conservation*, edited by A. E. Parkins and J. B. Whitaker, 1936. Also Galloway and associates, *Planning for America*, 1941. Quotations are from Renner's article. See also *Regional Factors in National Planning*, National Resources Committee, 1935.

various interests and their peculiar contributions into a uniform and integrated national programme", that is, the planner needs "a sub-national unit of area, which will furnish him a wieldy and manageable region". For this purpose the State is not a satisfactory unit. Very many problems, such as navigable waterways, are of an inter-state character, that is, they demand agreement between adjacent States rather than action by the federal government, which is very limited in sovereignty. Further, such problems as reafforestation and the development of hydro-electricity call for joint action by groups of States. Moreover, social and economic allegiances extend across State boundaries to form distinct socio-economic units based on history and present interests which demand recognition in questions of planning.

Many proposals have been put forward for a regional division of the United States which shall serve as a basis for the inter-State treatment of physical planning. These regions fall into five categories, arrangements involving groups of states (Fig. 72); the spheres of influence of the great metropolitan cities (Fig. 74); regions designed by federal bureaus for administrative convenience with easy access to a central administrative city; single-function areas (Fig. 73); and composite-function areas (Fig. 75). Each of the first four types of region has its obvious defects. Renner, in a general discussion of the problem, concludes that

> in order to have genuine regional planning, it must be based upon *a composite of factors and elements*, both physical and human. This type of region has been defined as an area characterized by general unity in its human ecology.[1]
>
> A major geographic principle is that in an area where resources are roughly uniform throughout, there is permitted the development of a general socio-economic homogeneity. The basis for determining the lineaments of major regionalism within the nation is inherent in such a principle. Indeed, this has to a certain extent already emerged spontaneously, as is revealed by the common use of such terms as the Middle West, the South, the East, the Pacific Northwest, and other endemic regional designations. Such regions, if their outlines be sufficiently flexible, would seem to be fairly satisfactory sub-national divisions for decentralized national planning. More-

[1] "When like thoughts and attitudes dominate entire groups of people, there emerges the dynamic force of public opinion or group thinking, which shapes and moulds the life of a region or a nation." In a measure, these evolve from "the human, economic and physical surroundings". See Helen Strong, "Regionalism: its Cultural Significance, *Economic Geography*, Vol. XII, 1936, pp. 392–410. This author then instances as such regions: Middle West, South, New England, Pacific Northwest, the West, California and the East.

over, they possess the added virtues of being readily identifiable, and of expressing already existing regional loyalty and consciousness to which planning can be harnessed.[1]

It is not possible, however, continues this writer, to define regions in advance of planning, for differing areas will be required for different purposes. Thus, regional planning headquarters will be established in those areas where there is a marked clustering of related resources and a marked consciousness of unity among the peoples. The boundaries of the region would not be fixed for all purposes, but would vary according to specific planning problems, though all would pivot on the same central capital centre. The boundaries would thus, in effect, be elastic.

A region for planning purposes within the framework of nation-wide planning in the United States, as stated in the report of a Special Committee to the National Resources Committee, should fulfil the following requirements :[2] It should be characterized by compactness and contiguity, by homogeneity, by " unity, organic interrelationship and cohesion ", by a major combination of resources (that is, it should be a natural economic unit), it should include whole problem areas and not partial areas, it should be a " total areal pattern of culture and works ", it should not cut across such patterns, it should conform to existing regional consciousness and sentiments, it should possess regional identity as evidenced by popular usage and by a popular name (such as New England or the Middle West), and lastly, it should be of a fairly large size.

The authors of this report to the National Resources Committee submitted a scheme of regions for nation-wide planning purposes. They considered every type of natural and administrative division of the country and discovered that very many of these regions coincided in their general extent and that they had a common core, although the boundaries were indeterminate and zonal. Using as a basis a great variety of criteria, twelve regions were established for " composite planning purposes ", and these were given a popular regional name which, however, is devoid of political meaning. These regions are shown on Fig. 75.

The characteristics of these geographical regions may be illustrated by reference to two clearly defined major planning

[1] Renner, on "National Regional Planning in Resource Use", in Parkins and Whitaker, *Our Natural Resources*, 1936.

[2] *Regional Factors in National Planning*, National Resources Committee, 1935.

regions, namely, New England and the Pacific *Northwest.* In both these regions Regional Planning Commissions were established in 1934, in accordance with suggestions from the National Planning Board, with the object of co-ordinating the planning of interstate problems. They have purely advisory functions. The New England area covers the six States east of the Hudson river with an area of 62,000 square miles and a population of just over eight millions.

It is the geographical region east of the Hudson valley. It is the historical region of the Yankee. It is the ethical region of the New England conscience and of Puritanism. It is an industrial region separate from all other industrial regions, a recreational region of rugged coast, tumbled mountains, crystal streams and lakes, sloping orchards and white pine forests. Movements such as the one that culminated in the organization of the New England Council show that New England is not too large to have a consciousness of community aims.[1]

There are, however, fundamental contrasts within this whole especially as between the north and the south, the latter in particular being overwhelmingly attracted to Boston and New York City.

Industrial southern New England is quite different from northern New England, and it has a close relation with the entire manufacturing region of the north-eastern States ; part of Connecticut is definitely associated with the New York metropolitan area more than with New England ; certain important land-use problem areas extend westward beyond the borders of Massachusetts and Vermont ; the great northern recreational belt is functionally related to the whole North Atlantic seaboard city complex.[2]

The Pacific Northwest, as the name indicates, lies in the extreme north-west of the United States on the Pacific slope, and its character as a physical and human unit is undoubted in both the scientific and popular senses, although the extent of the region from each point of view differs widely. The region finds varied definition. From the standpoint of regional planning the Pacific Northwest Commission defined it as including " not only the Columbia Basin, but Puget Sound in north-western Washington, and that part of the Missouri basin lying in eastern Montana, as well as the coastal areas fronting the Pacific in Oregon and Washington. All are inseparably linked economic-

[1] W. R. Greeley, " Regional and City Planning in New England ", in *New England's Prospect*, 1933, American Geographical Society, p. 406.

[2] *Regional Factors in National Planning*, p. 122.

ally and socially into one zone." [1] This whole area, embracing the States of Washington, Oregon, Idaho and Montana, has an area of 392,000 square miles and a population of three and a half millions.

But, again, the whole, while recognized by its authorities as a good planning unit, is geographically diverse, as well as being very large, and has three major divisions—the Pacific mountain and valley section, the Columbia Basin section (that is mainly agricultural), and Montana, characterized by wheat farming and live-stock ranching, with spots of irrigated agricultural land.

> The coastal section is dominated by the metropolitan influence of Portland and Seattle and their neighbours, while eastern Montana looks towards the Twin Cities. There are also marked political differences, the coastal section voting Republican, the "Inland Empire", centred on Spokane, voting Democratic, while Montana varies in its allegiance between the two major parties and has third party leanings. There is, however, one social factor tending for regionality in the Pacific Northwest. This is the unusual homogeneity of the population of the region, which is primarily Anglo-Saxon, with a considerable Scandinavian population in the port cities and in regions favourable to the fishing industries.[2]

The Pacific Northwest Commission, on behalf of the National Resources Committee, undertook a thorough investigation of the nature and degree of the homogeneity of the north-west as a basis for the regional treatment of common planning problems. The basis of regional homogeneity was sought in tests of "historical data relating to cultural habits and economic intercourse".

> It has studied a few of these factors, such as the distribution of lumbering in the four States, the distribution of different types of agriculture . . ., the traffic density as shown by studies of motor vehicle traffic . . ., the analysis of regional transportation facilities, and the correspondent bank relationships between cities and towns within the four States and with banking centres outside. The regional organization of the two great mail-order banks furnishes another clue to economic and commercial intercourse. Among possible culture tests of homogeneity, the study made use of available information concerning religious affiliation, newspaper distribution, and long-distance telephone messages between selected toll centres. It also studied . . . the votes cast on initiative and referendum measures in two elections separated by intervals of about 20 years. Lack of time and inaccessibility of other significant test data made

[1] *Pacific Northwest Regional Planning Commission, Progress Report,* February 1935, Portland, Oregon, p. 111, quoted in *Regional Factors in National Planning,* National Resources Committee, December 1935, p. 122.

[2] Ibid., quoted in *Regional Factors in National Planning,* p. 123.

it impossible to be sure of a precise boundary in Montana and in Idaho, but the net effect of the factors examined appears to be that Oregon and Washington, western Montana and all of Idaho except the south-east counties made up a unit which for general planning purposes might rule and constitute the Pacific Northwest region.[1]

The test of homogeneity for regional determination, from whatever angle we approach it, is in final analysis the behaviour of people. We need therefore to look for similarities of living habits and standards, similarities of knowledge and skills required to solve their economic difficulties (in which the conservation and utilization of natural resources loom very large), unity of religious outlooks, and expressions of feelings of regional unity. As incidental clues to some of these elements we shall look for physical (e.g. transportation) and institutional (e.g. banking) ties that act as canals for an intense social intercourse.[2]

The Pacific Northwest, as thus defined, is an economic and cultural entity that is suited for many purposes of Federal Government.

[1] *Regional Planning, Part I—Pacific Northwest,* National Resources Committee, May 1936, p. xii.

[2] Ibid., p. 100. Staff Report—Section III.

CHAPTER 13

THE INTERNATIONAL ASPECT OF REGIONALISM: CONCLUSION

It is important to realize the close relations that exist between the city-region and the State. " There are ", writes an American economist, " all kinds of regions. But the regionalism that is of greatest importance is metropolitan. Here we have an area inhabited by producers and consumers who from a radius usually of over a hundred miles look to one big centre for marketing their products and serving their supplies ", and this region " has grown to be a potential rival of the State ".[1]

The great city, as a centre of industry, commerce, culture and administration, and often as a great political capital, has grown up in the past, and especially in the last hundred years, through access to a unified political and economic unit and through access to international world-wide markets. It reaches the status of a super-metropolitan city, a " primate city ", to use Jefferson's term. Vienna and Constantinople, which functioned for centuries as the capitals of great States, the one the centre of the Austro-Hungarian Empire in the Danubian lands, the other the centre of the Turkish Empire in the Middle East and the Balkans, have suffered by the disintegration, for various reasons, of these empires and by the erection in the inter-war period of tariff barriers to prevent the free interchange of goods. The fate of Vienna is well known. It was a great city of two million people. Its tributary area was cut down to the size of Austria and it was severed from its established markets in Danubia. And it still houses nearly two million inhabitants. The difficulty of feeding its people, let alone giving them employment, is a familiar story. " The disrobing of Vienna and Constantinople are crimes against metropolitan regionalism which are bound to cost dear ", wrote an economist in 1929.[2]

" Cities do not grow of themselves. Country-sides set them up to do tasks that must be performed in central places ".[3]

[1] N. S. B. Gras, " Regionalism and Nationalism ", *Foreign Affairs*, Vol. VII, April, 1929, pp. 454–67.
[2] Gras, op. cit.
[3] M. Jefferson, " The Distribution of the World's City Folks ", *Geographical Review* Vol. XXI, 1931, pp. 446–65.

Turning to the large and densely populated States for a standard of measure, it is found that there is one city with over 100,000 people for every one to one and a quarter million people, and it is a fairly safe assumption that almost a million people in Europe and America are needed to support a city with over 100,000 inhabitants, remembering always that the proportion of big cities (as of the grand total of all urban population) to the total population of a country depends both on the economic structure of the country and upon the measure of its dependence upon outside markets for the service of its urban population. Modern big cities grew in the short period of urban growth from 1870 to 1914, and in Europe the changes of political frontiers in 1918 often cut them off from the countryside they had grown to serve. This applies not only to the lands encircling inland cities but also to the " hinterlands " of ports—the areas to and from which they distribute and collect supplies—with the result that a number of them remain centres severed from their limbs and, suffering from " surgical shock ", some have never recovered fully. Danzig is a port that has traditionally served the Vistula basin, the heart of Poland, behind it. But during the nineteenth century—from the partition of Poland down to 1919—it was entirely severed from this hinterland, because it lay in the belt of German territory that extended through West Prussia to East Prussia, south of which lay the Russian province of Congress Poland. The German-Russian frontier was a complete economic and cultural barrier. Poland was a backwater in Russia deliberately cut off from all communications across the German frontier. Danzig dwindled to a shadow of its former self, testimony to its greatness in the Middle Ages down to the end of the eighteenth century being its rows of great multi-storeyed warehouses on the waterfront. The Treaty of Versailles established the independence of Poland, but set up Danzig, which is an entirely German-speaking city, as a Free State, and thus made difficult the unrestricted use of Danzig as an outlet for the trade of Poland. The result was that Gdynia was established as a direct outlet on Polish soil. The two in effect serve as partners as sea-outlets for the trade of Poland.

The Baltic States have suffered similarly.[1] Their chief cities are far too large to serve such small and predominantly agricultural States. Riga, the chief port and city of Esthonia, has

[1] E. van Cleef, " East Baltic Ports and Boundaries with special reference to Königsberg ", *Geographical Review*, Vol. XXXV, 1945, pp. 257–72.

385,000 people in a State with just over one million. It grew to its disproportionate size as a port serving extensive areas in West Russia. " There were three or four times the number of people in West Russia who used and needed Riga, if analogies mean anything, though the economic countryside of a great city is not easy to delineate or define." [1] In other words, a small country is often not big enough to support a large city. Copenhagen, for instance, accounts alone for a fifth of the population of Denmark : it is as big as Sweden's three 100,000 cities put together, though Denmark has only half the population of Sweden. One is led to suspect that Copenhagen does indeed draw upon southern Sweden both as a source for its population and for its trade. Switzerland has four 100,000 cities, but no big dominant city, and the commercial capitals of other countries —especially Paris, Berlin, Milan, to say nothing of the ports of Genoa, Antwerp and Marseille—carry out for Switzerland many of the functions that would normally go to the dominant capital city.

There is another aspect to this question of the city in relation to political frontiers. The city, or, to be more correct, the urban tract, has been rapidly expanding geographically and economically during the past fifty years. This expansion is effected by the spread of houses, roads, railways, canals ; the movement of workers to and from factories ; the movement of daily supplies of milk, vegetables and meat (that is, perishable foodstuffs) ; the supply and control of public services in general (electricity, water supply, drainage, gas, telephones and other piped services) ; and the interdependence of factories on each other and on the warehouses, offices and exchanges in the city centre. All these space-relations unite a large area around the city into an effective natural geographical unit whose borders are very wide and vague, but whose city centres are the meeting-place of more and more of these relations. If a political frontier cuts through such a grouping it is a potential source of danger to the efficient functioning and organization of the whole and of its parts. If existing *before* the modern development took place, a compromise may have been found and the friction entailed in the existence of a frontier may be limited to an absolute minimum in the daily round and organization of all aspects of the life of the area. Such has been the case,

[1] M. Jefferson, " The Distribution of the World's City Folks ", *Geographical Review*, Vol. XXI, 1931, pp. 446–65.

for example, in the densely peopled industrial areas in Flanders on the borders of northern France and Belgium. This great textile-producing area lies for the most part on the French side, with its capital and greatest city in Lille, and its commercial and industrial centre in the combined towns of Roubaix–Tourcoing, but the populous areas across the frontier are akin to those on the French side in respect of their towns and countryside, in their economic activities and in their organization, and there is a great daily current of traffic and workers across the frontier. The opposite has been the case in Upper Silesia. This great coalfield lay, before 1919, almost entirely in German territory, and found its markets mainly in south-eastern Europe. Since 1919 it has been mainly in Poland, with relatively small sectors of it in Germany and Czechoslovakia. Its markets have been, in large measure, oriented towards Poland, although, owing to the competition of the German industrial areas of the middle Elbe basin and the Ruhr, its manufactured products, in the event of its full-scale development becoming comparable with that of the Ruhr, must find their chief markets in East Central Europe. Its unfavourable location, stretching across three States and in the extreme corners of two of them (Germany and Poland), and its lack of first-class waterways to connect it with the North and Baltic Seas and the eastern Alps, whence it must draw almost all the iron ores for large-scale development in the iron and steel industries, make the problem of the industrial development of Upper Silesia essentially an international one. Just as serious, however, is the problem of the organization of this extensive, straggling, and expanding urban area—the kind of problem that normally comes under the heading of "regional planning", but is here confounded and complicated by the existence of political frontiers. For efficiency of organization in all respects, Upper Silesia must function as a unit. It is not necessary to go into the details of this particular problem since they have been examined often and thoroughly. One of the aims of the League of Nations solution of this problem was to facilitate this free functioning and unity of organization and of daily movements of workers across the German-Polish frontier, which, as German propagandists have rightly stated, cuts arbitrarily right across the urban complex. The League of Nations control ended in the late 'thirties and the frontier difficulties were accentuated by the advent of the Nazi régime and its anti-Polish policy. The Silesian urban area, whose industrial development is in its infancy, has big prospects as the only large coalfield between the Ruhr (excluding the lignite field of central Germany) and the Donetz. It presents a dual problem—that of access to raw materials (iron ore) and markets (iron and steel goods and coal), and that of organizing the area as an economic and urban complex which must, in the interests of efficiency, override any potential barriers raised by political frontiers which cross it.

Finally, within the State, the city has emerged more and more in the last hundred years as the focus of activities of the towns, villages and countryside around it. Indeed, the growth of cities has proceeded so far that the really great city is too large from the standpoint of efficiency of service and of offering the best conditions for human living. Disintegration has already started through the operation of centrifugal forces, the shift of factories, institutions, houses and people, out from the centre—although hitherto in a haphazard fashion without any attempt at a " design for living ". But the future will see the continuance of this redistribution of urban functions and buildings from the congested centres, so that the city settlement area, and, beyond it, the city trade area, will emerge more and more as the effective grouping of real social and economic life, ready to be adopted in principle as the unit of democratic activity and land planning with the same conception as the city-state of antiquity.

The metropolitan region within the State is indeed a rival of the component States or Provinces, as in the U.S.A. and in Germany, and of the existing major political provinces (the direct descendants of the historical provinces) as in Britain (counties), France (*départements*), Italy (*compartimenti*), and of the *Provinces* in Prussia. Each of these political divisions is a unit within fixed political boundaries in respect of many aspects of social organization—such as government, administration and law—and, in the U.S.A., and Germany until recently, has enjoyed, in varying degree, the right of a sovereign State. But, the circulation unit, clearly defined as to its great city centres and their environs, often vague as to its limits, is the effective *de facto* unit of many of the most vital aspects of modern life, and has emerged as the natural (unplanned) framework of many activities. This fact is revealed by population distribution, circulation flows, the distribution of economic activities and interests, and of the multifarious private and government organizations of the State. Such groupings, therefore, form the best units in which to handle many aspects of the scientific study of society, since they have more in common than any other groupings of similar size. It is for this reason that in recent studies of social and economic trends this type of grouping is emphasized as the real unit of modern life ; and that in plans for reorganizing administrative units, from whatever point of view, the city and its region must be the basic and primary consideration.

The fundamental importance of the major regional groupings in the life of modern society is especially relevant to the problem of Germany.[1] Owing to the detailed treatment given to this subject in *The Regions of Germany* in this series, we have omitted consideration of regionalism in Germany in this book, but brief reference to the problem of the future of Germany is essential to the present theme. We have shown that Germany consists of eleven major natural provinces with two smaller units. These are Rhineland-Westphalia, centred on the Ruhr complex and Cologne ; Lower Saxony (*Niedersachsen*), centred on Hanover ; Rhine-Main, centred on the Frankfurt–Mainz–Wiesbaden complex; the Southwest, mainly Baden and Württemberg, centred on Stuttgart ; Bavaria ; Central Germany (*Mitteldeutschland*), embracing the middle Elbe basin ; Nordmark (Schleswig-Holstein and Mecklenburg) ; Pomerania ; East Prussia ; Brandenburg, centred on Berlin ; Silesia ; and the agglomerations of Hamburg and its tributary areas on the lower Elbe ; and Bremen with its similarly tributary areas on the lower Weser. Each of these units is a closely integrated whole, that often cuts right across the existing political frontiers inside Germany, and each is a recognized unit in the social and economic life of modern Germany.

Now, if we consider the present nebulous international political situation in Germany, it would appear to be certain that the eastern provinces of East Prussia, Pomerania, Silesia and Brandenburg east of the Oder are to be permanently amputated from Germany and annexed to Russia or Poland. This decision, that deprives the more thickly populated parts of western Germany of their chief surplus food-producing areas, must presumably be accepted, for good or ill, as a *fait accompli*. In the west, France is pressing for the political separation of an area that is vaguely defined as the Rhineland, eastwards as far as the Rhine, together with a tongue of territory in the Ruhr coalfield and industrial

[1] R. E. Dickinson, *The German Lebensraum*, Penguin Special, 1943, and *The Regions of Germany*, 1945.

area. Western Germany falls into three major provinces that lie astride the Rhine—Rhineland-Westphalia, Rhine-Main (or Hesse as it may alternatively be called) and the Southwest. The Ruhr cannot be severed from Cologne and its great lignite area, nor from the rest of north-western Germany. The Frankfurt–Mainz complex cannot be severed from the territories around it. The Southwest is a unit and, what is just as important, it has common interests and affiliations with the French provinces of Alsace and Lorraine, and with the Saar, that are based not only upon linguistic and cultural ties, but also on their common interests in the navigation of the Rhine. A second fact of fundamental importance is the essential economic interdependence of these three Rhineland provinces with each other and with Alsace-Lorraine, Switzerland, Belgium and Holland—which are comparable in area and population with the German provinces. This interdependence is based upon the interchange that is necessary between the coalfields of the Ruhr and Belgium and the vast iron-ore fields of French Lorraine. Since 1919, when Germany lost her Lorraine ores, that had been developed in the closest economic dependence with the Ruhr, the latter imported larger quantities of foreign ores up the Rhine and, to a lesser extent, through Emden by canal. The iron and steel industries on the coalfields of the Ruhr and central Belgium need Lorraine ores, and the iron and steel industries of Lorraine and the Saar need the coking coals of these areas, since those of the northern coalfield of France and imports are inadequate. Further, there is the common interest of western Germany, France, Switzerland, Belgium and the Netherlands in the navigation of the Rhine. Antwerp and Rotterdam are the chief ports for the whole of the Rhinelands, together with the river ports of Duisburg–Ruhrort, Mannheim–Ludwigshaven and Strasbourg. The essential economic interdependence of all these lands is indisputable and some kind of revival of Napoleon's *Rheinbund*, closely associated with the western powers, would seem to be the most feasible solution of the political future of Germany. But whatever the political fate of western Germany as a whole, the three German natural provinces should figure in the new political set-up. This is recognized in the American zone of occupation, as is evident in the formation of the new administrative unit of Hesse. The *territorial* separation of the narrow wedge of urban land popularly known as the Ruhr is impracticable (as mentioned on p. 251). International control of the war potential of the Ruhr industries is essential, but this need not involve territorial annexation and separation from the surrounding land.

The "rump" of Germany, lying between the Polish and Russian annexed territories in the east and the Rhinelands in the west, includes the natural provinces of Lower Saxony, the Nordmark, Pomerania, Brandenburg–Berlin, central Germany (*Mitteldeutschland*), and the ports of Bremen and Hamburg. Whatever political fate befalls this central sector, it is essential again that these provinces find expression in a new political framework. It may be noted that central Germany (see p. 252) is a vast and complicated industrial and agricultural complex, with great brown-coal or lignite fields, electricity and chemical plants and sugar-beet industries, while Saxony is a chief seat of the light engineering and textile industries. This whole complex is intimately related to Hamburg, Bremen, and Berlin. The capital normally derives the bulk of its electricity and fuel from this complex, and Hamburg in particular is its principal outlet overseas both by the river Elbe and by rail.

This mode of approach to the problem of the political future of Germany urges the recognition of the major natural geographical divisions as elaborated in this book. These divisions have been widely recognized in the organization and life of Germany between the wars and have been given much attention since the first days of the Weimar Republic. As in Britain, France and the United States, they are *de facto* units in the life and organization of the country. Recognition of these units would build on the trends of development and thought in Germany itself and would thus ensure continuity of development, based on the essential space relationships, rather than on the arbitrary boundaries and divisions imposed by power politics.

Finally, a word may be said concerning the planning of the city and its region.

The first draft of this book was ready in the early years of the war, before the complete destruction of large sections of great cities by area bombing. At this time, too, the public was not very alive to the needs and problems of planning the location and arrangement of buildings in order to form a fitting habitat for Man in Society. The destruction of our cities has rendered the business of reconstruction imperative and urgent. The need is far greater in German and in certain other Continental cities than in this country. In the United States, on the other hand, no bombs have fallen, but cities have grown at such a phenomenal rate through the growth of war-time industries that enormous housing estates have been erected that must inevitably give rise to precisely the same problems. These developments demand an acceleration of the business of structural analysis of our cities and the elucidation of principles for geographical planning of town and country. That such has been the case is evident from researches under government auspices during the war on both sides of the Atlantic.

We have emphasized the structure of the city, both as a built-up entity and in its regional relations. Let it not be supposed, however, that we hold the view that the large city, as a vast amorphous agglomeration, must inevitably continue to dominate the life and thought of our countryside. The city, in the fullest sense—and there are few cities that fill this ideal—must continue as the head of our civilization. It is, however, generally agreed that planning for the future should aim at reducing the size of the great urban agglomeration while improving and making more widely accessible the amenities of city civilization in town and country alike. Decentralization of industry and population, in the sense of a widespread redistribution far from the direct influence of the great city, was only an incipient development in the inter-war years—it should more accurately be described as deconcentration, since the shifts took place merely to the margins of the built-up areas and tended to expand the area of the urban tract. War-time evacuation of factories and commercial and administrative concerns was, on the other hand, real decentralization, real dispersal. But however far decentralization may go in the future and the urban agglomeration be reduced in size, the city as the fount of civilization, providing the best it has to offer, will and must remain. And such a trend will inevitably tend to increase the reality of the extensive city-region through the interdependence of its parts and their relations with the chief cities.

There are relatively few attempts in recent planning programmes to allow for real decentralization on an adequate scale from the urban areas and for a great reduction of their population. No city council can, or is likely to, write off in its plans several hundred thousand people as superfluous, to be housed somewhere other than in their own community sphere. All it can do is to make allowance for "overspills", and even this involves interminable wrangles with neighbouring local government authorities. Deliberate decentralization obviously lies beyond the powers of town and regional-town planning authorities, and can be effected only through the guidance and authority of a body with nation-wide powers. This objection has been raised, for instance, against the County of London Plan, and in the United States planned decentralization raises much more serious issues, for there the large city is much more impersonal and megalopolitan in character. The preconditions of urban reconstruction have been stated by Mr. Lewis Mumford as follows :

> (1) A National Population Policy, looking toward its stabilization, if not its increase, instead of permitting the threatened decrease of population to go on unchecked. (2) A Policy of Urban Land Utilization which will liquidate the present structure of urban values and permit large-scale reconstruction to be economically carried on, in a fashion favourable to family life and balanced communal relationships. (3) A National Policy of Industrial Decentralization along the lines laid down in the Barlow Report : a policy which will progressively move population out of London and other large centres until a net reproduction rate close to 1·0 is achieved. (4) Regional Administrative Units that will undertake the task of resettlement and building outside the existing municipal or county areas and will co-ordinate the work of the municipalities themselves.

The last will involve " the division of Britain into regional administrative areas and the creation of a new pattern of decentralized government and administration ".[1]

The aim of this book has been not to elaborate the theoretical aspects of these aims or principles, but to indicate something of the pattern and fabric of those spatial associations in western Europe and the United States upon which such planning and reconstruction should find its basic geographical foundations.

This whole field of study has now become of great importance as a basis for plans of reconstruction. Its investigation has made a tardy appearance in Britain and often seems to be regarded as new. So far our elaborate social surveys have neglected this aspect, with few exceptions, such as the survey of Merseyside. Civic and Regional Surveys have been dominated by the idea of Place, Work and Folk, which embraces all the relevant sciences on a footing of equality in a broad and invaluable philosophical concept, but fails, in the actual organization and prosecution of research in town or country, to canalize and direct investigation to specific problems in their spatial aspects with a clearly defined central objective. Nor, indeed, can it be expected that the

[1] Lewis Mumford, *The Plan of London County*, Rebuilding Britain Series, No. 12. Faber & Faber, 1945, p. 38.

architect or the professional planner can possibly cover the whole field of scientific investigation that should be preliminary to his formulation of planning proposals for houses, open spaces, new towns, roads and the like. There is, in other words, a very serious dearth of trained investigators in this whole field. The social sciences, particularly in the field of human ecology, have much leeway to make up, as investigations in the United States in the inter-war years so clearly demonstrate. There is great need for what Max Lock has called "civic diagnosis". The same may be said of "rural diagnosis", on the lines indicated by the recent survey prepared by Dr. Orwin and his collaborators. Moreover, there is no clear-cut division between one problem and another, and between one discipline and another. The fusion of disciplines in the common pursuit of particular problems is a characteristic of the scientific approach in our time, and while the specialist has his own field, he is constantly impinging on the fields of others. We have tried in these pages merely to indicate, for the consideration of social scientists, the contribution that geography can make in this field. For the study of modern society on the same lines as the anthropologist studies primitive societies, there is immediate need for institutes of research, attached to the Universities, with trained personnel, and working in close collaboration with the planners and builders as the common ground of the social sciences, and dedicated to what is, in effect, a virgin field of investigation. Without diagnosis of all aspects of society by the social scientist and the careful mapping of his data whenever relevant, the planner cannot provide for the optimum use of the land (witness some of the absurd results of zoning proposals in this country), the architect cannot build to suit the immediate and long-term needs of the people, and principles of planning, taking shape as the law of the land, can neither be worked out, nor put into effect, on a permanent and socially desirable basis.

INDEX

For Product Safety Concerns and Information please contact our EU representative GPSR@taylorandfrancis.com
Taylor & Francis Verlag GmbH, Kaufingerstraße 24, 80331 München, Germany

www.ingramcontent.com/pod-product-compliance
Lightning Source LLC
LaVergne TN
LVHW011111300926
849160LV00008B/348

* 9 7 8 0 4 1 5 8 6 0 4 5 1 *